Springer Series on Biofilms

More information about this series at http://www.springer.com/series/7142

Luis E. Chávez de Paz • Christine M. Sedgley •
Anil Kishen
Editors

The Root Canal Biofilm

Volume 9

Springer

Editors
Luis E. Chávez de Paz
The Swedish Academy for Advanced Clinical Dentistry
Gothenburg
Sweden

Christine M. Sedgley
Department of Endodontology
Oregon Health & Science University
Portland
Oregon
USA

Anil Kishen
Department of Clinical Sciences
Discipline of Endodontics
University of Toronto
Faculty of Dentistry
Toronto
Ontario
Canada

ISSN 1863-9607 ISSN 1863-9615 (electronic)
Springer Series on Biofilms
ISBN 978-3-662-47414-3 ISBN 978-3-662-47415-0 (eBook)
DOI 10.1007/978-3-662-47415-0

Library of Congress Control Number: 2015953861

Springer Heidelberg New York Dordrecht London

Printed on acid-free paper

Springer-Verlag GmbH Berlin Heidelberg is part of Springer Science+Business Media (www.springer.com)

Luis E. Chávez de Paz would like to dedicate his contribution to Patricia, Luis Diego, Nicolás, and Andrés, for all their encouragement and support.
Anil Kishen acknowledges Arunthathi, Abinav, and Aaryan for their encouragement and patience.
Christine Sedgley dedicates her contribution to Victor, her greatest supporter.

Preface

Biofilms are recognized as one of the earliest ecosystems on earth. They are composed of aggregates of microbial cells enclosed in a self-produced matrix adherent to a surface. Root canal biofilms are complex polymicrobial structures adherent to the root canal surface that are formed by microorganisms invading the pulpal space of teeth. Important histopathological studies published several decades ago first noted the presence of adherent cells on root canal surfaces. However, it was not until the introduction of advanced microscopy and molecular biology techniques that they were recognized to be the dominant form of microbial life in the root canal system. Similarly, it was only in the past decade that root canal infections were acknowledged to be biofilm infections. Subsequently, recent studies have shown that root canal biofilms are associated with persistent endodontic infections and as such are likely to be significant contributing factors determining the outcome of endodontic treatment.

Concerted efforts to study root canal biofilms have been made in the past decade resulting in the publication of observational and experimental studies that detail the morphology and biology of these structures in infected root canals. In addition to confirming that bacteria in root canals do not exist in free-floating planktonic states as previously assumed, this new information on root canal biofilm infections has provided an opportunity to reevaluate conventional clinical protocols and improve endodontic therapeutic measures.

The aim of this volume is to provide a current understanding of the basic scientific aspects of root canal biofilm biology within a clinically applicable context. This volume is divided into three sections. Part I discusses the basic biology of root canal biofilms and addresses key questions about the ecological and physiological aspects that play a role in the formation and resistance of biofilms in root canals (chapter "Ecology and Physiology of Root Canal Microbial Biofilm Communities"). The last two chapters of this section review the general mechanisms of biofilm adhesion (chapter "Molecular Principles of Adhesion and Biofilm Formation"), and the mechanisms of antimicrobial resistance in endodontic-related pathogens (chapter "Antimicrobial Resistance in Biofilm Communities"). In Part II, attention focuses on observational and experimental evidence of root canal microbial biofilms. Part II starts with an overview of observations of biofilms in root canals using scanning electron microscopy (chapter "The Use of Scanning Electron Microscopy (SEM) in Visualizing the Root Canal Biofilm"). Evidence for biofilm formation in histopathological preparations, and a review of novel molecular techniques to identify bacteria in biofilm populations in clinical samples, is provided in chapter "Bacterial Biofilms and Endodontic Disease: Histobacteriological and Molecular Exploration". Part II closes with a description of common experimental approaches utilized to study root canal biofilms including in vitro biofilm modeling techniques (chapter "Laboratory Models of Biofilms: Development and Assessment") and examines the challenges behind anatomic complexities in root canals as these may play a role in root canal disinfection (chapter "Root Canal Anatomy: Implications in Biofilm Disinfection"). The final section, Part III, considers how infections caused by root canal biofilms are clinically treated and review the implementation of novel anti-biofilm approaches. An overview of the outcome of persisting root canal biofilm infections and appropriate treatment options is first presented (chapter "Biofilm-Associated Infections in Root Canals: Treatment and Outcomes"). This is followed by an explanation of the influence of clinical irrigation techniques (chapter "Root Canal Irrigation") and the importance of inter-appointment medication on root canal biofilms (chapter "Inter-appointment Medi cation with Calcium Hydroxide in Routine Cases of Root Canal Therapy"). Finally, innovative methods and devices directed towards the removal of biofilms from root canals are discussed (chapter "Advanced Therapeutic Options to Disinfect Root Canals").

This volume will be of interest to a wide range of endodontics-related professionals, including basic microbiologists, clinical microbiologists, and clinicians, and should be useful to undergraduate, postgraduate, and postdoctoral scientists working at the frontier of a new understanding of the role of microbial biofilms in endodontic disease.

Gothenburg, Sweden — Luis E. Chávez de Paz DDS, MS, PhD
Portland, USA — Christine Sedgley MDS, MDSc, FRACDS, MRACDS(ENDO), PhD
Toronto, Canada — Anil Kishen
March 2015

Acknowledgements

It has been a fascinating journey since we started with the idea to edit a book dedicated to microbial biofilms formed in root canals. First of all we wish to thank all the authors for providing their excellent contributions, without their dedication and involvement in this project it wouldn't have been possible to complete it. We wish to give a special acknowledgement to Dr. William Costerton, former book series editor, for proposing that it is timely to publish a book on root canal biofilms.

The editors assembling the root canal biofilm project at a coffee shop at the Hynes Convention Center in Boston, during the American Association of Endodontists Annual Session, April 2012

Contents

Part I
General Biological Aspects

Ecology and Physiology of Root Canal Microbial Biofilm Communities

Luis E. Chávez de Paz and Philip D. Marsh

Abstract Microbial communities formed in root canals of teeth constitute the heart of the infected root canal ecosystem, and yet their establishment and development remains challenging to measure and predict. Identifying the ecological and physiological drivers of microbial community colonization, including resistance (insensitivity to disturbance) and resilience (the rate of recovery after disturbance), is important for understanding their response to antimicrobial treatment. This chapter will provide an overview of the ecological and physiological factors that are relevant for root canal microbial communities in terms of their establishment and endurance in root canal ecosystems. Initially, insights from ecological and physiological parameters that are useful for defining and measuring activities in root canal biofilm communities will be reviewed. The ecological progress of root canal infections will be discussed in terms of three ecological processes: (1) selection of successful root canal colonizers by habitat filtering, (2) selection of resistant bacteria to major disturbances in the environment (e.g., provoked by antimicrobial therapy in endodontics), and (3) resilience of the community after the disturbance. Finally, current methods for analyzing these ecological processes will be described, as these are key elements for identifying the biological features of individual microorganisms and of root canal microbial communities.

1 Introduction

Our current understanding of the microbiota of infected root canals is based on the findings from classical culture-based studies (Möller 1966; Bergenholtz 1974; Sundqvist 1976; Baumgartner and Falkler 1991) and in recent years from studies that have applied modern culture-independent molecular technologies (Munson

L.E. Chávez de Paz (✉)
Endodontics, The Swedish Academy for Advanced Clinical Dentistry, Gothenburg, Sweden
e-mail: luis.chavez.de.paz@gmail.com

P.D. Marsh
Division of Oral Biology, School of Dentistry, University of Leeds, and Public Health England, Porton Down, Salisbury, UK

L.E. Chávez de Paz et al. (eds.), *The Root Canal Biofilm*, Springer Series on Biofilms 9, DOI 10.1007/978-3-662-47415-0_1

et al. 2002; Spratt 2004; Rocas and Siqueira 2010; Chugal et al. 2011). The accumulated information from both these approaches has led to the characterization of species diversity in different clinical situations from the necrotic tooth to the chronically infected root-filled tooth. Notwithstanding this increase in valuable information on the identification of members from root canal microbial communities in different clinical scenarios, there are still significant limitations in explaining the fundamental ecological and physiological basis by which these microbial communities form in root canals. In particular, we still lack a clear understanding of the basis of ecology community-level functions and the potential physiological role that key members of the microbial communities play to maintain stability and structure after antimicrobial treatment. While studies based on 16S rRNA identification have characterized communities of bacteria in root canals with tens to hundreds of species (for a review see chapter "Bacterial Biofilms and Endodontic Disease: Histo-Bacteriological and Molecular Exploration"), it is usually not possible to experimentally establish which species actively take part in the community and perform pivotal functions. The development of root canal microbial communities might also depend on the nature of the primary root canal infection, as well as on environmental selection and physiological adaptation, the effects of which would be difficult to control or characterize under laboratory conditions.

Furthermore, the influence of each of the ecological drivers on the composition of a community established in root canals can vary according to the temporal scale of observation. For example, untreated necrotic root canals are found to be dominated by proteolytic anaerobic organisms, while treated root canals seem to be dominated by a less diverse community with species that can persist for long periods of time under harsh conditions, e.g., facultative anaerobic Gram-positive organisms (Sundqvist 1992, 1994; Figdor and Sundqvist 2007). The low numbers of such facultative anaerobic species detected in untreated necrotic cases using sequencing approaches create a further challenge in establishing the relation between community composition and dynamics from one clinical state to the other.

In this chapter, an ecological concept is presented, focusing on three main ecological processes. These will introduce an ecological and physiological interpretation of the role of microbial biofilm communities in the pathobiology of root canal infections. The first of these processes occurs after the invasion of bacteria from the oral cavity into the root canal, in which the root canal environment acts as a *habitat filter* to select for specific microorganisms. The second process occurs after or during root canal treatment. It will be proposed that the application of antibacterial solutions, dressings, etc., will cause a *simplification* in the root canal microbiota, where stressful environmental changes will select for more resistant microorganisms. Finally, the third process comprises the *resilience* of the remaining community, where the multiple ecological adaptive factors by which microorganisms will establish as a post-disturbance community are examined. Insights obtained from studying the ecology of microbial communities in root canals can be used to improve the management of endodontic infections.

2 Habitat Filtering: Selection of Root Canal Colonizers

A general consensus is that the growth and survival of microorganisms invading the pulpal space is controlled by a variety of environmental factors occurring at the time of the infection. These factors, of physical and chemical nature, constitute the habitat filter that will limit the growth of certain organisms compared to others (Fig. 1). To be able to define the factors included in this ecological filtering process, it is important to first define the status of the pulp at the time of the microbial invasion, i.e., the presence of a responsive or necrotized pulp. Clearly, the main difference between these two states of the pulp is the capacity to exert an inflammatory reaction in response to the bacterial invasion (Bergenholtz 2001). In the case of a responsive pulp that is exposed to the oral microbiota due to trauma as well as in a pulp that is undergoing an acute inflammatory reaction due to a deep carious process, the invading organisms must face an environment characterized by the infiltration of neutrophils. In this case, the chemical composition of the environment is represented by the tissue-destructive elements released by neutrophils, including oxygen radicals, lysosomal enzymes, and high concentrations of nitric oxide (NO).

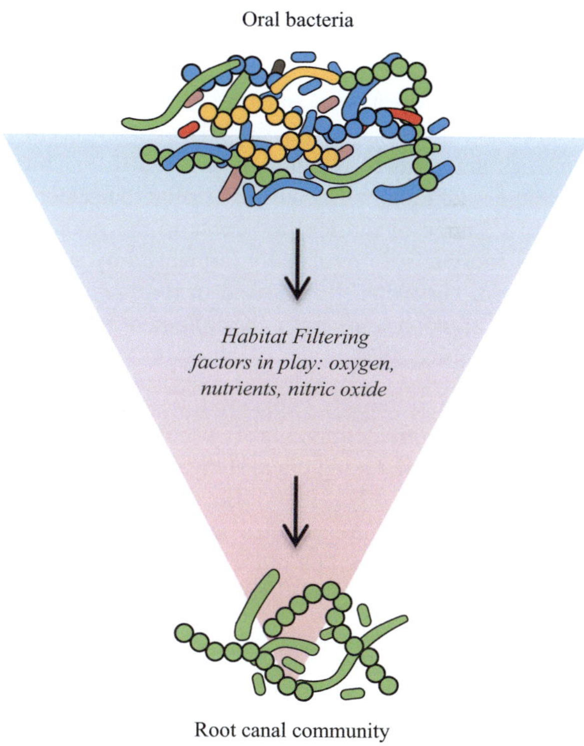

Fig. 1 Selection of root canal colonizers by habitat filtering. Schematic depiction of the habitat-filtering process, showing invading oral bacteria (*cells in colors*), the ecological filtering factors, and the successful root canal colonizers (*cells in green*). Oral bacteria invading the pulp chamber after exposure via caries, trauma, or periodontal disease are ecologically filtered by environmental factors such as oxygen, nutrients, and nitric oxide. The presence of nitric oxide in the pulp ecosystem is due to the infiltration of neutrophils during the inflammatory process in the pulp. Successful colonizers will constitute the root canal microflora

Nitric Oxide (NO) NO is a small, lipophilic, and freely diffusible radical that has strong cytotoxic properties due to its high reactivity. NO directly affects the activity of enzymes in bacteria by the reaction with bound free radicals or with metal ions (Kim et al. 2008; Zagryazhskaya et al. 2010; Pearl et al. 2012). NO has been found to affect bacterial respiration and amino acid biosynthesis, thereby causing cell growth arrest and suppression of DNA synthesis (Jyoti et al. 2014; Kolpen et al. 2014; Liu et al. 2015). Although the molecular interaction of NO with root canal bacteria has not been clarified, the ability of bacteria to adapt their phenotype in order to survive NO environments may be a crucial characteristic of oral microorganisms for colonization in the root canal ecosystem.

Oxygen Oxygen is the terminal electron acceptor in aerobic respiration that is by far the most efficient type of energy metabolism. Oxygen levels in the pulpal ecosystem may play an important role in selection and in determining functional interactions and spatial structures of root canal microbial communities. Studies on the dynamics of root canal infections have shown that the relative proportions of anaerobic microorganisms increase with time and that the facultative anaerobic bacteria are outnumbered when the canals have been infected for 3 months or more (Möller et al. 1981; Dahlán et al. 1987; Fabricius et al. 2006). In the infected root canal environment, there are concentration gradients in oxygen that can vary from low to complete anoxia. Although the oxygen gradients can be relatively stable over time, oxygen seems to be a major ecological factor in the root canal milieu and one that promotes the development of an anaerobic or microaerophilic microbiota (Sundqvist 1992, 1994).

Nutrients All organisms must scavenge nutrients and then coordinate central metabolism, monomer synthesis, and macromolecule polymerization for biomass synthesis and growth (Chubukov et al. 2014). Thus, one of the most important environmental factors that will determine the selection of root canal bacteria is the principal source of nutrient available to the microbiota for growth. The invading oral microorganisms are usually influenced by saliva, its components, and the diet of the host, but would be exposed in root canals primarily to serum constituents, including glycoproteins from the inflamed pulp and periapical tissues (Svensäter and Bergenholtz 2004).

The large and densely connected network of metabolites, enzymatic reactions, and regulatory interactions makes it challenging to understand the metabolic and regulatory network taking place at the time of colonization of the pulpal space in its totality (Sundqvist 1992, 1994). However, by means of specific laboratory models, it is possible to define in vitro individual regulatory circuits that will provide specific information on the nutritional demands of individual members or groups of the root canal microbial community (for a review see Sundqvist and Figdor 2003).

It is of great interest, however, that some of the actual molecular components and mechanisms that control the nutritional demands of root canal bacteria are determined by phenotypic adaptation to the environment. As the prime energy source of facultative anaerobic bacteria is carbohydrates, it is believed that a decrease in availability of carbohydrates in the root canal will limit the growth opportunities for these organisms (Sundqvist 1994; Figdor and Sundqvist 2007). However, bacteria from the oral cavity possess complementary patterns of glycosidase and protease activities and combine their complementary metabolic capabilities to degrade host glycoproteins in a synergistic manner (Bradshaw et al. 1994). In a process known as phenotypic switching, dual metabolic patterns found in some oral bacteria are proposed to play a role in the catabolism of complex glycoproteins from saliva (Wickström et al. 2009). For example, when the saccharolytic organism *S. oralis* was exposed to carbohydrate-deprived environments, this organism upregulated a number of proteolytic enzymes that help them to increase in numbers relative to other oral species (Beighton and Hayday 1986; Homer et al. 1990; Heinemann and Sauer 2010). This particular ability in *S. oralis* to digest protein could be considered as an advantage for their survival in the oral community at the times of carbohydrate famine. Phenotypic switching is an efficient strategy of bacteria to thrive in nutrient-limited environments, by a high frequency and reversible switch (ON/OFF) of the expression of one or more genes (Casadesus and Low 2013; Hammerschmidt et al. 2014). Phenotypic switching by oral bacteria in response to the availability of nutrients in the environment, e.g., in response to the lack of carbohydrates and presence of serum proteins, is an important characteristic that may explain the persistence of facultative anaerobic organisms in all clinical stages of the endodontic infection.

Studies of the root canal environment as a habitat filter will aid us to understand how oral microbes adapt their phenotypes in response to environmental changes to successfully colonize the root canal. Concentrating basic research on these adaptive mechanisms by, for example, using comparative genomic approaches to investigate habitat filtering, will help us relate changes in individual microbial genomes and the environments to which they are selected. In the future, this problem may be approached from a single-species genomics, carefully tested in a range of environmental conditions, and can be followed by the investigation of the complete genomes of representative multispecies communities obtained directly from root canal environments. Furthermore, with the advent of novel post-genomic techniques such as microbial metabolomics, the complete set of metabolites within a selected microorganism could be monitored, as well as their global outcome of interactions between its development processes and its environment (Takahashi et al. 2010). These results will finally help us elucidate the association between sequenced root canal microorganisms and the root canal habitat.

3 Disturbance and Selection of Resistant Bacteria

In ecology, disturbances are causal events that alter the immediate environment and have possible repercussions for a community of organisms (Gonzalez et al. 2011; Shade et al. 2012). Ecological disturbances may also directly alter a community by killing their members or change their relative abundances (Shade et al. 2012). Disturbances occur at various spatial and temporal scales with different frequencies, intensities, extents, and periodicities. Communities have nonlinear responses to disturbances that are mainly determined by their levels of resistance and resilience. Resistance is defined as the degree to which a community is insensitive to a disturbance (Ding and He 2010; Wardle and Jonsson 2014), and resilience is the rate at which a community returns to a pre-disturbance condition (see below resilience process). A related concept, sensitivity, is the inverse of resistance and defined as the degree of community change following a disturbance.

In endodontics, mechanical instrumentation in combination with chemical antimicrobial agents is a good example of an ecological disturbance. In this example, shear forces applied through direct contact of machine-driven files on the surfaces of the root canals aim to achieve physical removal of biofilm communities. Root canal biofilm control is further accomplished by the use of antimicrobials that aim to kill bacteria. Recently, microbiological research in endodontics has focused on evaluating the killing effect of chemicals with antimicrobial properties for disinfection (Kobayashi et al. 2014; Wang et al. 2014; Xhevdet et al. 2014). According to such studies, however, it is apparent that a portion of the microbial biofilm communities in root canals may tolerate and remain viable after treatment (see an schematic depiction of selection in Fig. 2). For example, a recent study showed that

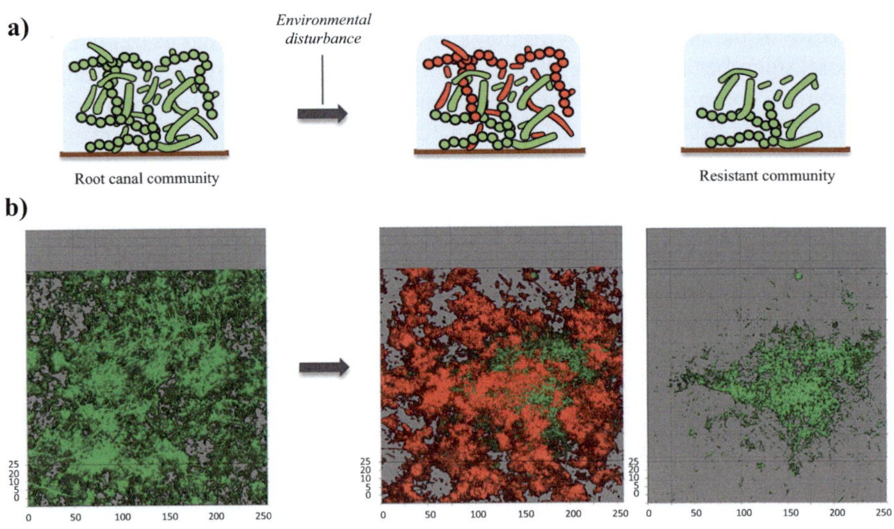

Fig. 2 Selection of resistant root canal bacteria. (**a**) Schematic depiction of the selective process in a root canal community by environmental disturbances. Environmental disturbances in endodontics are the shear forces by mechanical instrumentation, irrigation with antimicrobials, and intracanal medication. Affected cells are shown in red and the resistant bacteria selected after disturbance are shown in *light green*. (**b**) 3D reconstructions show microscopic structure of biofilms formed by oral bacteria grown on polystyrene surfaces and exposed to 2.5 % NaOCl and stained with the LIVE/DEAD stain (*green* (viable cells) and *red* (damaged cells)). The units on the axes are micrometers

it was not possible to eradicate wild strains of bacteria of endodontic origin grown in vitro using ampicillin, doxycycline, clindamycin, azithromycin, or metronidazole (Al-Ahmad et al. 2014).

Clinical studies have also confirmed the tolerance of selected members of microbial communities to endodontic procedures and that these organisms are likely to play a role in treatment failures (Engström et al. 1964; Gomes et al. 1996; Molander et al. 1998; Sundqvist et al. 1998; Sunde et al. 2002). Not surprisingly, bacteria that are tolerant to endodontic treatment are normal inhabitants of the oral cavity, with some exceptions like *E. faecalis* (Sedgley et al. 2004). Gram-positive facultative anaerobic bacteria from the genera *Streptococcus*, *Lactobacillus*, and *Actinomyces* are frequently recovered from root canals of teeth after treatment (Engström et al. 1964; Molander et al. 1998; Sundqvist et al. 1998; Chávez de Paz et al. 2003). The higher tolerance of Gram-positive bacteria may be related to different structural and physiological factors, for example, in cell-wall structure, innate resistance to antimicrobials, and phenotypic plasticity that allows them to adapt and endure harsh environmental conditions (Dessen et al. 2001; Berger-Bachi 2002). However, the information about the mechanisms involved specifically in tolerance to environmental disturbances in root canals is scarce. There have been few studies that have tried to model the influence of environmental conditions on microbial physiological responses and microbial community composition changes (Chávez de Paz 2007, 2012). In this section, the hypothesis will be introduced that a sudden change in the root canal environment will create conditions that are stressful for microorganisms and to which they are obliged to adapt for survival. Under such a selective ecological pressure, root canal microbes must have physiological adaptive mechanisms to survive and remain active in the face of this stress or they will die.

Adaptive Mechanisms of Resistance The main mechanisms of microbial resistance to survive disturbances in the environment rely on their ability to adapt their phenotype in the form of a rapid physiological response. In general, the cellular machinery of bacteria is prepared to change in response to various types of environmental threats such as shifts from aerobic to anaerobic conditions and rapid fluctuations of pH, temperature, and osmotic conditions (Bowden and Hamilton 1998; Marsh 2003). In the case of changes provoked by antimicrobials, immediate responses can often be achieved by regulation of the activities of preformed enzymes (Svensäter et al. 2001). In the case of mechanisms of tolerance by root canal bacteria, a recent study analyzed the survival of a selected group of root canal bacteria in biofilms under alkaline stress. In this study, it was observed that biofilm bacteria resisted by releasing specific enzymes out into the environment (Chávez de Paz et al. 2007). Cytoplasmic housekeeping enzymes, such as phosphocarrier HPr, the heat-shock chaperone DnaK, FBA, and GAPDH, were the most frequently identified proteins. Although the physiological role of these housekeeping enzymes outside the cell is presently unknown, most of these enzymes have also been found to be associated with the bacterial response to other similar environmental stresses such as acid challenge. Hence, it is not unreasonable to consider that the molecular mechanisms of stress response are orchestrated concomitantly from a main general stress response with the interplay of various regulatory processes taking place at the same time.

In the case of continual changes in the environment, for example, with the application of intracanal medicaments over a period of time, resistance may be also accomplished by alterations in the pattern of gene expression. This can be accomplished via an operon, where all related genes are located adjacent to each other in the chromosome and transcribed as a single transcript controlled by a single promoter site, or by means of regulatory units that utilize genes situated in different locations on the chromosome (Ghazaryan et al. 2014; Raivio 2014). Unfortunately, for most of the root canal microorganisms, we do not even have a minimal view of these fundamental molecular adaptive processes. Even in general medical microbiology there are actually not many microorganisms whose physiology is thoroughly understood and little is known about conditions prevailing in biofilms comprised of multispecies communities such as those in the root canals of teeth.

Recent studies on the whole genomes of a number of oral microorganisms have shown that there are more similarities than differences in the way bacteria handle stress (Jenkinson 2011; Zaura 2012; Wade 2013). As discussed above, an important feature of adaptation and survival of bacterial cells in stressful environments seems to be the expression of a range of proteins that promote the survival of the cells (Hamilton and Svensäter 1998). To understand how a single microbial cell is able to cope with an ecological disturbance within a multispecies community is an overwhelming challenge since adaptation or response to stress may take place at different levels in the community and vary in intensity among its members.

Tolerance to Antimicrobials by Biofilm Communities The physiology of a microbial community, like the one established in root canals, is certainly distinct from the physiology of individual members as the community lifestyle provides advantages compared to those of the component populations. In a multispecies community, the ranges of potential habitats for colonization are extended, resistance to stress and host defenses increase, and cooperative degradation of complex substrates can take place (Marsh 2003). Elucidating the physiologies of biofilm-associated communities is necessary for our understanding of infection and survival of bacteria in a changing environment.

In endodontics, there is an increasing interest in studying the effect of antimicrobials on multispecies biofilm communities. Studies have shown that mixed root canal microbial communities are variably "resistant" to disturbances, as measured by viability of the biofilm cells (Chávez de Paz 2012; Stojicic et al. 2013; Shrestha and Kishen 2014). Generally, these ex vivo studies that explored the effect of antimicrobials on multispecies communities are observational and typically involve large-scale antimicrobial disturbances or nutrient starvation (e.g., sodium hypochlorite, chlorhexidine, glucose starvation, etc.).

In a recent study, the phenotypic response of a multispecies biofilm model using four root canal bacterial isolates to the absence of glucose was determined (Chávez de Paz 2012). The results of this study showed a significant variation in the three-dimensional structure of the multispecies biofilms in response to the absence of glucose. In addition, physiological adaptation by members of the community to glucose depletion was observed. The metabolic activity was concentrated in the upper levels of the biofilms, while at lower levels, the metabolism of cells was considerably decreased. Subpopulations of species with high glycolytic demands, such as streptococci and lactobacilli, were found predominating in the upper levels of the biofilms. This distinct spatial organization in biofilms grown in the absence of glucose shows a clear reorganization of the community in order to satisfy their members' metabolism in order to enable the long-term persistence of the community. This result lends support to the hypothesis that the reorganization of subpopulations of cells in multispecies biofilms is also important for survival to stress from the environment (Shapiro 2007).

The results of these in vitro studies, however, suggest that we have still much to learn about the physiological adaptive mechanisms orchestrated by root canal microbial communities. In addition, only few studies have implemented multispecies models to investigate the compositional and functional responses to disturbance by a community of bacteria, which hinders more quantitative cross-system comparisons (see methodological review in chapter "Laboratory Models of Biofilms: Development and Assessment" and alternative ecological methods below). By developing laboratory experimentation on multispecies microbial communities, the implications of ecological disturbance screenings and their effect on root canal bacteria will advance. After these methodological systems are successfully established and results from different research groups are correspondingly replicated, the question arises as to how and at what level these artificial root canal multispecies communities (and data obtained from them) can be compared to their counterparts in the original environment.

Nevertheless, from classical studies on general biofilm biology, we know that the problem on the relative tolerance of bacteria, especially when growing as a biofilm, to antimicrobial agents is accounted for due to transport-based and physiology-based mechanisms or a combination (Mah and O'Toole 2001). Transport-based mechanisms indicate that the biofilms act as barrier to antibiotic/antimicrobial diffusion, although the main attributes of this mechanism rely on the features that govern transport rates and generate structural, chemical, and biological heterogeneity in biofilm communities (Stewart and Franklin 2008). Heterogeneity in biofilm communities is a result of the distinct metabolic activities of the cells that provoke different concentration gradients of nutrients and local chemical conditions (for a review in biofilm heterogeneity see (Stewart and Franklin 2008)).

Inherent Resistance to Ecological Disturbances Inherent resistance involves evolutionary selection of a growth form and a history strategy that allows a microbe to resist disturbances without having to induce specific mechanisms at the time of the disturbance. Developing such inherent resistance invariably involves physiological trade-offs that affect microbial function (Mah and O'Toole 2001). Among endodontically isolated organisms, for example, *E. faecalis* is thought to be more inherently resistant to alkaline stress than oxygen-sensitive Gram-negative bacteria.

The mechanisms behind the innate resistance of enterococci to alkaline pH are thought to include the activation of specific proton pumps and specific enzymatic systems and/or buffering devices that help to keep the internal pH neutral (Kayaoglu and Örstavik 2004). In a recent study, however, it was observed that in response to alkaline pH, a general transcriptional process including the expression of housekeeping genes, such as *dnaK* and GroEL, and the cytoskeletal molecule, *ftsZ*, took place in *E. faecalis* (Appelbe and Sedgley 2007). Thus, it would seem that a network of regulatory interactions including central cytoskeletal processes and expression of chaperones regulate the response of *E. faecalis* to alkaline stress (see chapter "Antimicrobial Resistance in Biofilm Communities"). Knowing the transcriptional regulatory network in this organism could aid in understanding central adaptive regulatory operations within a root canal biofilm community.

In conclusion, it is of importance to establish a stronger connection between microbial resistance to ecological disturbances and root canal ecology. With an enhanced understanding of microbial physiological responses to stress provoked by clinical procedures in endodontics, we will have a better understanding of the mechanisms employed by bacteria in response to antimicrobial therapies. Some questions that remain unanswered are, for example, how does physiological resistance to stress vary among microbial communities in root canals? How do those patterns of resistance relate to ecosystem-level consequences in response to stress?

4 Resilience of Root Canal Microbial Communities

This third ecological process addresses the concept of resilience of root canal microbial communities that have been selected by environmental disturbances (see above) and that have resisted antimicrobial endodontic treatment. The resilience of a microbial community focuses on its capacity to surmount ecological disturbances and still preserve viability and physiological function. But there is also another aspect of bacterial resilience that concerns the capacity for regrowth, reorganization, and development, which in the case of endodontic infections is essential for maintaining chronic inflammatory periapical lesions (Fig. 3).

In a resilient root canal microbial community, ecological disturbances may cause important physiological consequences in cells. For instance, it is likely that in the resilient community, stress-adapted cells may differentiate into low physiological states or dormancy. In these states of low metabolism, bacteria are implicitly driven to stasis to thrive in environments where nutrient resources may be scarce.

Dormancy and Adaptation to Starvation Bacteria under the stress of starvation have developed efficient adaptive regulatory reactions to shift their metabolic balance away from biosynthesis and reproduction, toward the acquisition of energy for essential biological functions (Matin 1992; Nyström 1999). Under nutrient limitation, bacteria rapidly reallocate cellular resources by stopping the synthesis of DNA, stable RNAs, ribosomal proteins, and membrane components (Potrykus

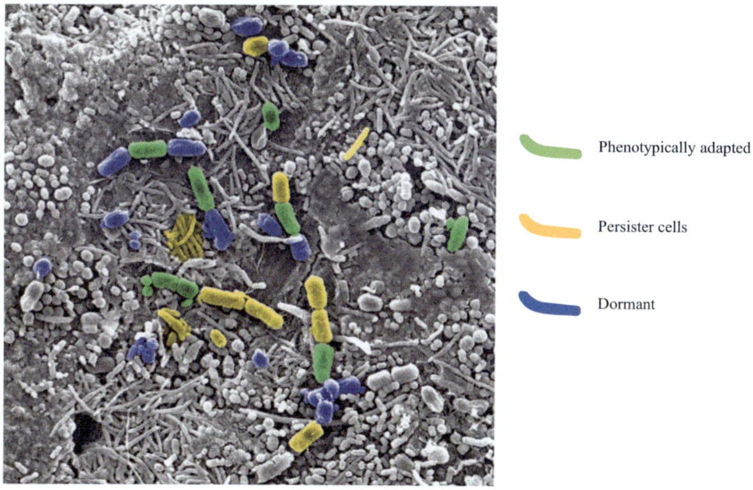

Fig. 3 Scanning electron microscopy (SEM) image of a resilient root canal microbial community in the apex of an infected root canal. The SEM image show *false colors of green, yellow*, and *blue* representing cells that are phenotypically adapted, persister, and dormant, respectively

and Cashel 2008). This effective responsive process to nutrient stress, termed "the stringent response," is characterized by the production of factors that are crucial for stress resistance, glycolysis, and amino acid synthesis (Dalebroux and Swanson 2012). The stringent response is accomplished in part by a massive switch in the transcription profile, coordinated by an effective alarmone system that includes the nucleotides guanosine tetraphosphate (ppGpp) and guanosine pentaphosphate (pppGpp). (p)ppGpp plays an important role in low-nutrient survival of *E. faecalis*, an organism that is known to withstand prolonged periods of starvation and remain viable in root-filled teeth for at least 12 months (Molander et al. 1998; Sundqvist et al. 1998). Furthermore, the alarmone system (p)ppGpp has also a profound effect on the ability of *E. faecalis* to form, develop, and maintain stable biofilms (Chávez de Paz et al. 2012). These improved understandings of the alarmone mechanisms underlying biofilm formation and survival by *E. faecalis* may facilitate the identification of pathways that could be targeted to control persistent infections by this organism.

From a physiological perspective, nutrient deprivation causes bacteria to reversibly switch to a state of metabolic arrest (dormancy) (Nyström 1999). In the dormant phenotype, bacteria will survive a wide range of environmental threats, in addition to deprivation of nutrients, such as temperature shifts and extreme pH changes, as well as exhibiting decreased sensitivity to antimicrobial agents (Stewart and Franklin 2008). When the nutrient supply is favorable again, the stress response is released and the bacteria resume metabolic activity and cell division. A large amount of RNA and protein appears to be degraded rapidly at the onset of starvation, which is believed to be part of a general stress response (GSR) that is connected to survival responses in changing environments like the oral cavity (Bowden and Hamilton 1998).

In some cases, the occurrence of persister cells has been identified. Persistence is a feature where bacteria that are phenotypically susceptible to antibiotics are not effectively eliminated upon exposure to high doses of those drugs. Persister cells do not have specific regulatory mechanisms of resistance, but they undergo general physiological changes, like diminishing their metabolism similar to a dormant state. The persistent phenotype is believed to be responsible for the recalcitrant nature and therapy unresponsiveness of several chronic infections (see chapters "Molecu lar Principles of Adhesion and Biofilm Formation" and "Antimicrobial Resistance in Biofilm Communities").

Metabolic Reactivation In cases with a chronic periapical infection that suddenly reactivates and causes an acute inflammatory response after many years, it is reasonable to assume that resilient dormant cells have "woken up" and resumed their metabolic activity to provoke acute periapical inflammation. Thus, from the metabolic perspective, the reactivation of dormant cells will render biofilm bacteria able to contribute to the persistence of inflammation. For example, a recent case report of a tooth that was adequately treated and showed no signs of disease revealed recurrent disease after 12 years. Histopathological analyses showed a heavy dentinal tubule infection surrounding the area of a lateral canal providing evidence on the persistence of an intra-radicular infection caused by bacteria possibly located in dentinal tubules (Vieira et al. 2012). This hypothesis on the metabolic reactivation of biofilm cells was tested in a recent study (Chávez de Paz et al. 2008). Biofilm cultures of oral isolates of *Streptococcus anginosus* and *Lactobacillus salivarius* were forced to enter a state of dormancy by exposing them to nutrient deprivation for 24 h in PBS buffer. After the starvation period the number of metabolically active cells decreased dramatically to zero and their cell membrane integrity was kept intact. Biofilm cells were then exposed to a "reactivation period" with fresh nutrients, but even after 96 h, the cultures were dominated by undamaged cells that were metabolically inactive. The data produced by this study showed that starved biofilm cells exhibit a slow physiological response and do not reactivate in short time periods even in the presence of fresh nutrients. This observation confirms the slower physiological response of biofilm cells, which may act as a strategic mechanism to resist further disturbances (Mah 2012).

In conclusion, global regulators of bacterial physiology are involved in microbial community resilience and have important roles in biofilm reorganization, virulence, and antibiotic resistance. All these molecular processes can be taken into consideration in the development of treatment strategies for bacterial infections resistant to conventional antimicrobial root canal treatment.

5 Methods to Analyze Microbial Ecosystems

In order to monitor physiological responses of organisms in communities and understand their relevance to resist and overcome environmental disturbances, it is necessary to build up a strong database of information describing the physiology of the participating organisms under controlled conditions. Many physiological properties of bacteria can be investigated by means of common microscopic tools and analytic strategies. Modern molecular tools offer approaches to in situ studies of specific physiological processes in the presence of essential nutrients or in disturbed environments (e.g., after application of antimicrobials). This section describes some of the most common microbiological methods to analyze microbial ecosystems in situ and under laboratory conditions.

Scanning Electron Microscopy (SEM) Modern techniques to analyze microbial ecosystems comprise a variety of traditional and modern microscopy techniques. For example, electron microscopy has provided a vast amount of information on the structure of microbial ecosystems. As discussed in chapter "The Use of Scanning Electron Microscopy (SEM) in Visualizing the Root Canal Biofilm," the use of SEM analysis has increased due to its rapidity and sensitivity to detect structural changes in microbial ecosystems. As seen in Fig. 3, imaging of the intra-radicular biofilm alongside a segment of an infected root canal by scanning electron microscopy clearly demonstrates the heterogeneous architecture of the oral biofilm. The biofilm is adherent adjacent to the dentinal tubule lumen and is characterized by cocci, filaments, and both yeast and hyphal cell-forming networks of extracellular matrix strands. In specific sections of the image, densely packed cells are accumulated surrounded by extracellular material. Although detailed qualitative information is obtained from these types of SEM images, other studies have showed limitations of the SEM technique. In 1994, Sutton et al. (1994) compared conventional scanning electron microscopy (SEM), low-temperature SEM, and electroscan wet mount SEM in monocultures of *S. crista* to expose large differences in the final grayscale image. It was observed that under natural conditions, extracellular polymeric substances (EPS) take over the resulting image, not yet allowing exploration of the cellular distribution in the biofilm below. The use of SEM to analyze oral biofilms and infected root canals is further reviewed in chapter "The Use of Scanning Electron Microscopy (SEM) in Visualizing the Root Canal Biofilm."

Confocal Scanning Laser Microscopy (CSLM) Confocal scanning laser microscopy (CSLM) has become the preferred technique to study the architecture of biofilms because it provides a powerful microscopy tool to analyze microbial communities in situ. Usually CSLM is applied with fluorescent probe techniques that take advantage of the optical geometry construction of the CSLM. The coherent light beams of CSLM have a very narrow depth of focus at the same time as all out-of-focus information is discarded. CSLM produces a series of narrow focal planes that are recorded at different depths throughout a three-dimensional sample (Neu et al. 2010). Subsequently, the single-plane images can be assembled using image-processing techniques to generate three-dimensional digitized images. These three-dimensional reconstructions of microbial biofilm sections allow the in-depth profile of a biofilm sample in situ. These techniques have helped reveal the highly heterogeneous structure of microbial biofilm.

Fluorescent Probes The above-discussed CSLM technique is usually applied in combination with fluorescent probes to discriminate between classes, genera, species, and also the viability of individual organisms present in the microbial ecosystem. Furthermore, chemical interactions within the biofilm can also be monitored. Common fluorescent probes include negative stains such as fluorescein, which provides a fluorescent background upon which the bacteria can be viewed as unstained cells. Other agents such as resazurin are used to distinguish between "live" and "dead" cells (Netuschil et al. 2014). Actively metabolizing cells reduce resazurin to a colorless nonfluorescent form, in contrast to dead cells, which maintain the fluorescent dye in their cytoplasm.

CSLM in combination with commercial fluorescent probes can aid in distinguishing between living and dead organisms within a microbial biofilm community. Although in traditional microbiology, a living cell could only be determined as one that can grow and reproduce, in the end developing a colony, with the advent of fluorescence microscopy, it is assumed that organisms capable of catalyzing fluorescent metabolites are metabolically active. For example, tetrazolium salts are markers that target oxidation/reduction reactions. In contrast, cell membrane integrity can be investigated with the widely used commercial agent, the LIVE/DEAD BacLight viability probe. In this technique, undamaged cells fluoresce green, whereas cells whose membrane structure is damaged (but not necessarily dead) fluoresce red (Fig. 2b).

Other more sophisticated fluorescent techniques include fluorophores that are linked to other agents in order to specify their target elements. For example, conjugated lectins can be used to determine the distribution of oligosaccharides in the biofilm matrix (Neu and Lawrence 2014). Monoclonal antibodies attached to fluorophores can also be used to determine the location of species within a biofilm (Chalmers et al. 2007). A more advanced technique is to use fluorophores attached to 16S rRNA oligonucleotide sequences in order to identify bacterial species in situ (see below).

In recent years, with the development of super-resolution microscopy, a suite of cutting edge microscopy methods that are able to surpass the resolution limits of light microscopy have dramatically improved both the localization and quantification of target molecules in single cells (Moraru and Amann 2012). In a pilot study, a combination of super-resolution microscopy and rRNA targeted oligonucleotide probing provided the subcellular localization of ribosomes in *E. coli* (Moraru and Amann 2012). It was observed that ribosomes were localized surrounding the central nucleoid and that some of the cells have two distinct nucleoids. The inter-nucleoid rRNA indicated the position of the division septum, most probably following rRNA localization along the cell membrane during the division processes. This highly advanced technique could be used to allow the tracking of ribosome-associated changes in activity levels and subcellular localization at the single-cell level in complex microbial communities. These would give insights into variations occurring across community members and after different environmental conditions.

Fluorescence In Situ Hybridization (FISH) The microbial compositions of biofilm communities, such as those growing in the oral cavity or root canal of teeth, are generally diverse. Thus, it is imperative that in situ determination of the different species present and their distribution in a three-dimensional space are accomplished for subsequent analysis and interpretations. Figure 4 depicts the main methodological approach for identification of relevant organisms in biofilm communities by fluorescence in situ hybridization (FISH). FISH allows for the simultaneous detection of phylogenetically different bacteria. This method detects bacteria at the species, genus, and family levels, and FISH with oligonucleotide probes based on ribosomal RNA (rRNA) specifically identifies targeted bacteria (Brileya et al. 2014).

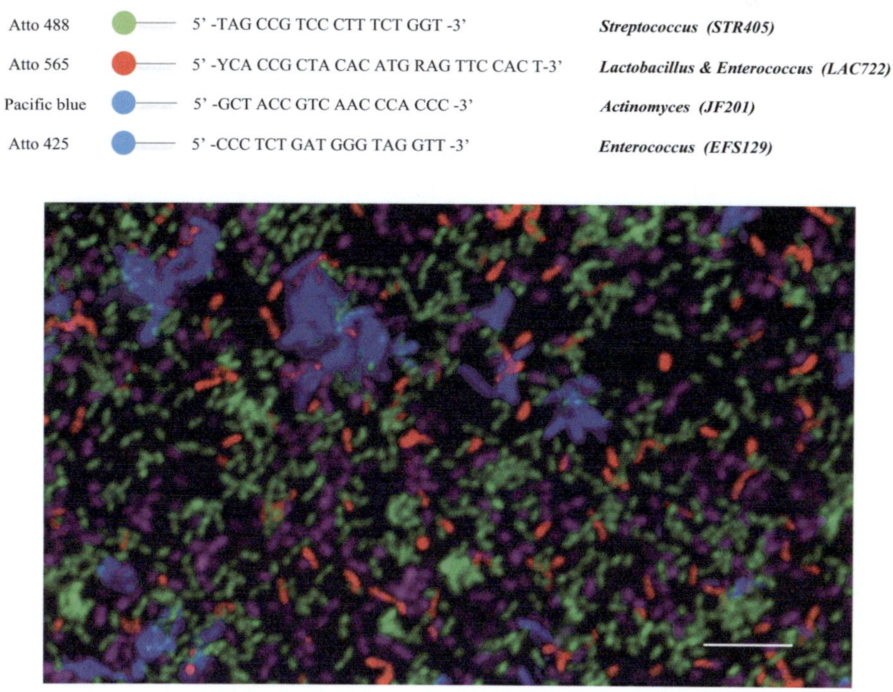

Fig. 4 Identification of bacteria by fluorescence in situ hybridization (FISH). 16S rRNA fluorescent probes used to identify four root canal bacteria on multispecies biofilms cultures, *Streptococcus gordonii* (*green*), *Lactobacillus salivarius* (*red*), *Actinomyces naeslundii* (*blue*), and *Enterococcus faecalis* (*violet*). The red probe for *Lactobacillus* targeted also *Enterococcus*; hence, the addition of an additional blue probe for *Enterococcus* shows labeling results in *violet* fluorescence. Bar = 10 μm

With the aid of the FISH technique, the identification of individual bacterial cells within a community will represent an important advantage in order to understand the organization of microbial ecosystems. Oligonucleotide probes that are designed to target specific regions of the 16S rRNA gene are then labeled with specific fluorescent dyes. Different probes can be generated: for example, one that recognizes the *Lactobacillus*, another for *Streptococcus*, and successively more specific probes for particular groups of bacteria right down to individual species. Figure 4 illustrates an example of four-species biofilm targeted with a cocktail of fluorescent oligonucleotide probes to detect *L. salivarius* (red), *S. gordonii* (green), *A. naeslundii* (blue), and *E. faecalis* (violet) to map the diversity of a root canal microbial population.

In conclusion, with the FISH technique, a nondestructive identification of a complex microbial population could be accomplished.

6 Concluding Remarks

Understanding the adaptive mechanisms and implications of resistance in root canal microbial biofilm communities depends on research into the ecological and physiological processes occurring in the root canal ecosystem. This era is an exciting time for microbial ecology research because the complete genomes of many oral pathogens have been sequenced and are available for analysis on diverse laboratory setups. It is therefore possible now to investigate regulatory genes in root canal bacteria, including those needed to establish and adapt to different environmental disturbances, so investigators will soon be able to analyze and monitor the responsiveness of bacteria to environmental threats, e.g., antimicrobials used in endodontics. The availability of replication origins, chromosome ends, and many of the genes for DNA, RNA, and protein synthesis will contribute to studies of basic physiological responses during colonization, resistance, and resilience, as well as providing access to some of the key elements in gene regulation and root canal biofilm formation. The near future should see progress toward a clearer understanding of how interspecies interactions lead to the coordination of physiological events in microbial communities.

Acknowledgments The helpful suggestions of G. Bergenholtz are greatly appreciated.

References

Al-Ahmad A, Ameen H, Pelz K, Karygianni L, Wittmer A, Anderson AC, Spitzmuller B, Hellwig E (2014) Antibiotic resistance and capacity for biofilm formation of different bacteria isolated from endodontic infections associated with root-filled teeth. J Endod 40(2):223–230

Appelbe OK, Sedgley CM (2007) Effects of prolonged exposure to alkaline pH on *Enterococcus faecalis* survival and specific gene transcripts. Oral Microbiol Immunol 22(3):169–174

Baumgartner JC, Falkler WA Jr (1991) Bacteria in the apical 5 mm of infected root canals. J Endod 17(8):380–383

Beighton D, Hayday H (1986) The influence of diet on the growth of streptococcal bacteria on the molar teeth of monkeys (*Macaca fascicularis*). Arch Oral Biol 31(7):449–454

Bergenholtz G (1974) Micro-organisms from necrotic pulp of traumatized teeth. Odontol Revy 25 (4):347–358

Bergenholtz G (2001) Factors in pulpal repair after oral exposure. Adv Dent Res 15:84

Berger-Bachi B (2002) Resistance mechanisms of gram-positive bacteria. Int J Med Microbiol 292 (1):27–35

Bowden GH, Hamilton IR (1998) Survival of oral bacteria. Crit Rev Oral Biol Med 9(1):54–85

Bradshaw DJ, Homer KA, Marsh PD, Beighton D (1994) Metabolic cooperation in oral microbial communities during growth on mucin. Microbiology 140(Pt 12):3407–3412

Brileya KA, Camilleri LB, Fields MW (2014) 3D-fluorescence in situ hybridization of intact, anaerobic biofilm. Methods Mol Biol 1151:189–197

Casadesus J, Low DA (2013) Programmed heterogeneity: epigenetic mechanisms in bacteria. J Biol Chem 288(20):13929–13935

Chalmers NI, Palmer RJ Jr, Du-Thumm L, Sullivan R, Shi W, Kolenbrander PE (2007) Use of quantum dot luminescent probes to achieve single-cell resolution of human oral bacteria in biofilms. Appl Environ Microbiol 73(2):630–636

Chávez de Paz LE (2007) Redefining the persistent infection in root canals: possible role of biofilm communities. J Endod 33(6):652–662

Chávez de Paz LE (2012) Development of a multispecies biofilm community by four root canal bacteria. J Endod 38(3):318–323

Chávez de Paz LE, Dahlén G, Molander A, Möller A, Bergenholtz G (2003) Bacteria recovered from teeth with apical periodontitis after antimicrobial endodontic treatment. Int Endod J 36 (7):500–508

Chávez de Paz LE, Bergenholtz G, Dahlén G, Svensäter G (2007) Response to alkaline stress by root canal bacteria in biofilms. Int Endod J 40(5):344–355

Chávez de Paz LE, Hamilton IR, Svensäter G (2008) Oral bacteria in biofilms exhibit slow reactivation from nutrient deprivation. Microbiology 154(Pt 7):1927–1938

Chávez de Paz LE, Lemos JA, Wickström C, Sedgley CM (2012) Role of (p)ppGpp in biofilm formation by *Enterococcus faecalis*. Appl Environ Microbiol 78(5):1627–1630

Chubukov V, Gerosa L, Kochanowski K, Sauer U (2014) Coordination of microbial metabolism. Nat Rev Microbiol 12(5):327–340

Chugal N, Wang JK, Wang R, He X, Kang M, Li J, Zhou X, Shi W, Lux R (2011) Molecular characterization of the microbial flora residing at the apical portion of infected root canals of human teeth. J Endod 37(10):1359–1364

Dahlén G, Fabricius L, Holm SE, Möller A (1987) Interactions within a collection of eight bacterial strains isolated from a monkey dental root canal. Oral Microbiol Immunol 2 (4):164–170

Dalebroux ZD, Swanson MS (2012) ppGpp: magic beyond RNA polymerase. Nat Rev Microbiol 10(3):203–212

Dessen A, Di Guilmi AM, Vernet T, Dideberg O (2001) Molecular mechanisms of antibiotic resistance in gram-positive pathogens. Curr Drug Targets Infect Disord 1(1):63–77

Ding C, He J (2010) Effect of antibiotics in the environment on microbial populations. Appl Microbiol Biotechnol 87(3):925–941

Engström B, Hård AF, Segerstad L, Ramström G, Frostell G (1964) Correlation of positive cultures with the prognosis for root canal treatments. Odontol Revy 15:257–270

Fabricius L, Dahlén G, Sundqvist G, Happonen RP, Möller AJ (2006) Influence of residual bacteria on periapical tissue healing after chemomechanical treatment and root filling of experimentally infected monkey teeth. Eur J Oral Sci 114(4):278–285

Figdor D, Sundqvist G (2007) A big role for the very small–understanding the endodontic microbial flora. Aust Dent J 52(1 Suppl):S38–S51
Ghazaryan L, Tonoyan L, Ashhab AA, Soares MI, Gillor O (2014) The role of stress in colicin regulation. Arch Microbiol 196(11):753–764
Gomes BP, Lilley JD, Drucker DB (1996) Variations in the susceptibilities of components of the endodontic microflora to biomechanical procedures. Int Endod J 29(4):235–241
Gonzalez A, Clemente JC, Shade A, Metcalf JL, Song S, Prithiviraj B, Palmer BE, Knight R (2011) Our microbial selves: what ecology can teach us. EMBO Rep 12(8):775–784
Hamilton IR, Svensäter G (1998) Acid-regulated proteins induced by Streptococcus mutans and other oral bacteria during acid shock. Oral Microbiol Immunol 13(5):292–300
Hammerschmidt K, Rose CJ, Kerr B, Rainey PB (2014) Life cycles, fitness decoupling and the evolution of multicellularity. Nature 515(7525):75–79
Heinemann M, Sauer U (2010) Systems biology of microbial metabolism. Curr Opin Microbiol 13 (3):337–343
Homer KA, Manji F, Beighton D (1990) Inhibition of protease activities of periodontopathic bacteria by extracts of plants used in Kenya as chewing sticks (mswaki). Arch Oral Biol 35 (6):421–424
Jenkinson HF (2011) Beyond the oral microbiome. Environ Microbiol 13(12):3077–3087
Jyoti A, Singh AK, Dubey M, Kumar S, Saluja R, Keshari RS, Verma A, Chandra T, Kumar A, Bajpai VK, Barthwal MK, Dikshit M (2014) Interaction of inducible nitric oxide synthase with rac2 regulates reactive oxygen and nitrogen species generation in the human neutrophil phagosomes: implication in microbial killing. Antioxid Redox Signal 20(3):417–431
Kayaoglu G, Örstavik D (2004) Virulence factors of *Enterococcus faecalis*: relationship to endodontic disease. Crit Rev Oral Biol Med 15(5):308–320
Kim HA, Kim SH, Ko HM, Choi JH, Kim KJ, Oh SH, Cho KO, Choi IW, Im SY (2008) Nitric oxide plays a key role in the platelet-activating factor-induced enhancement of resistance against systemic candidiasis. Immunology 124(3):428–435
Kobayashi Y, Hayashi M, Yoshino F, Tamura M, Yoshida A, Ibi H, Lee MC, Ochiai K, Ogiso B (2014) Passive ultrasonic irrigation in the presence of a low concentration of hydrogen peroxide enhances hydroxyl radical generation and bactericidal effect against Enterococcus faecalis. J Oral Sci 56(1):35–39
Kolpen M, Bjarnsholt T, Moser C, Hansen CR, Rickelt LF, Kuhl M, Hempel C, Pressler T, Hoiby N, Jensen PO (2014) Nitric oxide production by polymorphonuclear leucocytes in infected cystic fibrosis sputum consumes oxygen. Clin Exp Immunol 177(1):310–319
Liu H, Liang X, Wang D, Zhang H, Liu L, Chen H, Li Y, Duan Q, Xie K (2015) Combination therapy with nitric oxide and molecular hydrogen in a murine model of acute lung injury. Shock 43(5):504–511
Mah TF (2012) Biofilm-specific antibiotic resistance. Future Microbiol 7(9):1061–1072
Mah TF, O'Toole GA (2001) Mechanisms of biofilm resistance to antimicrobial agents. Trends Microbiol 9(1):34–39
Marsh PD (2003) Are dental diseases examples of ecological catastrophes? Microbiology 149 (Pt 2):279–294
Matin A (1992) Physiology, molecular biology and applications of the bacterial starvation response. Soc Appl Bacteriol Symp Ser 21:49S–57S
Molander A, Reit C, Dahlén G, Kvist T (1998) Microbiological status of root-filled teeth with apical periodontitis. Int Endod J 31(1):1–7
Möller ÅJR (1966) Microbiological examination of root canals and periapical tissues of human teeth. Methodological studies. Odontol Tidskr 74(5):Suppl:1–380
Möller AJ, Fabricius L, Dahlén G, Öhman AE, Heyden G (1981) Influence on periapical tissues of indigenous oral bacteria and necrotic pulp tissue in monkeys. Scand J Dent Res 89(6):475–484
Moraru C, Amann R (2012) Crystal ball: fluorescence in situ hybridization in the age of super-resolution microscopy. Syst Appl Microbiol 35(8):549–552

Munson MA, Pitt-Ford T, Chong B, Weightman A, Wade WG (2002) Molecular and cultural analysis of the microflora associated with endodontic infections. J Dent Res 81(11):761–766
Netuschil L, Auschill TM, Sculean A, Arweiler NB (2014) Confusion over live/dead stainings for the detection of vital microorganisms in oral biofilms–which stain is suitable? BMC Oral Health 14:2
Neu TR, Lawrence JR (2014) Investigation of microbial biofilm structure by laser scanning microscopy. Adv Biochem Eng Biotechnol 146:1–51
Neu TR, Manz B, Volke F, Dynes JJ, Hitchcock AP, Lawrence JR (2010) Advanced imaging techniques for assessment of structure, composition and function in biofilm systems. FEMS Microbiol Ecol 72(1):1–21
Nyström T (1999) Starvation, cessation of growth and bacterial aging. Curr Opin Microbiol 2 (2):214–219
Pearl JE, Torrado E, Tighe M, Fountain JJ, Solache A, Strutt T, Swain S, Appelberg R, Cooper AM (2012) Nitric oxide inhibits the accumulation of CD4 + CD44hiTbet + CD69lo T cells in mycobacterial infection. Eur J Immunol 42(12):3267–3279
Potrykus K, Cashel M (2008) (p)ppGpp: still magical? Annu Rev Microbiol 62:35–51
Raivio TL (2014) Everything old is new again: an update on current research on the Cpx envelope stress response. Biochim Biophys Acta 1843(8):1529–1541
Rocas IN, Siqueira JF Jr (2010) Identification of bacteria enduring endodontic treatment procedures by a combined reverse transcriptase-polymerase chain reaction and reverse-capture checkerboard approach. J Endod 36(1):45–52
Sedgley CM, Lennan SL, Clewell DB (2004) Prevalence, phenotype and genotype of oral enterococci. Oral Microbiol Immunol 19(2):95–101
Shade A, Peter H, Allison SD, Baho DL, Berga M, Burgmann H, Huber DH, Langenheder S, Lennon JT, Martiny JB, Matulich KL, Schmidt TM, Handelsman J (2012) Fundamentals of microbial community resistance and resilience. Front Microbiol 3:417
Shapiro JA (2007) Bacteria are small but not stupid: cognition, natural genetic engineering and socio-bacteriology. Stud Hist Philos Biol Biomed Sci 38(4):807–819
Shrestha A, Kishen A (2014) Antibiofilm efficacy of photosensitizer-functionalized bioactive nanoparticles on multispecies biofilm. J Endod 40(10):1604–1610
Spratt DA (2004) Molecular analysis in endodontic infections. Endod Topics 9
Stewart PS, Franklin MJ (2008) Physiological heterogeneity in biofilms. Nat Rev Microbiol 6 (3):199–210
Stojicic S, Shen Y, Haapasalo M (2013) Effect of the source of biofilm bacteria, level of biofilm maturation, and type of disinfecting agent on the susceptibility of biofilm bacteria to antibacterial agents. J Endod 39(4):473–477
Sunde PT, Olsen I, Debelian GJ, Tronstad L (2002) Microbiota of periapical lesions refractory to endodontic therapy. J Endod 28(4):304–310
Sundqvist G (1976) Bacteriological studies of necrotic dental pulps. Umeå University, Umeå, Sweden
Sundqvist G (1992) Ecology of the root canal flora. J Endod 18(9):427–430
Sundqvist G (1994) Taxonomy, ecology, and pathogenicity of the root canal flora. Oral Surg Oral Med Oral Pathol 78(4):522–530
Sundqvist G, Figdor D (2003) Life as an endodontic pathogen. Endod Top 6(1):3–28
Sundqvist G, Figdor D, Persson S, Sjögren U (1998) Microbiologic analysis of teeth with failed endodontic treatment and the outcome of conservative re-treatment. Oral Surg Oral Med Oral Pathol Oral Radiol Endod 85(1):86–93
Sutton NA, Hughes N, Handley PS (1994) A comparison of conventional SEM techniques, low temperature SEM and the electroscan wet scanning electron microscope to study the structure of a biofilm of Streptococcus crista CR3. J Appl Bacteriol 76(5):448–454
Svensäter G, Bergenholtz G (2004) Biofilms in endodontic infections. Endod Topics 9:27–36
Svensäter G, Welin J, Wilkins JC, Beighton D, Hamilton IR (2001) Protein expression by planktonic and biofilm cells of *Streptococcus mutans*. FEMS Microbiol Lett 205(1):139–146

Takahashi N, Washio J, Mayanagi G (2010) Metabolomics of supragingival plaque and oral bacteria. J Dent Res 89(12):1383–1388
Vieira AR, Siqueira JF Jr, Ricucci D, Lopes WS (2012) Dentinal tubule infection as the cause of recurrent disease and late endodontic treatment failure: a case report. J Endod 38(2):250–254
Wade WG (2013) The oral microbiome in health and disease. Pharmacol Res 69(1):137–143
Wang Z, Shen Y, Haapasalo M (2014) Dentin extends the antibacterial effect of endodontic sealers against Enterococcus faecalis biofilms. J Endod 40(4):505–508
Wardle DA, Jonsson M (2014) Long-term resilience of above- and below ground ecosystem components among contrasting ecosystems. Ecology 95(7):1836–1849
Wickström C, Hamilton IR, Svensäter G (2009) Differential metabolic activity by dental plaque bacteria in association with two preparations of MUC5B mucins in solution and in biofilms. Microbiology 155(Pt 1):53–60
Xhevdet A, Stubljar D, Kriznar I, Jukic T, Skvarc M, Veranic P, Ihan A (2014) The disinfecting efficacy of root canals with laser photodynamic therapy. J Lasers Med Sci 5(1):19–26
Zagryazhskaya AN, Lindner SC, Grishina ZV, Galkina SI, Steinhilber D, Sud'ina GF (2010) Nitric oxide mediates distinct effects of various LPS chemotypes on phagocytosis and leukotriene synthesis in human neutrophils. Int J Biochem Cell Biol 42(6):921–931
Zaura E (2012) Next-generation sequencing approaches to understanding the oral microbiome. Adv Dent Res 24(2):81–85

Molecular Principles of Adhesion and Biofilm Formation

Jens Kreth and Mark C. Herzberg

Abstract Oral bacteria are responsible for oral health and disease, including caries, periodontal disease, and endodontic infections. The development of oral diseases is intimately linked with the ability of oral bacteria to form and reside in an adherent multispecies consortium named biofilm. The oral biofilm provides a protective environment for the bacterial community and its formation is a genetically controlled process. In this chapter, we present a general overview of developmental mechanisms employed by individual members of the oral biofilm. The species composition of the oral biofilm and the oral microbiome is discussed historically and in the context of newly developed next-generation sequencing techniques. Furthermore, biofilm-specific regulatory mechanisms and phenotypic traits are explained to provide the reader with a comprehensive overview of oral biofilm formation and its role in health and disease.

1 Introduction

1.1 What Are Biofilms?

Microbial communities are commonly referred to as biofilms (Costerton et al. 1995). These communities are found associated with humans, generally on the skin or mucous membranes, but can also be found in natural (e.g., rivers and streams or soil) and artificial environments (e.g., on the surfaces of the places where we live and work). In general, biofilms connote the lifestyles of aggregated, sessile, or attached microbes in any environment and contrast free-floating, planktonic

J. Kreth (✉)
Department of Microbiology and Immunology, University of Oklahoma Health Sciences Center, Oklahoma City, OK, USA
e-mail: Jens-Kreth@ouhsc.edu

M.C. Herzberg
Department of Diagnostic and Biological Sciences, School of Dentistry, Mucosal and Vaccine Research Center, Minneapolis VA Medical Center, University of Minnesota, Minneapolis, MN, USA

L.E. Chávez de Paz et al. (eds.), *The Root Canal Biofilm*, Springer Series on Biofilms 9, DOI 10.1007/978-3-662-47415-0_2

counterparts. The definition of biofilms has changed over time to include relevant new discoveries in biofilm research and to appreciate their structural and developmental complexity. An early definition as microbial aggregates attached to a living or nonliving surface embedded within extracellular polymeric substances (EPS) of bacterial origin has been extended to include aggregated cell masses floating in a liquid phase and cell aggregates in air–liquid interfaces.

One of the defining steps in biofilm development is the production of extracellular polysaccharides (EPS) (Flemming and Wingender 2010). In nature, what we term EPS actually consists of bacterial polysaccharides, proteins, nucleic acids, and lipids. The EPS contributes to the architecture of the biofilm community. In medically relevant biofilms, host-derived components play an important role in the initiation of biofilm development and should be considered a part of the EPS. For example, a conditioning saliva-derived film (the acquired salivary pellicle) is essential for the attachment of the initial oral biofilm forming bacteria (Hannig et al. 2005). Hall-Stoodley et al. therefore suggested that biofilm with EPS exists as "aggregated, microbial cells surrounded by a polymeric self-produced matrix, which may contain host components" (Hall-Stoodley et al. 2012).

1.2 Why Do Biofilms Form?

Human microbes of medical interest live predominantly in biofilms. The microorganisms dwelling in biofilm communities are estimated to cause about 80 % of infections (Costerton et al. 1999; Costerton 2001), suggesting that virulence is favored for the biofilm residents. Most human-associated microbial species are highly adapted to a specific body site; residing in a biofilm avoids dislocation to a less favorable environment. By residing in sessile communities, microorganisms are less likely to face eradication. For example, to optimize retention on selected oral surfaces, members of oral biofilms have developed mechanisms to optimize binding to specific cell and tissue sites (Zhang et al 2005). The successive integration of new members into the initial biofilm is also promoted by specific cell surface receptors to facilitate species–species aggregation (Jakubovics et al. 2014). To enable growth in the selected oral niche, the colonizing microflora can effectively metabolize salivary components.

The biofilm community also protects its ecological niche against invading, nonresident species that would otherwise overrun the space. In a process called colonization resistance, this community-based interference or antagonism prevents integration of *Pseudomonas aeruginosa*, for example, into human salivary microbial biofilms (He et al. 2011; van der Waaij et al. 1971). The protective biofilm environment also extends to the host–biofilm interface. The innate and adaptive arms of the immune system more effectively eliminate planktonic cells than microorganisms in biofilms. Several mechanisms can be in play. For example, in biofilms *Staphylococcus epidermidis* cells resist deposition of the antimicrobial complement component C3b and immunoglobulin G (IgG) on cell surfaces, thus diminishing opsonization required for phagocyte-mediated killing (Kristian et al. 2008). Similarly, *Staphylococcus aureus* cells in biofilms resist macrophage phagocytosis by circumventing bacterial recognition pathways mediated by toll-like receptors TLR2 and TLR9 (Thurlow et al. 2011). Both TLRs usually recognize bacterial components (pathogen-associated molecular patterns, PAMPs), which are expressed on cells in the biofilm, but appear masked by the presence of the EPS. Interestingly, initial biofilm formation by *Pseudomonas aeruginosa* is facilitated by the presence of human neutrophils through *P. aeruginosa* attachment to neutrophil-derived actin and DNA (Walker et al. 2005), further illustrating that adaptation to the protective, anti-phagocytic biofilm environment sustains viability and long-term persistence.

The biofilm can likely modulate the host-immune response depending on the species composition. *Porphyromonas gingivalis*, a member of the subgingival biofilm community associated with the development of periodontal disease, can downregulate specific immune mediators. For example, the presence of *P. gingivalis* in a ten species in vitro biofilm model was required to downregulate proinflammatory interleukin-Iβ and the NLRP3 inflammasome (Belibasakis et al. 2012), which are required for the effective elimination of bacteria by the host (Taxman et al. 2010). *P. gingivalis* is therefore suggested to use this strategy to manipulate the local inflammatory immune response and evade host surveillance with the ultimate benefit of survival at the host–biofilm interface (Bostanci and Belibasakis 2012).

Microorganisms residing in a biofilm community also enjoy greater resistance against antimicrobials. To combat infecting microbes in biofilms, conventional antibiotics are required at 10- to 1000-fold greater concentrations. Bacteria are also able to respond to the presence of antibiotics like methicillin by forming biofilms, as shown for *S. aureus* and several other species (Kaplan 2011), indicating a specific mechanistic behavior of bacterial cells encountering potentially life-threatening conditions. Similarly, microbes encounter daily challenges from the host innate immunity. Antimicrobial peptides are produced by the oral mucosa to target bacteria residing or passing through the oral cavity (Diamond et al. 2008; Gorr 2012). Found in saliva, antimicrobial peptides are less effective against biofilms than planktonic cells (Helmerhorst et al. 1999; Mazda et al. 2012; Wei et al. 2006).

1.2.1 Formation of Diffusion Barrier and Adsorbant Surface

Several factors influence biofilm susceptibility to antimicrobials as described in the following section. The EPS can form both a diffusion barrier and affinity matrix, partitioning the microbes from an antimicrobial compound or peptide. As an affinity matrix, the EPS actively binds antimicrobial substances to limit penetration into the biofilm. By slowing and limiting the diffusion of the antimicrobial, EPS reduces the effective local concentration reaching the viable cells in the biofilm community. The structure of the EPS affects the rate of diffusion of antimicrobials;

diffusion would be more limiting as the size of the antimicrobials increases. For example, fluorescent probes of varying molecular weights—surrogates for antimicrobial compounds—penetrated into a preformed in vitro biofilm consortium with different efficiencies. The diffusion limitation reflected molecular sieving, which could be predicted by the molecular weight of the fluorescent probes. The pore diameter for the particular biofilm EPS was estimated to be between 2.6 and 4.6 nm (Thurnheer et al. 2003). The pore size of EPS is expected to vary with the microbial species composition in the biofilm and the structure of the synthesized EPS, but little is known. In some conditions, diffusion limitation is not achieved by the EPS. For example, the antibiotics vancomycin and rifampin can effectively penetrate the biofilms of *S. epidermidis*, but fail to eradicate the biofilm-dwelling bacteria (Dunne et al. 1993), suggesting another mechanism responsible for the reduced susceptibility.

Biofilms, including the EPS and the compact colonies of cells, also limit diffusion of components required for the growth of the resident cells and removal of secreted metabolic end products. The EPS and cell colonies function generally as a constraint on diffusion and as a molecular sieve. As shown for ex vivo oral biofilms, oxygen availability is limited in deeper parts of the biofilm (von Ohle et al. 2010), repressing the respiratory activity of oral biofilm bacteria (Nguyen et al. 2002). The cells grow slower because of suboptimal conditions for metabolic activity. The microbial heterogeneity in segments of the biofilm community is both a cause and consequence of regional differences in metabolic activity. Heterogeneous metabolic activity in biofilms leads to reduced cellular content of RNA and proteins in some regions of the biofilm, while growth occurs elsewhere (Sternberg et al. 1999; Xu et al. 1998). Regions with slow growth may show greater antibiotic resistance. Mature subgingival ex vivo biofilms with limited nutrient supply are less susceptible to chlorhexidine and other antibiotics when compared to newly formed, metabolically active biofilms (Sedlacek and Walker 2007; Shen et al. 2011).

1.2.2 Biofilm-Specific Development of Genetic Resistance

During biofilm development, specific traits can be expressed that confer antibiotic resistance, which is not observed in planktonic cells. For example, a glucosyltransferase (encoded by *ndvB*) in *P. aeruginosa* is responsible for the production of cyclic periplasmatic glucans. The expression of *ndvB* is specific for biofilm cells and seems to be absent in planktonic cells. Inactivation of *ndvB* leads to the loss of high-level, biofilm-specific antibiotic resistance (Mah et al. 2003). The cyclic glucans produced by NdvB can interact with antibiotics, thus sequestering antibiotics away from their cellular targets (Beaudoin et al. 2012; Mah et al. 2003). Similarly, during biofilm development, cells of *S. aureus* and *Salmonella enterica* serovar Typhimurium upregulate specific multidrug efflux pumps that transport antimicrobial compounds out the cell (He and Ahn 2011). *Candida albicans* also upregulates drug efflux pumps during biofilm formation, which may increase resistance to antifungal components during oral candidiasis (Ramage et al. 2002).

1.2.3 Emergence of Persister Cells

Bacterial persistence during antibiotic treatment can be attributed to persister cells (Bigger 1944). Persister cells are a small metabolically inactive, dormant subpopulation found in biofilm and planktonic cultures (Balaban 2011; Lewis 2010). Their minimal physiological activity facilitates extreme tolerance against antimicrobial treatments. In contrast, resistant bacteria acquire either a mutation or encode a specific gene conferring antibiotic resistance. The persistent state is not passed on to the offspring. Once the antibiotic challenge to the population is removed, the persister cells resume metabolic activity and repopulate the infected area, and new persister cells can appear.

The biofilm can also shield inhabitants from clearance by the immune system, contributing to the pathogenesis of chronic infections such as cystic fibrosis and tuberculosis (Allison et al. 2011). Immune cells and antibodies can attack the outermost surface of the biofilm, but bacterial cells within are protected. Antimicrobial therapy can also select for increased occurrence of persisters, as shown for *C. albicans* isolated from biofilms of oral candidiasis patients (Lafleur et al. 2010). For detailed information about the genetic regulation of persistence, see reference Lewis (2010). Persisting bacteria and fungi can cause recurrence of infection once antibiotic treatment concludes.

Biofilms, therefore, show greater resistance to antibiotics and immune defense mechanisms due to several mechanisms, which may have evolved to protect the community. The outermost layers of the biofilm form both an antibiotic diffusion barrier and adsorbent. Since human-associated biofilms are typically polymicrobial consortia, each resident species can create its own microenvironment. Bacterial species differ in their susceptibility to antibiotic treatment. Under selective pressure of antibiotic treatment, the less susceptible species will tend to survive. In concert with the selective advantages provided by formation and growth in biofilms, infections associated with biofilms tend to be treated ineffectively by antibiotics.

Growth of microbes in biofilm communities offers other advantages. During starvation conditions, survival of bacteria generally favors species residing in a community. Indeed, several species form biofilms to mitigate starvation conditions. Within biofilms, the lower metabolic activity might help cells to survive times of insufficient carbohydrates, nitrogen, and phosphorus in the oral biofilm. In general, microbes in biofilms are more resistant to stress (Coenye 2010), which can include nutrient deprivation, changes in oxygen tension, or extremes of pH and temperature.

In a biofilm community, resident cells enjoy intercellular communication. Microbial communication is crucial for concerted gene regulation as a response to environmental changes. The close proximity of cells in the biofilm seems to create an ideal environment to talk. A common form of communication between bacteria is the production of signaling molecules, which can be sensed by neighbors. The reduced diffusion in biofilms facilitates a localized critical increase of signaling molecules, which can trigger a corresponding response in the microenvironment. Intercellular communication is important for general stress adaptation and community development. Intercellular communication is discussed in detail in Sects. 5 and 6.

1.3 Challenges for Biofilms in the Oral Environment

Among the human-associated bacterial communities, oral biofilms reside in an anatomical site that encounters diverse environmental challenges. These challenges include perturbations from the external environment and microbial growth control provided within the oral cavity by the innate arm of the immune system. The oral cavity is bathed in saliva, which contains innate immune effector molecules that originate in the salivary glands and in the mucosal epithelial cells that line the oral surfaces. The net effect of the challenges to biofilm communities will vary in the different niches found on the oral surfaces and within tissue folds and crevices. The oral mucosal epithelium is a continuously shedding and renewing tissue. Cells containing oral microorganisms are shed and replaced with new sterile cells, which rapidly bind and are invaded by oral microorganisms. The cycle of shedding, renewal, and reinfection repeats continuously.

Located in close proximity to the oral epithelium, teeth are non-shedding surfaces. Tooth eruption and entry into the oral cavity breach the covering mucosal epithelium. As the teeth erupt, the gingiva, a band of keratinized squamous mucosal epithelium, surrounds and attaches to the tooth surface and forms a penetration barrier. As the erupting tooth and the surrounding epithelium mature, a gingival crevice forms between the tooth and the attached gingiva. The gingival crevice is bathed with a specialized fluid called gingival crevice fluid, which arises as a serum transudate that percolates from the connective tissues through the thin crevicular epithelium. The oral surfaces are more generally bathed in saliva. The composition of bathing saliva varies in intraoral locales based on proximity to the different major and minor salivary glands. Oxygen tension, temperature, and humidity can also differ because of proximity to ambient air. For microbes, the mouth contains, therefore, an infinite number of microenvironments, each one contiguous with its neighbors. The distinct ecological determinants tend to select for survival and growth of certain microbial species while excluding others. Therefore, the bacterial composition differs for the subgingival and supragingival biofilms, which also differ from the species composition of the tongue.

Oral biofilms are also challenged frequently by sudden environmental changes due to host behavior. During host food intake, low nutrient availability for biofilms can be replaced by relative overabundance. Masticated foods and lactic acid released mainly by oral streptococci and actinomyces after carbohydrate fermentation cause sudden changes in pH. Additional physical and chemical stresses to the biofilm communities can be caused by sudden changes in temperature, osmolarity, and the mastication process during and after food ingestion. To accommodate to the stresses, the oral biofilm communities have become genetically diverse.

In individual subjects, the species richness is estimated to be 250 to 300 different species (Keijser et al. 2008; Zaura et al. 2009). The species diversity creates intrinsic competition for space and nutrients. Nonetheless, the oral biofilms maintain homeostasis as a vital part of overall human oral health.

2 Species Composition of the Human Oral Biofilm

The first ever description of bacteria by Antonie Philips van Leeuwenhoek was based on his own dental plaque samples. Visualized through his primitive compound microscope, the initial drawings reported in 1683 suggested that the oral biofilm is a multispecies consortium (Hall 1989). As technology has advanced in the last several decades, our knowledge of the species composition of oral biofilms has exploded in healthy subjects and others with oral diseases. Our understanding of disease development has increased and potential pathogenic species have been associated with caries and periodontal disease.

A new concept of disease development has been developed (Beighton 2005; Kleinberg 2002; Marsh 2003; Takahashi and Nyvad 2008; van Houte 1994). Today, the events leading to caries and periodontal disease are better explained by shifts in bacterial ecology and disturbances in biofilm homeostasis than the presence of a specific pathogen. The ecological plaque theory as defined by Marsh suggests that environmental pressures like low pH cause a shift in the bacterial ecology of the dental biofilm from health-promoting bacteria to primarily acidogenic, aciduric, and cariogenic species (Marsh 2003, 2009). Furthermore, the polymicrobial synergy and dysbiosis (PSD) model acknowledge that the presence and biofilm-specific activities of "keystone" species like *Porphyromonas gingivalis* and *Tannerella forsythus* are required to fully promote the pathogenic potential of the biofilm (Hajishengallis and Lamont 2012). Healthy microbial homeostasis is usually promoted by the interplay between host behavior, host defense mechanisms, and biofilm intrinsic mechanisms. Excessive intake of fermentable carbohydrates can perturb dental plaque homeostasis by lowering the pH. Higher proportions of conditional pathogens are selected, whereas under healthy conditions the composition of the dental plaque would not cause any problems for the host.

More than 300 years after the first visualization of the life-forms in dental plaque, new techniques and approaches are used to study oral microbial composition. Improvements in sensitivity and the development of high-throughput sequencing allow us to estimate that the oral biofilms contain up to 19,000 species/phylotypes (Keijser et al. 2008). The complexity of oral biofilms is far greater than previously realized. Reflecting the complexity of these communities, the composition of oral biofilms is now called the microbiome. By use of informatics approaches and continued technological advances, new understanding of the oral microbiome in health and disease and breakthrough discoveries are to be expected.

2.1 Composition of Human Oral Biofilms: Pre-microbiome Era

The microbial composition of dental plaque has been characterized using culture-dependent and culture-independent approaches (Aas et al. 2005, 2008; Becker et al. 2002; Kroes et al. 1999; Kumar et al. 2003; Paster et al. 2001, 2006). The culture-independent method uses amplification of the species-specific 16S rRNA gene, followed by cloning of the respective amplicons and sequencing. When plaque was sampled from five subjects at nine different healthy intraoral sites using this method, 141 predominant species were detected (Aas et al. 2005). The genus *Streptococcus* is the most abundant, including the species *Streptococcus sanguinis*, *S. gordonii*, *S. mitis*, *S. oralis*, and *S. salivarius* (Aas et al. 2005). The streptococci constitute over 80 % of initial biofilm formers and 20 % of mature plaque biofilms (Rosan and Lamont 2000). Because of their interactions with other genera and species, the presence of certain species of streptococci specifies the initial species succession of successful biofilm formation as discussed below. Among the five subjects, several genera were identified to be in common including *Streptococcus*, *Veillonella*, *Granulicatella*, and *Gemella* (Aas et al. 2005). The nine intraoral plaque sites each showed distinct microflora. Interestingly, *Streptococcus mitis* was isolated from all intraoral sites and was present in all 5 individuals. Consistent with the absence of clinically detectable dental diseases, "red complex" species associated with periodontal disease were not found including *Porphyromonas gingivalis*, *Treponema denticola*, and *Tannerella forsythia* (Socransky et al. 1998). Similarly cariogenic species like *S. mutans*, *Lactobacillus* spp., and *Bifidobacterium* spp. were not isolated (Aas et al. 2005).

Using a more sensitive method in species detection, 16S rRNA amplification, cloning, and sequencing, the bacterial diversity of the oral cavity was determined in 10 healthy subjects (Bik et al. 2010). Overall, 247 species-level phylotypes were identified, including "red complex" species in three of the subjects. Periodontal pathogens seem to be present in subjects with periodontal health, although at low levels. All 10 subjects shared a core set of 15 bacterial genera, with subject-specific differences at the species and strain level (Bik et al. 2010). Several species were only found in 1 or 2 subjects, suggesting that the interindividual environment, behavior, and genetic background contribute to the species composition of the oral biofilm. These studies provided important contributions to our understanding of the overall composition of normal bacterial flora of the oral cavity.

To understand bacterial ecology during caries development, the microbial composition of healthy enamel in 30 subjects was compared with caries lesions of increasing severity in 30 subjects with childhood caries (Becker et al. 2002). The 23 most prominent bacterial species were investigated. As expected, cariogenic *S. mutans* was identified in relatively high abundance in subjects with clinical signs of caries development, and significantly fewer were recovered on sound enamel surfaces and in subjects without any sign of caries. *Actinomyces gerencseriae* and *Bifidobacterium* spp. were also more prevalent in caries than in healthy sites.

Conversely, *S. sanguinis*, *S. gordonii*, and *Actinomyces* serotype II were associated with intact, caries-free teeth and lower caries prevalence. Caries-associated species like *S. mutans* and *Lactobacillus* spp. increase in numbers during caries development (Torlakovic et al. 2012). With more sophisticated analytical methods, however, the species composition is further defined and includes other previously not detected species (Torlakovic et al. 2012). These data indicate that the species composition of oral biofilms varies from health to disease and fully support the ecological plaque theory.

2.2 Composition of Human Oral Biofilms: Microbiome Era

The Human Microbiome Project (HMP) Consortium seeks to provide a comprehensive overview of human microbial communities sampled on 18 sites of the healthy body (Consortium 2012a) (and http://www.hmpdacc.org). In contrast to the limited genus and species identification of methods in the pre-microbiome era, at least 300 bacterial reference genomes will be sequenced. The HMP uses a deep-sequencing approach and provides an in-depth analysis of the available data. New species can be identified and the overall metabolic pathways can be constructed when shared by the sampled body site and the inhabiting biofilm community. The sampling protocols and processing are precisely defined allowing for direct comparison of results from the samples from each subject (Consortium 2012a).

What have we learned about the oral biofilm community so far from the HMP? Nearly 5000 simultaneous samples were collected from 15 (male) to 18 (female) body sites from 242 healthy adult men and women. The human ecosystem is now known to be comprised of more than 10,000 microbial species; 81–99 % of the genera have been identified. With the first milestone announcement on June 13, 2012, several of the many concurrent publications report on the oral biofilm community (Consortium 2012b; Rho et al. 2012; Segata et al. 2012; Wu et al. 2012). When compared to the other sampled sites, the oral communities were among the most diverse (Huse et al. 2012). At the genus and species levels, tooth-associated communities were moderately distinct from other oral surfaces although some genera were widely present in most sites (Consortium 2012b; Huse et al. 2012). The unique composition of each site during health appeared stable. Genera with pathogenic members were well represented among this disease-free cohort (Segata et al. 2012). Although the microbial composition differed from site to site between buccal mucosa, supragingival plaque, and tongue dorsum, core metabolic processes were expressed in most environments (Consortium 2012b). In each environment, different consortia of species performed the core metabolic processes. Refinements in the analysis of available data sets, notwithstanding, core metabolic functions can be provided to the biofilm by previously unrecognized functional and ecological biodiversity in different oral sites and dramatically different distributions of oral species among individuals (Eren et al. 2014).

Clearly, the HMB project has given us a high-resolution overview of the species composition of the human oral biofilm with novel insights into the microbial collaborations needed for full metabolic capabilities of the community. The appreciation of the diversity at different oral sites begs the question of how colonization and metabolic activity reflects the health or disease potential of the biofilm in caries, for example. To address this question, it is feasible to determine the functional gene expression of dental plaque bacteria from different oral sites using RNA-seq. The actual metabolically active proportion of the consortium can be identified in terms of gene expression at any given oral site. RNA generation requires active metabolism and inactive or dead bacteria would be excluded from such an analysis. This meta-transcriptomic approach has been used only recently, but the actively transcribed gene sets and the resulting metabolic profile of the consortia were clearly defined (Benitez-Paez et al. 2014; Peterson et al. 2014). By ascribing functions to genes expressed in dental plaque and other oral communities, activities can be characterized. Further exciting discoveries will increase our understanding of oral biofilms at a different level.

3 First Steps in the Development of Biofilms

The crucial steps in the initial development of biofilms have been well documented. For example, the spatiotemporal pattern of early oral biofilm formation has been traced in the human host using specific removable appliances harboring dental enamel chips (Diaz et al. 2006). After 4 and 8 h of chip placement, oral streptococci were the predominant initially colonizing species. Indeed, the initial colonizers generally belong to the genus *Streptococcus*; some species are constant members representing a core group of initial biofilm formers (Diaz et al. 2006; Rosan and Lamont 2000).

Initial attachment is considered a reversible process. Initial or pioneer colonizers adhere and detach from the tooth surface, and the cycle repeats until permanent attachment is achieved. This irreversible attachment stage is typically mediated by high-affinity binding between proteins and polysaccharides on the bacterial cell surface to specific receptors on the tooth or cell. Within the bacteria, adhesion is controlled by a genetic program associated with a change in the expression of biofilm-related genes (Cowan et al. 1987; Hasty et al. 1992; Nobbs et al. 2009).

Biofilm development progresses with the formation of small micro-colonies of streptococci and a few non-streptococci (Diaz et al. 2006). This newly formed biofilm matures into a complex mature biofilm community by intrinsic growth and recruitment of other species into the developing biofilm. The growth and accrual of a complex community of microbial cells is accompanied by EPS production.

To discuss the molecular principles of biofilm formation, we will concentrate on oral streptococci as the best-studied oral bacterial species and initial biofilm former.

3.1 Attachment to the Tooth Surfaces

3.1.1 Salivary Pellicle Adhesion

The molecular principles of adhesion are based on successive steps. First, oral streptococci adhere to macromolecular complexes found on the saliva-coated tooth or mucosal cells. This salivary film is commonly called pellicle. Although oral streptococci are able to adhere directly to hydroxyapatite (Tanaka et al. 1996), the major mineral found in dental enamel, the initial attachment process involves the pellicle since the teeth and oral surfaces are consistently bathed with saliva and naked enamel occurs rarely in the oral cavity. All evidence indicates that the streptococcal surface interacts with salivary components including salivary proteins.

Salivary proteins solubilized in the liquid or gel fractions of saliva are able to attach avidly to the tooth surface via negatively charged residues and electrostatic interactions through the hydrophilic regions; when tightly bound to hydroxyapatite, hydrophobic domains buried within the three-dimensional structure of the salivary proteins can become exposed and facilitate additional interactions (Lamkin and Oppenheim 1993; Lindh 2002). This adsorbed complex of proteins and other macromolecules constitutes the acquired salivary pellicle, which immediately forms on the emerging tooth or after tooth cleaning by bathing in the constant flow of saliva. Components in saliva and in the pellicle known to interact with the adhering microorganisms include proline-rich proteins, alpha-amylase, secretory IgA, mucin glycoproteins, and glycoprotein (gp) 340 (for comprehensive reviews, see Jenkinson and Lamont 1997, Nobbs et al. 2009, Nobbs et al. 2011).

Binding to salivary pellicle proteins is mediated by protein–protein or protein–carbohydrate interactions between the pellicle surface and the streptococcal surface. Amylase is the most abundant salivary protein and is present in the salivary pellicle and dental plaque (Aguirre et al. 1987; Orstavik and Kraus 1973). Amylase can complex with sIgA in the salivary pellicle to form a binding receptor for certain oral streptococci (Gong et al. 2000). Several oral streptococci bind amylase alone. *S. gordonii* and *S. mitis* encode specific amylase binding proteins (Li et al. 2002a; Vorrasi et al. 2010). Amylase binding protein B, AbpB, contributes to biofilm formation based upon the inability of an AbpB mutant to colonize starch-fed rats (Tanzer et al. 2003). For initial biofilm formation, binding to amylase is of greater physiological significance than the physical attachment process. After streptococcal binding, amylase retains about 50 % of its enzymatic function to catalyze the hydrolysis of α-1,4-glucosidic linkages in dietary starch (Scannapieco et al. 1990). Ingestion of starch enzymatically produces glucose, maltose, and maltodextrins in close proximity to the streptococcal surface. These hydrolysis products of starch can be immediately transported into streptococci by carbohydrate-specific transporters. Adhesion is therefore linked to a metabolic advantage, which illustrates the complex nature of the oral biofilm.

Several other surface adhesins have been identified in oral streptococci mediating the binding to salivary components, including SsaB, FimA, Hsa, GspB, SspB, and SpaA (Holmes et al. 1998; Nobbs et al. 2011). *Streptococcus gordonii* adhesins Hsa and GspB recognize carbohydrate moieties of glycosylated salivary components. Both proteins mediate adhesion to the salivary pellicle through lectin-like recognition of sialic-acid-containing salivary mucin MG2 and salivary agglutinin. It is not surprising that oral streptococci have evolved several surface proteins with redundant function to ensure proper binding to their specific niche in the oral cavity. The oral cavity contains the only ecological niches allowing long-term survival of oral streptococci and these microorganisms must adhere or die. Interestingly, the genetic regulatory circuit of adhesin expression can compensate for imbalances in surface protein display, suggesting a fail-safe mechanism of adherence (unpublished data).

3.2 *Binding to Cellular Components*

Streptococcal surface proteins also bind the bacteria to host cellular components and salivary proteins that form a pellicle on mucosal epithelial cells. About 80 % of the oral cavity surface consists of soft tissue, providing a large area for bacterial attachment (Nobbs et al. 2011). Attachment to oral epithelial cells requires multiple *S. gordonii* adhesin proteins (Davies et al. 2009). Certain oral streptococci bind to mammalian fibronectin, a structural glycoprotein involved in cell–matrix interactions (Labat-Robert 2012; Nobbs et al. 2011), which facilitates adhesion to cells (Okahashi et al. 2010). *S. gordonii* Hsa mediates binding to sialic acid, which is found on the N-linked glycans of fibronectin and other glycoproteins. Binding to fibronectin is probably an important biological target evolutionarily since *S. gordonii* adhesins CshA and CshB also bind to fibronectin peptide domains present in the backbone. Similarly, the adhesins SspA and SspB also recognize fibronectin (Jakubovics et al. 2009).

Oral bacteria employ other host components as binding receptors including the integrin family of cell surface receptors responsible for cell–cell attachment (Nobbs et al. 2009, 2011). For example, the *S. mutans* multi-ligand antigen I/II family adhesin SpaP mediates interaction with $\alpha5\beta1$ integrins (Engels-Deutsch et al. 2011), other host components, and bacterial cells of other species. Antigen I/II also mediates binding to fibronectin, collagen, salivary glycoproteins, glycoprotein 340, platelets, integrins, and *A. naeslundii* (Brady et al. 2010). To confer different binding capabilities, heterologous protein targets can bind distinct protein domains in the antigen I/II protein. For example, the alanine-rich repeat domain (A-region) found on the N-terminus binds to collagen type I. The C-terminal region is responsible for interactions with *P. gingivalis* (Brady et al. 2010), whereas the central variable domain (V-region) interacts with surface ligands of *Actinomyces naeslundii*. The A-region and the V-region each interact with salivary glycoproteins. The proline-rich repeated domain (P-region) plays a role in antigen I/II protein folding and self-aggregation.

The periodontal pathogen *P. gingivalis* expresses cell surface fimbriae, encoded by *fimA*, which is critical for interactions with host cell membrane components. FimA mediates binding to epithelial cells through the interaction with extracellular matrix proteins like fibrinogen, vitronectin, and type I collagen (Murakami et al. 1996; Naito and Gibbons 1988; Nakamura et al. 1999). In a survey to connect the periodontal health status to the prevalence of specific *fimA* genotypes, a correlation was observed between *fimA* types II and IV and periodontitis patients. Type II had a prevalence of 75 % and is most likely a contributing factor to the etiology of periodontal disease (Amano et al. 2000; Amano 2003).

Interestingly, bacterial interactions with the host not only benefit colonization, but it also stimulates cell signaling and immune response. For example, *S. mutans* binding to epithelial surface receptors in the oral cavity induces the synthesis and release of proinflammatory cytokines like interleukin-6 and interleukin-8, which promote neutrophil chemotaxis and degranulation (Engels-Deutsch et al. 2003). Binding also induces expression of epithelial surface molecules such as ICAM-1 and VCAM-1 (Vernier-Georgenthum et al. 1998), which facilitate transendothelial migration of neutrophils during the inflammatory response. In contrast, *P. gingivalis* binding to cell surfaces dampens or interferes with the host-immune response and may also promote periodontal tissue destruction. *P. gingivalis* fimbriae trap ECM proteins like fibronectin and vitronectin, thereby interfering with integrin-mediated signal transduction. As a consequence, the periodontal pathogen appears to subvert the response to repair and heal compromised tissue (Scragg et al. 1999).

Therefore, surface adhesin proteins on oral bacteria do not only ensure binding to oral surfaces during biofilm development but also contribute to the virulence of the respective species and influence the capacity of the host to respond to the presence of the biofilm communities.

3.3 EPS Production

After initial attachment of bacterial cells, the next step in biofilm formation is defined by the production of EPS. The best investigated EPS in the oral cavity is glucans produced by streptococcal glucosyltransferases (Gtfs) from monomeric sugars. If dietary sucrose is available in excess, glucosyltransferase activity is intimately linked to caries development. Several oral species express Gtfs, but *S. mutans* appears to synthesize the majority of the glucans in oral biofilms. All using sucrose as substrate, three different Gtfs (GtfBCD) have been identified in *S. mutans*. GtfBCD each synthesize a chemically distinct glucan. GtfB synthesizes primarily insoluble α-1,3-linked glucans, GtfC produces a mixture of soluble (α-1,6-linked glucans) and insoluble glucans, and GtfD forms predominantly soluble glucans. Of special interest is the ability of the secreted enzymes to associate with the salivary pellicle. GtfC and GtfD can both bind to the pellicle, GtfD through the association with alpha-amylase. The *S. gordonii* glucosyltransferase GtfD also binds to alpha-amylase. GtfB remains bound to the streptococcal surface and is able to bind to surfaces of other bacterial species. On bacterial surfaces and the pellicle, the adsorbed Gtfs actively synthesize the respective glucan polymers, effectively rendering non-Gtf encoding species into glucan producers (Koo et al. 2013).

Adherence of streptococci to the preformed glucan polymers is facilitated by specific *S. mutans* glucan-binding proteins, mainly GbpC and GbpB. GbpC and GbpB bind cells of *S. mutans* to the mesh of glucan polymers and associated proteins forming an EPS superstructure important for morphogenesis and the 3D architecture of the oral biofilm. Subsequently, other species are able to bind to the glucans and increase the species richness of the biofilm.

The role of *S. mutans* in the oral biofilm has mainly been investigated in the context of caries development. The role of *S. mutans* in healthy conditions, however, has not been defined. *S. mutans* is found in low abundance in healthy subjects, and yet its ability to provide a mesh-like structure for attachment of several species might contribute to initial biofilm development for healthy communities. The transition from a healthy to a pathogenic cariogenic biofilm community would occur if excess dietary sucrose selects for increased abundance of *S. mutans*.

Another EPS component present in oral biofilms is extracellular DNA (eDNA), but its role in oral biofilm formation is less well studied. Several oral streptococci, including *S. mutans* (Klein et al. 2010), are able to release DNA into the extracellular environment through an autolytic process, which contributes to biofilm formation. When treated with DNase I to hydrolyze extracellular DNA, *S. mutans* biofilms grown in the presence of sucrose and starch showed a significant reduction in biomass. Growth in sucrose and starch upregulates the autolysin *lytT* gene, which correlates with presence of eDNA. Another mechanism of DNA release in *S. mutans* involves a newly identified auto-active bacteriocin, which induces cell death and release of eDNA from the bacteriocin producer (Perry et al. 2009).

Release of eDNA appears to be a highly conserved activity and other streptococci have evolved corresponding mechanisms. *S. gordonii* and *S. sanguinis* produce eDNA during aerobic growth closely associated with the production of H_2O_2 (Kreth et al. 2008). Both species release high molecular weight DNA of chromosomal origin (Kreth et al. 2009). Deletion of the gene encoding pyruvate oxidase dramatically reduced H_2O_2 production and eDNA release (Kreth et al. 2009). When H_2O_2 production is limited by growth in oxygen-limited conditions, eDNA is also reduced (Zheng et al. 2011a, b). Unlike *S. mutans*, *S. gordonii* and *S. sanguinis* only require H_2O_2 to induce the DNA release process. In *S. gordonii*, however, the major autolysin AtlS can also be involved in DNA release (Liu and Burne 2011). Deletion of AtlS in *S. gordonii* prevented autolysis and production of eDNA under aerobic conditions (Liu and Burne 2011). In contrast, under anaerobic conditions, eDNA release can be induced by H_2O_2 addition without any obvious bacterial cell lysis (Itzek et al. 2011). Streptococci may, therefore, have several mechanisms to trigger eDNA release in response to different internal and/or external stimuli.

The generation of eDNA as a consequence of aerobic growth in a H_2O_2-dependent mechanism is consistent with the roles of *S. gordonii* and *S. sanguinis* as initial biofilm colonizers. Once the biofilm establishes and matures with *S. sanguinis* and *S. gordonii* as permanent residents, the oxygen tension inside the biofilm declines, and these species would cease to produce H_2O_2 and eDNA. At that time point in biofilm formation, the contribution of eDNA to the EPS matrix is complete.

Overall, the binding capacities of oral streptococci to the salivary pellicle, host cellular components, and other members of the oral biofilm community allow for the colonization of the oral cavity, spanning the hard surfaces of teeth and the oral mucosa. Oral biofilms are, therefore, bacterial habitats with complex host–biofilm interactions.

4 Bacterial Interactions Leading to Species Succession in Biofilms: Co-aggregation and Adhesion

Oral biofilm formation is driven by individual species that adhere initially to non-colonized surfaces. On these surfaces, the initial colonizers show clonal growth. Other species are recruited, often by co-aggregation, or serendipitously interact to form a mature biofilm community. Concurrently, co-aggregation of genetically distinct bacteria can occur in saliva and multispecies complexes can then intercalate into the maturing biofilm. Co-aggregation during biofilm formation follows a specific colonization pattern based on the compatible co-aggregation pairings between bacterial species. Co-aggregation pairs have been well investigated in vitro and in situ over the years leading to a model of spatiotemporal tooth colonization reviewed in numerous excellent articles.

In biofilms, different species reside and grow in close proximity. The localization of colonies of different species is rooted in the mutual benefits each derives on the metabolic and cell signal levels. Hence, co-aggregation of streptococci, occurring through interactions between lectin-like adhesin with receptor polysaccharide on the partner species, facilitates intergeneric cell–cell communication. Bringing heterologous species into close proximity also facilitates metabolic collaborations. For example, veillonellae are known to co-aggregate with streptococci and actinomyces. In biofilms, the three species create a complex metabolic relationship. *Veillonella* spp. are unable to metabolize carbohydrates but readily ferment organic acids like lactic acid. The common end product of streptococcal and actinomyces metabolism is lactic acid, which is used by *Veillonella* spp. for growth. At the same time removal of lactic acid prevents acidification. Acidification would inhibit the growth of the lactic acid producer itself.

The production of H_2O_2 by oral streptococci is another important factor in this ménage à trois. Some species are producers of H_2O_2, whereas others are sensitive to the oxidative stress. Certain oral streptococci have the ability to produce H_2O_2 that selects for the integration of compatible species into the developing biofilm community since new members must be able to cope with oxidative stress. *A. naeslundii* is inhibited by H_2O_2 produced by *S. gordonii*, but in co-aggregation cultures both species grow together in close proximity (Jakubovics et al. 2008a, b). *S. gordonii* in coculture grows more readily than *A. naeslundii*, becoming the dominant species at a ratio of about 9 to 1. Why do *S. gordonii* need *A. naeslundii* in the community? *A. naeslundii* provides the H_2O_2-degrading enzyme catalase, which can reduce oxidative damage to surface proteins inflicted by H_2O_2 produced by *S. gordonii* (Jakubovics et al. 2008b). This example of interspecies cooperation compliments *S. gordonii*'s own mechanism to maintain the reduced state and function of the adhesins by using the oxidative repair activity of methionine sulfoxide reductase, MsrA (Lei et al. 2011). The low ratio of *A. naeslundii* to *S. gordonii* would, therefore, allow for sufficient production of H_2O_2 to inhibit susceptible species competing for the same niche. Like *A. naeslundii,* some strains of Veillonella also produce catalase, which would also protect *S. gordonii*. The interactions between streptococci, veillonellae, and actinomyces occur in vivo given the spatial proximity of the three species in human plaque samples (Valm et al. 2011), suggesting that the mutualisms actually occur in nature.

A striking and seemingly antagonistic relationship exists between *S. gordonii* and *P. gingivalis*. Both are aggregation partners and *P. gingivalis* can be found together during initial biofilm formation at 8 h (Diaz et al. 2006). During initial plaque formation, the environment shifts from oxygen-rich to microaerophilic. Yet, *P. gingivalis* is a strictly anaerobic species (Naito et al. 2008). During initial biofilm development, *P. gingivalis* is susceptible to oxidative stress resulting from the relatively high oxygen tension produced by inspired air and H_2O_2 produced by *S. gordonii*.

The initial interaction between oral streptococci and *P. gingivalis* is most likely mediated by *P. gingivalis* FimA and the streptococcal surface located glyceraldehyde-3-phosphate dehydrogenase (Maeda et al. 2004a, b, c). The binding increases in strength through the subsequent interaction between *S. gordonii* SspAB with the *P. gingivalis* short fimbrial protein subunit, Mfa (Demuth et al. 2001). During biofilm community development with *P. gingivalis*, several required *S. gordonii* genes were identified in a genetic screen (Kuboniwa et al. 2006). Among them is *spxB*, which encodes the pyruvate oxidase responsible for the production of H_2O_2. Why is this *S. gordonii* gene important for the interaction with *P. gingivalis* when the gene product generates a toxic by-product? One important feature could be the metabolic function of the pyruvate oxidase. The metabolic activity of streptococcal cells appears to create an anaerobic environment by rapidly consuming oxygen. As a consequence, H_2O_2 production declines. It is tempting to speculate that the initial association of *P. gingivalis* with *S. gordonii* is not accompanied by growth of *P. gingivalis*; but once the anaerobic environment is created, *P. gingivalis* can proliferate.

The initial in vivo oral biofilm microbial community is not restricted to *P. gingivalis* and *S. gordonii* but includes other streptococci and species like *Veillonella* and *Actinomyces*. Over time, the complexity of the community increases. The interplay between H_2O_2 and catalase production appears to be important in spatial and temporal development of the community.

Biofilm development through selective integration of species into existing microbial communities also reflects concurrent exclusion of certain other species. For example, *S. cristatus* downregulates the expression of the *P. gingivalis fimA* gene, which encodes the surface protein responsible for the attachment to other bacteria (Xie et al. 2000). *S. cristatus* communicates with *P. gingivalis* to exclude this species from the microenvironment. The downregulation of *fimA* requires initial contact of the two organisms and appears to depend on the presence of *S. cristatus* arginine deiminase (Lin et al. 2008). Hence, *S. cristatus* may have the ability to shape its own neighborhood by selecting its interacting partners. It is not hard to imagine how this ability could influence the acquisition of later successive colonizers in the developing biofilm.

5 Regulation of Adhesion and Biofilm Formation

When compared to planktonic conditions (when cells grow unattached to a surface), biofilm cells differentially regulate approximately 10 % of the bacterial genome, suggesting biofilm-specific regulation of gene expression (Rendueles and Ghigo 2012). For example, *S. mutans* UA159 cells in biofilms regulate about 12 % of its genome with about 139 genes activated and 104 genes repressed when compared to planktonic cells (Shemesh et al. 2007). Interestingly, most of the genes significantly downregulated were from cells in the depth of the biofilm close to the attachment surface. The selectively downregulated genes were largely involved in attachment, suggesting that those genes are involved in the transition from the planktonic to sessile phenotype, but that these genes are downregulated when no longer required for biofilm attachment or maturation (Shemesh et al. 2007). The initial step of biofilm formation and the subsequent maturation of the biofilm community are directed, therefore, by a genetically regulated developmental program.

The biofilm-specific regulation of genes seems to be specific for each species. During the transition into the biofilm phenotype, the periodontal pathogen *Aggregatibacter actinomycetemcomitans* does not downregulate any genes but upregulates about 355 open reading frames (ORF), including a large set of ribosomal genes and potential virulence genes. In contrast, *P. gingivalis* upregulated only 26 ORFs and downregulated 193 ORFs, most encoding "hypothetical" proteins (Frias-Lopez and Duran-Pinedo 2012).

The decision of planktonic cells to attach to a surface depends on environmental conditions (Petrova and Sauer 2012). Well-known environmental triggers include macrophages co-localized with planktonic bacteria and nutritional deficits (discussed above). Generally, the response to environmental signals is mediated by receptors complexed with response regulators (transcription factors). Typically these signaling systems are termed two-component systems (TCSs). Ca^{2+} is a common environmental cue and *S. mutans* regulates calcium-dependent biofilm formation by signaling through the CiaXRH three-component system. CiaX is a small, secreted peptide with a calcium-binding (SD) domain (He et al. 2008). Mutation of CiaX diminished *S. mutans* biofilm formation in vitro, probably because calcium is required to initiate biofilm developmental pathways.

Biofilm formation involves several other signal transduction systems. The TCS BfrAB is indispensible for *S. gordonii* single-species biofilm development (Zhang et al. 2004). By binding to a specific DNA consensus sequence, the response regulator BfrA directly regulates the expression of several genes, including the two ABC transporters encoded by the *bfrCD* and *bfrEFG* operons (Zhang et al. 2009). Interestingly, the DNA binding domain of *S. gordonii* BfrA is able to bind to the homologous operons identified in *S. sanguinis*, suggesting that BfrAB controls biofilm development in *S. sanguinis* as well. Two other genes controlled by BfrAB in *S. gordonii*, *bfrC* and *bfrG*, are also required for dual-species biofilm maturation with *P. gingivalis* (Kuboniwa et al. 2006). Hence, in vivo biofilm development requires coordinated signaling pathways to ensure adhesion and interaction with other species.

In general, TCSs appear to be pleiotropic regulators. Planktonic and biofilm phenotypes are influenced by more than one TCS, suggesting vigorous cross talk between individual TCSs to enable the biofilm lifestyle. For example, the TCSs VicRK, HdrRM, and CiaXRH all affect *S. mutans* biofilm formation, competence, and virulence. The competence regulatory TCS ComED is also involved in biofilm development and is discussed later in Sect. 6.

The transition from planktonic to sessile cells leading to the irreversible attachment in several species has been associated with the level of the ubiquitous intracellular messenger signaling molecule bis-(3'-5')-cyclic di-GMP (c-di-GMP) (Jonas et al. 2009). The production of c-di-GMP is under the control of diguanylate cyclases (DGCs) and phosphodiesterases (PDEs). c-di-GMP is synthesized from GTP by DGCs and degraded by PDEs (Mills et al. 2011). High c-di-GMP levels promote adhesion by the production of adhesive extracellular matrix components including fimbriae in *S. typhimurium* and *P. aeruginosa* (Kader et al. 2006; Kulasekara et al. 2005). Low intracellular levels of c-di-GMP promote motility behavior like swimming, swarming, and twitching motility (Simm et al. 2004). Inactivation of a DGC ortholog (*gcp*) in *S. mutans* also leads to impaired biofilm formation. Gcp encodes the only protein in *S. mutans* with DGC activity (Yan et al. 2010).

The oral streptococci need to adhere or become extinct. These species are not particularly invasive of soft tissues, and so to survive they clearly invest in mechanisms that facilitate adhesion to surfaces. Indeed these species may possess a fail-safe mechanism to ensure that the protein adhesins are always expressed sufficiently to withstand blockade or loss of one or more of these crucial surface proteins. For example, *S. gordonii* overexpresses two alternative, well-studied adhesins when one or more major adhesins (SspAB) are mutated and deleted (Zhang et al. 2005). Furthermore, when surface adhesins cannot be properly presented on the cell wall by deletion of the enzyme, Sortase A, virtually all of the protein adhesins are overexpressed (Nobbs et al. 2007a). Hence, the cells appear to be genetically programmed to provide an ample abundance of surface adhesin proteins to compensate for any loss of expression or defect in processing, and the importance of adhesion and biofilm formation to the survival of these species is evident from the multiple mechanisms that sustain and regulate these functions.

6 Ecological Aspects of Oral Biofilm Development

6.1 Communication and the Consequences

Bacteria use small diffusible molecules or peptides as signals for cell-to-cell communication. Communication can be intra- and interspecies specific and aid in the coordination of bacterial population behavior as needed in biofilm formation and community development. During early biofilm formation on the tooth surface, the sparse colonization and small clusters of cell aggregates attached to the salivary pellicle influence the ability of members of the biofilm community to communicate with each other. Only cells in the immediate vicinity of a communicating cell can interact with diffusible signals or metabolites. The distance from the target cell receptors (i.e., TCSs) and concentration of the diffusible signaling molecules are crucial limiting parameters in cell-to-cell communication. For example, the importance of spatial clustering on the efficiency of diffusible signals has been suggested by mathematical modeling (Hense et al. 2007). Communication can also be influenced by intrinsic distortion in the biofilm. For example, the important *S. mutans* communication peptide CSP (see below), which regulates stress adaptation, biofilm formation, and competence development, is degraded in dual-species biofilms with *S. gordonii*. Degradation is mediated by challisin, a secreted protease. Since a challisin homologue is encoded in the genome of the early colonizer *S. sanguinis*, other oral streptococci might use similar mechanisms.

Metabolic cooperativity between members of the oral biofilm also requires communication. For example, *S. gordonii* amylase gene showed increased expression when cells grew in close proximity to *V. atypica* in a saliva-conditioned flow cell (Egland et al. 2004). *V. atypica* benefits from induction of amylase gene expression in *S. gordonii*. Amylase degrades starch to monomeric glucose, which can be fermented by *S. gordonii* to lactic acid, promoting growth of *V. atypica*. This cooperative interaction is only effective over short distances, suggesting the action of a diffusible substance (Egland et al. 2004).

Diffusible signals are also required for dual-species biofilm formation of *S. gordonii* with *P. gingivalis* and *S. oralis* with *A. naeslundii* (McNab et al. 2003; Rickard et al. 2006). Biofilm development for these pairs of species depends on the signal autoinducer 2 (AI-2). AI-2 is a cell-density-dependent signaling molecule produced by many bacterial species. Although AI-2 shows species–species structural differences, many species can sense and respond to AI-2. Therefore, AI-2 is considered as a nonspecific signaling molecule for inter-bacterial communication (McNab and Lamont 2003). Accordingly, *S. gordonii* and *S. oralis* AI-2 mutants were not able to form a dual-species biofilm with *P. gingivalis* or *A. naeslundii* (McNab et al. 2003). In the natural setting of the oral biofilm, the metabolic signaling of AI-2 is likely to influence the response of multiple species and the structure of the developing oral biofilm.

6.2 Interspecies Antagonism

The oral biofilm is a competitive environment. Even during initial biofilm formation when colonization space is ample, competition exists. For example, the initial colonizers S. *sanguinis* and *S. gordonii* express specific cell surface adhesin molecules with similar binding specificities. Hence, they compete to bind similar sites in the salivary pellicle (Nobbs et al. 2007b). Yet, *S. sanguinis* has a greater natural prevalence in plaque and saliva, but fails to outcompete *S. gordonii completly*. *S. gordonii* apparently survives because it competes to adhere to the saliva-coated tooth more effectively with *S. sanguinis* than any other tested oral streptococci (Liljemark et al. 1979, 1981). The competitive advantage of *S. gordonii* was attributable to expression of the sialic-acid-binding protein Hsa based upon mutational analysis (Nobbs et al. 2007b). This surface adhesin enables *S. gordonii* to successfully compete for binding to the tooth surface with the genetically similar, but more abundant *S. sanguinis*. Hence, the efficiency of adherence to certain salivary components is crucial in the competitive oral environment.

Several oral bacteria can also express antimicrobial peptides (bacteriocins) to antagonize competing species. Interspecies antagonism of *S. sanguinis* or *S. gordonii* with *S. mutans* has been modeled with a specific deferred antagonism assay, and the outcome is determined by the sequence of colonization, nutritional availability, and environmental pressures (Kreth et al. 2005a, 2008). All three streptococcal species produce small chromosomally encoded bacteriocins (Deng et al. 2004; Fujimura and Nakamura 1979; Schlegel and Slade 1972). Bacteriocins of *S. mutans* have been well characterized (84, 99), and a recent review highlights the sophisticated regulation of their production (Merritt and Qi 2012; Qi et al. 2000, 2001). The production of certain bacteriocins by *S. mutans* is controlled in a cell-density-dependent manner by the ComCDE system that controls competence development and biofilm formation (Heng et al. 2007; Kreth et al. 2005b). One of the distinctive features of bacteriocins is the target range of susceptible bacteria. The bacteriocins produced by *S. mutans* are able to inhibit *S. sanguinis* and *S. gordonii* among other oral streptococci (Kreth et al. 2005a). Conversely, the bacteriocins produced by *S. gordonii* and *S. sanguinis* do not target *S. mutans*. Nonetheless, *S. gordonii* and *S. sanguinis* are able to inhibit the growth of *S. mutans* by generating an alternative antimicrobial compound, H_2O_2 (Kreth et al. 2005a), as described above in Sect. 4.

During initial colonization when the cell density is relatively low, H_2O_2 production might be ecologically advantageous relative to bacteriocin production, which requires high cell density to trigger bacteriocin gene expression. H_2O_2 production is oxygen dependent (Kreth et al. 2008), and the oxygen tension in saliva is sufficient to allow for aerobic respiration and hydrogen peroxide production during initial colonization (Marquis 1995). In addition, both *S. sanguinis* and *S. gordonii* are only weakly inhibited by their own H_2O_2 production. Once the biofilm reaches a critical thickness and cell density, diffusion becomes limited and the oxygen tension declines. The decline in oxygen tension might decrease H_2O_2 production to a non-inhibiting level. Under these biofilm conditions, *S. mutans* might use bacteriocins to inhibit *S. sanguinis* and *S. gordonii*, since bacteriocin production is oxygen independent (Kreth et al. 2008). In high-cell-density conditions, *S. mutans* becomes an aggressive competitor for *S. sanguinis* and *S. gordonii*, which may explain why it is able to initiate a shift in composition to the community to create a caries-promoting environment.

Bacteriocins target the membrane of the susceptible species and cause leakage of the cell contents (Oppegard et al. 2007). To effectively compete with *S. sanguinis* and *S. gordonii*, *S. mutans* coordinates bacteriocin production with another important cell function, competence development (see below) (Kreth et al. 2005b; van der Ploeg 2005). Competence is the natural ability of some oral streptococci to incorporate extracellular DNA from the environment (Cvitkovitch 2001). Under the control of the *S. mutans* competence system, mutacin IV is produced before inducing the expression of genes required for DNA uptake. In this way, *S. mutans* uses mutacin IV to kill the target organism and release the cellular contents, including DNA. Subsequently, *S. mutans* cells become competent and take up the released DNA (Kreth et al. 2005b). Although not beneficial for the lysed bacterial species, this interspecies genetic exchange tends to promote evolutionary survival of traits of the target species in the daughter cells of competent *S. mutans*.

6.3 Competence and Genetic Exchange

Competence-dependent genetic exchange requires availability of eDNA and the ability of the competent species to take up DNA from the environment. The development of competence is signaled by the accumulation of a species-specific competence peptide, which is produced and secreted into the environment (competence-stimulating peptide or CSP). CSP serves as a signal to induce competence in the producing population and serves as a form of intraspecies (or interspecies) communication (Johnsborg and Havarstein 2009).

Each streptococcal species produces a unique CSP (Whatmore et al. 1999). Competence development facilitates cell-density-dependent cell regulation but might also reflect other environmental conditions like flow rates and diffusion from the producing species. Competence development follows a basic architecture in most streptococci. Competence development is encoded in the *comCDE* operon, including CSP and the TCS ComDE, which senses and provides transcriptional response to the environmental peptide. Secreted CSP binds to the membrane-spanning ComD sensor kinase (a receptor). Binding of CSP induces a conformational change in ComD, which activates the intracellular kinase domain and promotes subsequent ComD autophosphorylation (P-ComD). ComE is the response regulator and the recipient of the transferred phosphoryl group from P-ComD. P-ComE is able to bind to chromosomal DNA to an imperfect direct repeat (ComE-box) (Ween et al. 1999). This specific binding motif (ComE-box) is found in the promoter region of several genes considered to be early competence genes. The early genes include the *comCDE* operon itself and the *comAB* operon. The *comAB* operon encodes the CSP-specific transporter ComAB, which processes and exports the pre-peptide CSP as the CSP. The secreted CSP, therefore, signals an increase in its own expression and production, secretion, and sensing via ComDE (Claverys et al. 2006; Johnsborg et al. 2007).

Another early competence gene, the gene encoding the alternative sigma subunit ComX of the DNA-dependent RNA polymerase, is transcribed. ComX is essential for competence development and regulates other genes required for the synthesis of the DNA uptake system and recombination. ComX recognizes a unique DNA element named *cin*-box important for the activation of late competence genes (Lee and Morrison 1999).

Competence development in *S. sanguinis* and *S. gordonii* seems to follow this blueprint first described in *S. pneumoniae*. After CSP is added to non-competent cell cultures, *S. gordonii* develops transient competence peaking at 10 to 20 min (Vickerman et al. 2007). After 5 min of exposure, the initial responsive genes are *comCDE*, *comAB*, and the *comX* homologue *comR*. Regulated by ComR, late genes are expressed including genes encoding proteins required for uptake and DNA recombination. Competence declines about 30 to 40 min after CSP addition (Vickerman et al. 2007).

In *S. sanguinis*, the ComAB homologues are not encoded to transport and process CSP. The pre-peptide lacks the double glycine processing site in the N-terminal leader sequence as found in ComAB processed CSPs of other streptococci (Rodriguez et al. 2011; Xu et al. 2007). The usage of DNA microarray technology enabled the identification of other genes not initially identified in the competence developmental circuit. Addition of CSP to *S. sanguinis* induced about 122 genes and downregulated about 83, compared to *S. gordonii* with 162 induced and 89 downregulated genes. However, the number of genes belonging to the early competence genes was considerably lower in *S. sanguinis* with only five genes compared to 28 for *S. pneumoniae* and 35 for *S. gordonii*. The group of early genes in *S. sanguinis* contained *comDE*, *comX*, and two uncharacterized open reading frames, suggesting that this constitutes a minimal set of early competence genes (Rodriguez et al. 2011; Xu et al. 2007).

The high-cell-density oral biofilm seems to be a favorable environment for genetic exchange (Li et al. 2002b). Several active mechanisms are employed by oral bacteria to ensure the availability of eDNA. *S. gordonii* and *S. sanguinis* release homologous and heterologous DNA in a H_2O_2-dependent process, and *S. mutans* is able to release its own DNA and DNA from closely related species in a bacteriocin-dependent process (Kreth et al. 2005b, 2008, 2009; Perry et al. 2009). The biofilm matrix contains eDNA, which might be a source for competent bacteria (Flemming and Wingender 2010). Oral bacteria are able to take up DNA during competence development, including the abundant oral biofilm genus *Streptococcus* and the periodontal pathogen *P. gingivalis*. Interestingly, a recent study with *P. gingivalis* demonstrated that natural competence is the dominant form of chromosomal DNA transfer in this perio-pathogen (Tribble et al. 2012). Indeed, eDNA is found in *P. gingivalis* biofilms, further indicating that this EPS compound might have multiple functions during biofilm development. Since competence development is density dependent, the high-cell-density conditions in biofilms favor genetic exchange.

S. mutans, *S. sanguinis*, and *S. gordonii* each show a competence-influenced biofilm phenotype (Bizzini et al. 2006; Li et al. 2002b; Zhu et al. 2011). Although the biofilm community seems to be better protected from outside environmental perturbation than planktonic cells, the biofilm cannot afford to be evolutionarily idle. Competence allows for a long-term adaptation due to the incorporation of new genetic traits via horizontal gene transfer (Roberts and Kreth 2014). Not only is horizontal gene transfer important in the evolution of bacterial genomes, it also allows bacteria to repair DNA damage via homologous recombination, a mechanism to swap and replace damaged segments of DNA with correct versions. Uptake of eDNA, therefore, appears to promote species diversity and adaptation using information stored in the DNA of any targeted bacterial species in the oral cavity.

S. sanguinis and *S. gordonii* control the expression of several other genes not involved in the uptake of DNA in a CSP-dependent manner. For example, contact with antibiotics can induce competence development. Consequently, it is now widely accepted that the competence system is part of a general stress response for oral streptococci to adapt to changing environmental conditions.

Acknowledgments Research in the authors' labs that is referred to in this chapter was supported by NIH-NIDCR R01DE021726, R00DE018400 and R03DE022601 (JK) and R01DE08590 (MCH).

References

Aas JA, Paster BJ, Stokes LN, Olsen I, Dewhirst FE (2005) Defining the normal bacterial flora of the oral cavity. J Clin Microbiol 43:5721–5732

Aas JA, Griffen AL, Dardis SR, Lee AM, Olsen I, Dewhirst FE, Leys EJ, Paster BJ (2008) Bacteria of dental caries in primary and permanent teeth in children and young adults. J Clin Microbiol 46:1407–1417

Aguirre A, Levine MJ, Cohen RE, Tabak LA (1987) Immunochemical quantitation of alpha-amylase and secretory IgA in parotid saliva from people of various ages. Arch Oral Biol 32:297–301

Allison KR, Brynildsen MP, Collins JJ (2011) Heterogeneous bacterial persisters and engineering approaches to eliminate them. Curr Opin Microbiol 14:593–598

Amano A (2003) Molecular interaction of *Porphyromonas gingivalis* with host cells: implication for the microbial pathogenesis of periodontal disease. J Periodontol 74:90–96
Amano A, Kuboniwa M, Nakagawa I, Akiyama S, Morisaki I, Hamada S (2000) Prevalence of specific genotypes of *Porphyromonas gingivalis fimA* and periodontal health status. J Dent Res 79:1664–1668
Balaban NQ (2011) Persistence: mechanisms for triggering and enhancing phenotypic variability. Curr Opin Genet Dev 21:768–775
Beaudoin T, Zhang L, Hinz AJ, Parr CJ, Mah TF (2012) The biofilm-specific antibiotic resistance gene *ndvB* is important for expression of ethanol oxidation genes in *Pseudomonas aeruginosa* biofilms. J Bacteriol 194:3128–3136
Becker MR, Paster BJ, Leys EJ, Moeschberger ML, Kenyon SG, Galvin JL, Boches SK, Dewhirst FE, Griffen AL (2002) Molecular analysis of bacterial species associated with childhood caries. J Clin Microbiol 40:1001–1009
Beighton D (2005) The complex oral microflora of high-risk individuals and groups and its role in the caries process. Community Dent Oral Epidemiol 33:248–255
Belibasakis GN, Guggenheim B, Bostanci N (2012) Down-regulation of NLRP3 inflammasome in gingival fibroblasts by subgingival biofilms: involvement of *Porphyromonas gingivalis*. Innate Immun 19(1):3–9
Benitez-Paez A, Belda-Ferre P, Simon-Soro A, Mira A (2014) Microbiota diversity and gene expression dynamics in human oral biofilms. BMC Genomics 15:311
Bigger JW (1944) Treatment of staphylococcal infections with penicillin by intermittent sterilization. Lancet 2:497–500
Bik EM, Long CD, Armitage GC et al (2010) Bacterial diversity in the oral cavity. ISME J. 4 (8):962–974
Bizzini A, Beggah-Moller S, Moreillon P, Entenza JM (2006) Lack of in vitro biofilm formation does not attenuate the virulence of *Streptococcus gordonii* in experimental endocarditis. FEMS Immunol Med Microbiol 48:419–423
Bostanci N, Belibasakis GN (2012) *Porphyromonas gingivalis*: an invasive and evasive opportunistic oral pathogen. FEMS Microbiol Lett 333(1):1–9
Brady LJ, Maddocks SE, Larson MR, Forsgren N, Persson K, Deivanayagam CC, Jenkinson HF (2010) The changing faces of Streptococcus antigen I/II polypeptide family adhesins. Mol Microbiol 77:276–286
Claverys JP, Prudhomme M, Martin B (2006) Induction of competence regulons as a general response to stress in Gram-positive bacteria. Annu Rev Microbiol 60:451–475
Coenye T (2010) Response of sessile cells to stress: from changes in gene expression to phenotypic adaptation. FEMS Immunol Med Microbiol 59:239–252
Consortium HMP (2012a) A framework for human microbiome research. Nature 486:215–221
Consortium HMP (2012b) Structure, function and diversity of the healthy human microbiome. Nature 486:207–214
Costerton JW (2001) Cystic fibrosis pathogenesis and the role of biofilms in persistent infection. Trends Microbiol 9:50–52
Costerton JW, Lewandowski Z, Caldwell DE, Korber DR, Lappin-Scott HM (1995) Microbial biofilms. Annu Rev Microbiol 49:711–745
Costerton JW, Stewart PS, Greenberg EP (1999) Bacterial biofilms: a common cause of persistent infections. Science 284:1318–1322
Cowan MM, Taylor KG, Doyle RJ (1987) Energetics of the initial phase of adhesion of *Streptococcus sanguis* to hydroxylapatite. J Bacteriol 169:2995–3000
Cvitkovitch DG (2001) Genetic competence and transformation in oral streptococci. Crit Rev Oral Biol Med 12:217–243
Davies JR, Svensater G, Herzberg MC (2009) Identification of novel LPXTG-linked surface proteins from *Streptococcus gordonii*. Microbiology 155:1977–1988

Demuth DR, Irvine DC, Costerton JW, Cook GS, Lamont RJ (2001) Discrete protein determinant directs the species-specific adherence of *Porphyromonas gingivalis* to oral streptococci. Infect Immun 69:5736–5741

Deng H, Ding Y, Fu MD, Xiao XR, Liu J, Zhou T (2004) Purification and characterization of sanguicin–a bacteriocin produced by *Streptococcus sanguis*. Sichuan Da Xue Xue Bao Yi Xue Ban 35:555–558

Diamond G, Beckloff N, Ryan LK (2008) Host defense peptides in the oral cavity and the lung: similarities and differences. J Dent Res 87:915–927

Diaz PI, Chalmers NI, Rickard AH, Kong C, Milburn CL, Palmer RJ Jr, Kolenbrander PE (2006) Molecular characterization of subject-specific oral microflora during initial colonization of enamel. Appl Environ Microbiol 72:2837–2848

Dunne WM Jr, Mason EO Jr, Kaplan SL (1993) Diffusion of rifampin and vancomycin through a *Staphylococcus epidermidis* biofilm. Antimicrob Agents Chemother 37:2522–2526

Egland PG, Palmer RJ Jr, Kolenbrander PE (2004) Interspecies communication in *Streptococcus gordonii-Veillonella atypica* biofilms: signaling in flow conditions requires juxtaposition. Proc Natl Acad Sci USA 101:16917–16922

Engels-Deutsch M, Pini A, Yamashita Y, Shibata Y, Haikel Y, Scholler-Guinard M, Klein JP (2003) Insertional inactivation of *pac* and *rmlB* genes reduces the release of tumor necrosis factor alpha, interleukin-6, and interleukin-8 induced by *Streptococcus mutans* in monocytic, dental pulp, and periodontal ligament cells. Infect Immun 71:5169–5177

Engels-Deutsch M, Rizk S, Haikel Y (2011) *Streptococcus mutans* antigen I/II binds to alpha5beta1 integrins via its A-domain and increases beta1 integrins expression on periodontal ligament fibroblast cells. Arch Oral Biol 56:22–28

Eren AM, Borisy GG, Huse SM, Mark Welch JL (2014) Oligotyping analysis of the human oral microbiome. Proc Natl Acad Sci USA 111:E2875–E2884

Flemming HC, Wingender J (2010) The biofilm matrix. Nat Rev Microbiol 8:623–633

Frias-Lopez J, Duran-Pinedo A (2012) Effect of periodontal pathogens on the metatranscriptome of a healthy multispecies biofilm model. J Bacteriol 194:2082–2095

Fujimura S, Nakamura T (1979) Sanguicin, a bacteriocin of oral *Streptococcus sanguis*. Antimicrob Agents Chemother 16:262–265

Gong K, Mailloux L, Herzberg MC (2000) Salivary film expresses a complex, macromolecular binding site for *Streptococcus sanguis*. J Biol Chem 275:8970–8974

Gorr SU (2012) Antimicrobial peptides in periodontal innate defense. Front Oral Biol 15:84–98

Hajishengallis G, Lamont RJ (2012) Beyond the red complex and into more complexity: the polymicrobial synergy and dysbiosis (PSD) model of periodontal disease etiology. Mol Oral Microbiol 27(6):409–419

Hall AR (1989) The Leeuwenhoek lecture, 1988. Antoni van Leeuwenhoek 1632–1723. Notes Rec R Soc Lond 43:249–273

Hall-Stoodley L, Stoodley P, Kathju S, Hoiby N, Moser C, William Costerton J, Moter A, Bjarnsholt T (2012) Towards diagnostic guidelines for biofilm-associated infections. FEMS Immunol Med Microbiol 65:127–145

Hannig C, Hannig M, Attin T (2005) Enzymes in the acquired enamel pellicle. Eur J Oral Sci 113:2–13

Hasty DL, Ofek I, Courtney HS, Doyle RJ (1992) Multiple adhesins of streptococci. Infect Immun 60:2147–2152

He X, Ahn J (2011) Differential gene expression in planktonic and biofilm cells of multiple antibiotic-resistant *Salmonella* Typhimurium and *Staphylococcus aureus*. FEMS Microbiol Lett 325:180–188

He X, Wu C, Yarbrough D, Sim L, Niu G, Merritt J, Shi W, Qi F (2008) The *cia* operon of *Streptococcus mutans* encodes a unique component required for calcium-mediated autoregulation. Mol Microbiol 70:112–126

He X, Hu W, He J, Guo L, Lux R, Shi W (2011) Community-based interference against integration of *Pseudomonas aeruginosa* into human salivary microbial biofilm. Mol Oral Microbiol 26:337–352

Helmerhorst EJ, Hodgson R, van't Hof W, Veerman EC, Allison C, Nieuw Amerongen AV (1999) The effects of histatin-derived basic antimicrobial peptides on oral biofilms. J Dent Res 78:1245–1250

Heng NC, Tagg JR, Tompkins GR (2007) Competence-dependent bacteriocin production by *Streptococcus gordonii* DL1 (Challis). J Bacteriol 189:1468–1472

Hense BA, Kuttler C, Muller J, Rothballer M, Hartmann A, Kreft JU (2007) Does efficiency sensing unify diffusion and quorum sensing? Nat Rev Microbiol 5:230–239

Holmes AR, Gilbert C, Wells JM, Jenkinson HF (1998) Binding properties of *Streptococcus gordonii* SspA and SspB (antigen I/II family) polypeptides expressed on the cell surface of *Lactococcus lactis* MG1363. Infect Immun 66:4633–4639

Huse SM, Ye Y, Zhou Y, Fodor AA (2012) A core human microbiome as viewed through 16S rRNA sequence clusters. PLoS ONE 7, e34242

Itzek A, Zheng L, Chen Z, Merritt J, Kreth J (2011) Hydrogen peroxide-dependent DNA release and transfer of antibiotic resistance genes in *Streptococcus gordonii*. J Bacteriol 193:6912–6922

Jakubovics NS, Gill SR, Iobst SE, Vickerman MM, Kolenbrander PE (2008a) Regulation of gene expression in a mixed-genus community: stabilized arginine biosynthesis in *Streptococcus gordonii* by coaggregation with Actinomyces naeslundii. J Bacteriol 190:3646–3657

Jakubovics NS, Gill SR, Vickerman MM, Kolenbrander PE (2008b) Role of hydrogen peroxide in competition and cooperation between *Streptococcus gordonii* and *Actinomyces naeslundii*. FEMS Microbiol Ecol 66:637–644

Jakubovics NS, Brittan JL, Dutton LC, Jenkinson HF (2009) Multiple adhesin proteins on the cell surface of *Streptococcus gordonii* are involved in adhesion to human fibronectin. Microbiology 155:3572–3580

Jakubovics NS, Yassin SA, Rickard AH (2014) Community interactions of oral streptococci. Adv Appl Microbiol 87:43–110

Jenkinson HF, Lamont RJ (1997) Streptococcal adhesion and colonization. Crit Rev Oral Biol Med 8:175–200

Johnsborg O, Havarstein LS (2009) Regulation of natural genetic transformation and acquisition of transforming DNA in *Streptococcus pneumoniae*. FEMS Microbiol Rev 33:627–642

Johnsborg O, Eldholm V, Havarstein LS (2007) Natural genetic transformation: prevalence, mechanisms and function. Res Microbiol 158:767–778

Jonas K, Melefors O, Romling U (2009) Regulation of c-di-GMP metabolism in biofilms. Future Microbiol 4:341–358

Kader A, Simm R, Gerstel U, Morr M, Romling U (2006) Hierarchical involvement of various GGDEF domain proteins in rdar morphotype development of *Salmonella enterica* serovar Typhimurium. Mol Microbiol 60:602–616

Kaplan JB (2011) Antibiotic-induced biofilm formation. Int J Artif Organs 34:737–751

Keijser BJ, Zaura E, Huse SM, van der Vossen JM, Schuren FH, Montijn RC, ten Cate JM, Crielaard W (2008) Pyrosequencing analysis of the oral microflora of healthy adults. J Dent Res 87:1016–1020

Klein MI, DeBaz L, Agidi S, Lee H, Xie G, Lin AH, Hamaker BR, Lemos JA, Koo H (2010) Dynamics of *Streptococcus mutans* transcriptome in response to starch and sucrose during biofilm development. PLoS ONE 5, e13478

Kleinberg I (2002) A mixed-bacteria ecological approach to understanding the role of the oral bacteria in dental caries causation: an alternative to *Streptococcus mutans* and the specific-plaque hypothesis. Crit Rev Oral Biol Med 13:108–125

Koo H, Falsetta ML, Klein MI (2013) The exopolysaccharide matrix: a virulence determinant of cariogenic biofilm. J Dent Res 92:1065–1073

Kreth J, Merritt J, Shi W, Qi F (2005a) Competition and coexistence between *Streptococcus mutans* and *Streptococcus sanguinis* in the dental biofilm. J Bacteriol 187:7193–7203

Kreth J, Merritt J, Shi W, Qi F (2005b) Co-ordinated bacteriocin production and competence development: a possible mechanism for taking up DNA from neighbouring species. Mol Microbiol 57:392–404

Kreth J, Zhang Y, Herzberg MC (2008) Streptococcal antagonism in oral biofilms: *Streptococcus sanguinis* and *Streptococcus gordonii* interference with *Streptococcus mutans*. J Bacteriol 190:4632–4640

Kreth J, Vu H, Zhang Y, Herzberg MC (2009) Characterization of hydrogen peroxide-induced DNA release by *Streptococcus sanguinis* and *Streptococcus gordonii*. J Bacteriol 191:6281–6291

Kristian SA, Birkenstock TA, Sauder U, Mack D, Gotz F, Landmann R (2008) Biofilm formation induces C3a release and protects *Staphylococcus epidermidis* from IgG and complement deposition and from neutrophil-dependent killing. J Infect Dis 197:1028–1035

Kroes I, Lepp PW, Relman DA (1999) Bacterial diversity within the human subgingival crevice. Proc Natl Acad Sci USA 96:14547–14552

Kuboniwa M, Tribble GD, James CE, Kilic AO, Tao L, Herzberg MC, Shizukuishi S, Lamont RJ (2006) *Streptococcus gordonii* utilizes several distinct gene functions to recruit *Porphyromonas gingivalis* into a mixed community. Mol Microbiol 60:121–139

Kulasekara HD, Ventre I, Kulasekara BR, Lazdunski A, Filloux A, Lory S (2005) A novel two-component system controls the expression of *Pseudomonas aeruginosa* fimbrial cup genes. Mol Microbiol 55:368–380

Kumar PS, Griffen AL, Barton JA, Paster BJ, Moeschberger ML, Leys EJ (2003) New bacterial species associated with chronic periodontitis. J Dent Res 82:338–344

Labat-Robert J (2012) Cell-Matrix interactions, the role of fibronectin and integrins. A survey Pathol Biol (Paris) 60:15–19

Lafleur MD, Qi Q, Lewis K (2010) Patients with long-term oral carriage harbor high-persister mutants of *Candida albicans*. Antimicrob Agents Chemother 54:39–44

Lamkin MS, Oppenheim FG (1993) Structural features of salivary function. Crit Rev Oral Biol Med 4:251–259

Lee MS, Morrison DA (1999) Identification of a new regulator in *Streptococcus pneumoniae* linking quorum sensing to competence for genetic transformation. J Bacteriol 181:5004–5016

Lei Y, Zhang Y, Guenther BD, Kreth J, Herzberg MC (2011) Mechanism of adhesion maintenance by methionine sulphoxide reductase in *Streptococcus gordonii*. Mol Microbiol 80:726–738

Lewis K (2010) Persister cells. Annu Rev Microbiol 64:357–372

Li L, Tanzer JM, Scannapieco FA (2002a) Identification and analysis of the amylase-binding protein B (AbpB) and gene (*abpB*) from *Streptococcus gordonii*. FEMS Microbiol Lett 212:151–157

Li YH, Tang N, Aspiras MB, Lau PC, Lee JH, Ellen RP, Cvitkovitch DG (2002b) A quorum-sensing signaling system essential for genetic competence in *Streptococcus mutans* is involved in biofilm formation. J Bacteriol 184:2699–2708

Liljemark WF, Bloomquist CG, Ofstehage JC (1979) Aggregation and adherence of *Streptococcus sanguis*: role of human salivary immunoglobulin A. Infect Immun 26:1104–1110

Liljemark WF, Bloomquist CG, Germaine GR (1981) Effect of bacterial aggregation on the adherence of oral streptococci to hydroxyapatite. Infect Immun 31:935–941

Lin X, Lamont RJ, Wu J, Xie H (2008) Role of differential expression of streptococcal arginine deiminase in inhibition of *fimA* expression in *Porphyromonas gingivalis*. J Bacteriol 190:4367–4371

Lindh L (2002) On the adsorption behaviour of saliva and purified salivary proteins at solid/liquid interfaces. Swed Dent J Suppl 1–57

Liu Y, Burne RA (2011) The major autolysin of *Streptococcus gordonii* is subject to complex regulation and modulates stress tolerance, biofilm formation, and extracellular-DNA release. J Bacteriol 193:2826–2837

Maeda K, Nagata H, Kuboniwa M, Kataoka K, Nishida N, Tanaka M, Shizukuishi S (2004a) Characterization of binding of *Streptococcus oralis* glyceraldehyde-3-phosphate dehydrogenase to *Porphyromonas gingivalis* major fimbriae. Infect Immun 72:5475–5477

Maeda K, Nagata H, Nonaka A, Kataoka K, Tanaka M, Shizukuishi S (2004b) Oral streptococcal glyceraldehyde-3-phosphate dehydrogenase mediates interaction with *Porphyromonas gingivalis* fimbriae. Microbes Infect 6:1163–1170

Maeda K, Nagata H, Yamamoto Y, Tanaka M, Tanaka J, Minamino N, Shizukuishi S (2004c) Glyceraldehyde-3-phosphate dehydrogenase of *Streptococcus oralis* functions as a coadhesin for *Porphyromonas gingivalis* major fimbriae. Infect Immun 72:1341–1348

Mah TF, Pitts B, Pellock B, Walker GC, Stewart PS, O'Toole GA (2003) A genetic basis for *Pseudomonas aeruginosa* biofilm antibiotic resistance. Nature 426:306–310

Marquis RE (1995) Oxygen metabolism, oxidative stress and acid–base physiology of dental plaque biofilms. J Ind Microbiol 15:198–207

Marsh PD (2003) Are dental diseases examples of ecological catastrophes? Microbiology 149:279–294

Marsh PD (2009) Dental plaque as a biofilm: the significance of pH in health and caries. Compend Contin Educ Dent 30:76–78, 80, 83–77; quiz 88, 90

Mazda Y, Kawada-Matsuo M, Kanbara K, Oogai Y, Shibata Y, Yamashita Y, Miyawaki S, Komatsuzawa H (2012) Association of CiaRH with resistance of *Streptococcus mutans* to antimicrobial peptides in biofilms. Mol Oral Microbiol 27:124–135

McNab R, Lamont RJ (2003) Microbial dinner-party conversations: the role of LuxS in interspecies communication. J Med Microbiol 52:541–545

McNab R, Ford SK, El-Sabaeny A, Barbieri B, Cook GS, Lamont RJ (2003) LuxS-based signaling in *Streptococcus gordonii*: autoinducer 2 controls carbohydrate metabolism and biofilm formation with *Porphyromonas gingivalis*. J Bacteriol 185:274–284

Merritt J, Qi F (2012) The mutacins of *Streptococcus mutans*: regulation and ecology. Mol Oral Microbiol 27:57–69

Mills E, Pultz IS, Kulasekara HD, Miller SI (2011) The bacterial second messenger c-di-GMP: mechanisms of signalling. Cell Microbiol 13:1122–1129

Murakami Y, Iwahashi H, Yasuda H, Umemoto T, Namikawa I, Kitano S, Hanazawa S (1996) *Porphyromonas gingivalis* fimbrillin is one of the fibronectin-binding proteins. Infect Immun 64:2571–2576

Naito Y, Gibbons RJ (1988) Attachment of *Bacteroides gingivalis* to collagenous substrata. J Dent Res 67:1075–1080

Naito M, Hirakawa H, Yamashita A et al (2008) Determination of the genome sequence of *Porphyromonas gingivalis* strain ATCC 33277 and genomic comparison with strain W83 revealed extensive genome rearrangements in *P. gingivalis*. DNA Res 15:215–225

Nakamura T, Amano A, Nakagawa I, Hamada S (1999) Specific interactions between *Porphyromonas gingivalis* fimbriae and human extracellular matrix proteins. FEMS Microbiol Lett 175:267–272

Nguyen PT, Abranches J, Phan TN, Marquis RE (2002) Repressed respiration of oral streptococci grown in biofilms. Curr Microbiol 44:262–266

Nobbs AH, Vajna RM, Johnson JR, Zhang Y, Erlandsen SL, Oli MW, Kreth J, Brady LJ, Herzberg MC (2007a) Consequences of a sortase A mutation in *Streptococcus gordonii*. Microbiology 153:4088–4097

Nobbs AH, Zhang Y, Khammanivong A, Herzberg MC (2007b) *Streptococcus gordonii* Hsa environmentally constrains competitive binding by *Streptococcus sanguinis* to saliva-coated hydroxyapatite. J Bacteriol 189:3106–3114

Nobbs AH, Lamont RJ, Jenkinson HF (2009) Streptococcus adherence and colonization. Microbiol Mol Biol Rev 73:407–450, Table of Contents

Nobbs AH, Jenkinson HF, Jakubovics NS (2011) Stick to your gums: mechanisms of oral microbial adherence. J Dent Res 90:1271–1278

Okahashi N, Nakata M, Sakurai A et al (2010) Pili of oral *Streptococcus sanguinis* bind to fibronectin and contribute to cell adhesion. Biochem Biophys Res Commun 391:1192–1196

Oppegard C, Rogne P, Emanuelsen L, Kristiansen PE, Fimland G, Nissen-Meyer J (2007) The two-peptide class II bacteriocins: structure, production, and mode of action. J Mol Microbiol Biotechnol 13:210–219

Orstavik D, Kraus FW (1973) The acquired pellicle: immunofluorescent demonstration of specific proteins. J Oral Pathol 2:68–76

Paster BJ, Boches SK, Galvin JL, Ericson RE, Lau CN, Levanos VA, Sahasrabudhe A, Dewhirst FE (2001) Bacterial diversity in human subgingival plaque. J Bacteriol 183:3770–3783

Paster BJ, Olsen I, Aas JA, Dewhirst FE (2006) The breadth of bacterial diversity in the human periodontal pocket and other oral sites. Periodontol 2000(42):80–87

Perry JA, Jones MB, Peterson SN, Cvitkovitch DG, Levesque CM (2009) Peptide alarmone signalling triggers an auto-active bacteriocin necessary for genetic competence. Mol Microbiol 72:905–917

Peterson SN, Meissner T, Su AI, Snesrud E, Ong AC, Schork NJ, Bretz WA (2014) Functional expression of dental plaque microbiota. Front Cell Infect Microbiol 4:108

Petrova OE, Sauer K (2012) Sticky situations: key components that control bacterial surface attachment. J Bacteriol 194:2413–2425

Qi F, Chen P, Caufield PW (2000) Purification and biochemical characterization of mutacin I from the group I strain of *Streptococcus mutans*, CH43, and genetic analysis of mutacin I biosynthesis genes. Appl Environ Microbiol 66:3221–3229

Qi F, Chen P, Caufield PW (2001) The group I strain of *Streptococcus mutans*, UA140, produces both the lantibiotic mutacin I and a nonlantibiotic bacteriocin, mutacin IV. Appl Environ Microbiol 67:15–21

Ramage G, Bachmann S, Patterson TF, Wickes BL, Lopez-Ribot JL (2002) Investigation of multidrug efflux pumps in relation to fluconazole resistance in *Candida albicans* biofilms. J Antimicrob Chemother 49:973–980

Rendueles O, Ghigo JM (2012) Multi-species biofilms: how to avoid unfriendly neighbors. FEMS Microbiol Rev 36(5):972–989

Rho M, Wu YW, Tang H, Doak TG, Ye Y (2012) Diverse CRISPRs evolving in human microbiomes. PLoS Genet 8, e1002441

Rickard AH, Palmer RJ Jr, Blehert DS, Campagna SR, Semmelhack MF, Egland PG, Bassler BL, Kolenbrander PE (2006) Autoinducer 2: a concentration-dependent signal for mutualistic bacterial biofilm growth. Mol Microbiol 60:1446–1456

Roberts AP, Kreth J (2014) The impact of horizontal gene transfer on the adaptive ability of the human oral microbiome. Front Cell Infect Microbiol 4:124

Rodriguez AM, Callahan JE, Fawcett P, Ge X, Xu P, Kitten T (2011) Physiological and molecular characterization of genetic competence in *Streptococcus sanguinis*. Mol Oral Microbiol 26:99–116

Rosan B, Lamont RJ (2000) Dental plaque formation. Microbes Infect 2:1599–1607

Scannapieco FA, Bhandary K, Ramasubbu N, Levine MJ (1990) Structural relationship between the enzymatic and streptococcal binding sites of human salivary alpha-amylase. Biochem Biophys Res Commun 173:1109–1115

Schlegel R, Slade HD (1972) Bacteriocin production by transformable group H streptococci. J Bacteriol 112:824–829

Scragg MA, Cannon SJ, Rangarajan M, Williams DM, Curtis MA (1999) Targeted disruption of fibronectin-integrin interactions in human gingival fibroblasts by the RI protease of *Porphyromonas gingivalis* W50. Infect Immun 67:1837–1843

Sedlacek MJ, Walker C (2007) Antibiotic resistance in an *in vitro* subgingival biofilm model. Oral Microbiol Immunol 22:333–339

Segata N, Haake SK, Mannon P, Lemon KP, Waldron L, Gevers D, Huttenhower C, Izard J (2012) Composition of the adult digestive tract bacterial microbiome based on seven mouth surfaces, tonsils, throat and stool samples. Genome Biol 13:R42

Shemesh M, Tam A, Steinberg D (2007) Differential gene expression profiling of Streptococcus mutans cultured under biofilm and planktonic conditions. Microbiology 153:1307–1317

Shen Y, Stojicic S, Haapasalo M (2011) Antimicrobial efficacy of chlorhexidine against bacteria in biofilms at different stages of development. J Endod 37:657–661

Simm R, Morr M, Kader A, Nimtz M, Romling U (2004) GGDEF and EAL domains inversely regulate cyclic di-GMP levels and transition from sessility to motility. Mol Microbiol 53:1123–1134

Socransky SS, Haffajee AD, Cugini MA, Smith C, Kent RL Jr (1998) Microbial complexes in subgingival plaque. J Clin Periodontol 25:134–144

Sternberg C, Christensen BB, Johansen T, Toftgaard Nielsen A, Andersen JB, Givskov M, Molin S (1999) Distribution of bacterial growth activity in flow-chamber biofilms. Appl Environ Microbiol 65:4108–4117

Takahashi N, Nyvad B (2008) Caries ecology revisited: microbial dynamics and the caries process. Caries Res 42:409–418

Tanaka H, Ebara S, Otsuka K, Hayashi K (1996) Adsorption of saliva-coated and plain streptococcal cells to the surfaces of hydroxyapatite beads. Arch Oral Biol 41:505–508

Tanzer JM, Grant L, Thompson A, Li L, Rogers JD, Haase EM, Scannapieco FA (2003) Amylase-binding proteins A (AbpA) and B (AbpB) differentially affect colonization of rats' teeth by *Streptococcus gordonii*. Microbiology 149:2653–2660

Taxman DJ, Huang MT, Ting JP (2010) Inflammasome inhibition as a pathogenic stealth mechanism. Cell Host Microbe 8:7–11

Thurlow LR, Hanke ML, Fritz T et al (2011) *Staphylococcus aureus* biofilms prevent macrophage phagocytosis and attenuate inflammation in vivo. J Immunol 186:6585–6596

Thurnheer T, Gmur R, Shapiro S, Guggenheim B (2003) Mass transport of macromolecules within an in vitro model of supragingival plaque. Appl Environ Microbiol 69:1702–1709

Torlakovic L, Klepac-Ceraj V, Ogaard B, Cotton SL, Paster BJ, Olsen I (2012) Microbial community succession on developing lesions on human enamel. J Oral Microbiol 4

Tribble GD, Rigney TW, Dao DH, Wong CT, Kerr JE, Taylor BE, Pacha S, Kaplan HB (2012) Natural competence is a major mechanism for horizontal DNA transfer in the oral pathogen *Porphyromonas gingivalis*. MBio 3

Valm AM, Mark Welch JL, Rieken CW, Hasegawa Y, Sogin ML, Oldenbourg R, Dewhirst FE, Borisy GG (2011) Systems-level analysis of microbial community organization through combinatorial labeling and spectral imaging. Proc Natl Acad Sci USA 108:4152–4157

van der Ploeg JR (2005) Regulation of bacteriocin production in *Streptococcus mutans* by the quorum-sensing system required for development of genetic competence. J Bacteriol 187:3980–3989

van der Waaij D, Berghuis-de Vries JM, Lekkerkerk L-v (1971) Colonization resistance of the digestive tract in conventional and antibiotic-treated mice. J Hyg (Lond) 69:405–411

van Houte J (1994) Role of micro-organisms in caries etiology. J Dent Res 73:672–681

Vernier-Georgenthum A, al-Okla S, Gourieux B, Klein JP, Wachsmann D (1998) Protein I/II of oral viridans streptococci increases expression of adhesion molecules on endothelial cells and promotes transendothelial migration of neutrophils in vitro. Cell Immunol 187:145–150

Vickerman MM, Iobst S, Jesionowski AM, Gill SR (2007) Genome-wide transcriptional changes in *Streptococcus gordonii* in response to competence signaling peptide. J Bacteriol 189:7799–7807

von Ohle C, Gieseke A, Nistico L, Decker EM, DeBeer D, Stoodley P (2010) Real-time micro-sensor measurement of local metabolic activities in *ex vivo* dental biofilms exposed to sucrose and treated with chlorhexidine. Appl Environ Microbiol 76:2326–2334

Vorrasi J, Chaudhuri B, Haase EM, Scannapieco FA (2010) Identification and characterization of amylase-binding protein C from *Streptococcus mitis* NS51. Mol Oral Microbiol 25:150–156

Walker TS, Tomlin KL, Worthen GS et al (2005) Enhanced *Pseudomonas aeruginosa* biofilm development mediated by human neutrophils. Infect Immun 73:3693–3701

Ween O, Gaustad P, Havarstein LS (1999) Identification of DNA binding sites for ComE, a key regulator of natural competence in *Streptococcus pneumoniae*. Mol Microbiol 33:817–827

Wei GX, Campagna AN, Bobek LA (2006) Effect of MUC7 peptides on the growth of bacteria and on *Streptococcus mutans* biofilm. J Antimicrob Chemother 57:1100–1109

Whatmore AM, Barcus VA, Dowson CG (1999) Genetic diversity of the streptococcal competence (*com*) gene locus. J Bacteriol 181:3144–3154

Wu YW, Rho M, Doak TG, Ye Y (2012) Oral spirochetes implicated in dental diseases are widespread in normal human subjects and carry extremely diverse integron gene cassettes. Appl Environ Microbiol 78:5288–5296

Xie H, Cook GS, Costerton JW, Bruce G, Rose TM, Lamont RJ (2000) Intergeneric communication in dental plaque biofilms. J Bacteriol 182:7067–7069

Xu KD, Stewart PS, Xia F, Huang CT, McFeters GA (1998) Spatial physiological heterogeneity in *Pseudomonas aeruginosa* biofilm is determined by oxygen availability. Appl Environ Microbiol 64:4035–4039

Xu P, Alves JM, Kitten T et al (2007) Genome of the opportunistic pathogen *Streptococcus sanguinis*. J Bacteriol 189:3166–3175

Yan W, Qu T, Zhao H, Su L, Yu Q, Gao J, Wu B (2010) The effect of c-di-GMP (3'-5'-cyclic diguanylic acid) on the biofilm formation and adherence of *Streptococcus mutans*. Microbiol Res 165:87–96

Zaura E, Keijser BJ, Huse SM, Crielaard W (2009) Defining the healthy "core microbiome" of oral microbial communities. BMC Microbiol 9:259

Zhang Y, Lei Y, Khammanivong A, Herzberg MC (2004) Identification of a novel two-component system in *Streptococcus gordonii* V288 involved in biofilm formation. Infect Immun 72:3489–3494

Zhang Y, Lei Y, Nobbs A, Khammanivong A, Herzberg MC (2005) Inactivation of *Streptococcus gordonii* SspAB alters expression of multiple adhesin genes. Infect Immun 73:3351–3357

Zhang Y, Whiteley M, Kreth J, Lei Y, Khammanivong A, Evavold JN, Fan J, Herzberg MC (2009) The two-component system BfrAB regulates expression of ABC transporters in *Streptococcus gordonii* and *Streptococcus sanguinis*. Microbiology 155:165–173

Zheng L, Itzek A, Chen Z, Kreth J (2011a) Environmental influences on competitive hydrogen peroxide production in *Streptococcus gordonii*. Appl Environ Microbiol 77:4318–4328

Zheng LY, Itzek A, Chen ZY, Kreth J (2011b) Oxygen dependent pyruvate oxidase expression and production in *Streptococcus sanguinis*. Int J Oral Sci 3:82–89

Zhu L, Zhang Y, Fan J, Herzberg MC, Kreth J (2011) Characterization of competence and biofilm development of a *Streptococcus sanguinis* endocarditis isolate. Mol Oral Microbiol 26:117–126

Antimicrobial Resistance in Biofilm Communities

Christine Sedgley and Gary Dunny

Abstract Biofilms are composed of microcolonies encased in an extracellular polymeric substance (EPS) matrix. Wide-ranging differences exist between the biofilm and planktonic states in growth, structure, behavior, and physiology, all of which can have profound effects on their susceptibility to antimicrobials. Other factors that can contribute to the decreased susceptibility of biofilm microorganisms to antimicrobial agents include provision of a physical barrier to antimicrobial agents by the EPS matrix, facilitation of horizontal gene transfer (HGT) of DNA trapped within the extracellular matrix, quorum sensing and stress responses resulting in the recruitment and expression of resistance determinants such as multidrug resistance efflux pumps, the presence of persister cells that survive antibiotic treatment, and metabolic heterogeneity throughout the biofilm resulting in slow growth and protection against antibiotics active on rapidly growing bacteria. While further work is needed to fully understand antimicrobial resistance in biofilm communities, including the multispecies biofilms found in root canal infections, the accumulative effects of various processes, rather than individual involvement, are likely to be important. It is clear that much remains to be learned about the critical events in the development of antimicrobial resistance in biofilm communities.

1 Introduction

Biofilms are composed of microcolonies encased in an extracellular polymeric substance (EPS) matrix. They vary greatly among different species, strains, and in different environments. Much has been discovered about the structure of medically important biofilms from detailed evaluation of pathogens such as

C. Sedgley (✉)
Department of Endodontology, School of Dentistry, Oregon Health & Science University, Portland, OR, USA
e-mail: sedgley@ohsu.edu

G. Dunny
Department of Microbiology, University of Minnesota, Minneapolis, MN, USA

L.E. Chávez de Paz et al. (eds.), *The Root Canal Biofilm*, Springer Series on Biofilms 9, DOI 10.1007/978-3-662-47415-0_3

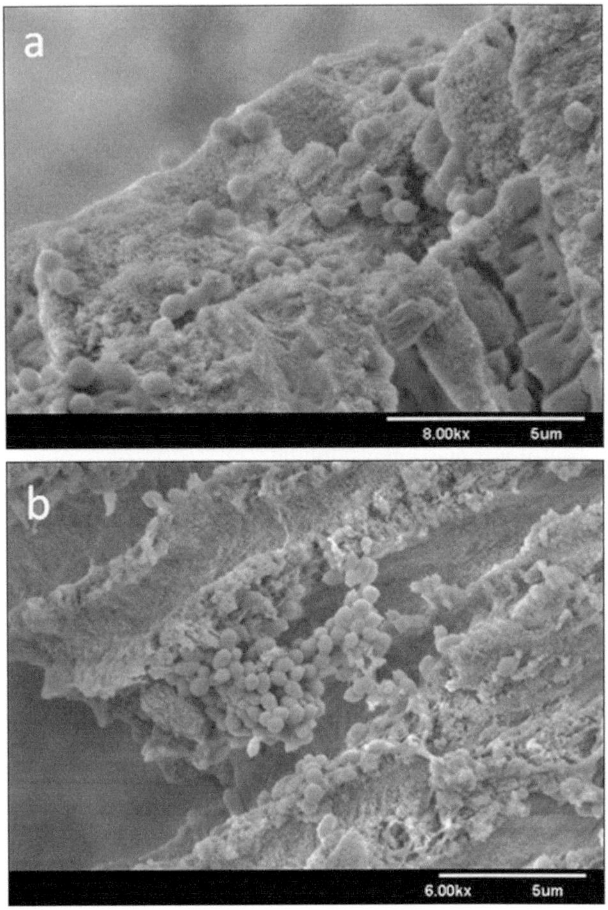

Fig. 1 Scanning electron micrographs showing biofilm (**a**) on the root canal wall (**b**) and within dentinal tubules

Pseudomonas aeruginosa, *Escherichia coli*, and *Staphylococcus* spp. and mixed species biofilms from oral plaque (Kolenbrander et al. 2010; Elias and Banin 2012; Rendueles and Ghigo 2012) and endodontic infections (Ricucci et al. 2009; Ricucci and Siqueira 2010; Carr et al. 2009) (Fig. 1).

Compared to growth in planktonic conditions, biofilm growth has been shown to increase resistance to multiple antibiotics in many diverse species (Hoiby et al. 2010; Mah and O'Toole 2001). For example, the antibiotic susceptibilities of *E. coli*, *S. aureus*, and *P. aeruginosa* were up to 1000-fold higher for biofilm populations compared to planktonic forms in minimal biofilm eradication concentration (MBEC) assays (Ceri et al. 1999). The resistance to antimicrobial agents by biofilms has resulted in considerable difficulties in the clinical management of recalcitrant infections seen in patients with infected medical indwelling devices and cystic fibrosis (CF) patients with *P. aeruginosa* lung infections (Hoiby et al. 2010; Costerton et al. 1999; Stewart and Costerton 2001). In addition, resistance to antimicrobial agents by biofilms has been described as a major obstacle to successful endodontic treatment (Siqueira et al. 2010). This chapter will discuss factors and mechanisms that contribute to the decreased susceptibility of biofilm communities to antimicrobial agents.

2 Factors and Mechanisms Involved in Antimicrobial Resistance in Biofilms

Based on the detection of antibiotic resistance genes in bacteria from 4 million years ago, "antibiotic resistance is natural, ancient, and hard wired in the microbial pangenome" (Bhullar et al. 2012). This section provides examples of how microbial cells in biofilms can utilize a variety of factors and mechanisms for survival under challenging conditions that include exposure to antimicrobial agents (Mah and O'Toole 2001; Fux et al. 2005; Hoiby et al. 2010; Lewis 2010; Roberts and Mullany 2010; Poole 2012; Stewart and Franklin 2008). While further work is needed to fully understand these processes, their accumulative effects, rather than individual involvement, are likely to contribute to increased resistance in biofilm communities (Zhang and Mah 2008).

2.1 EPS Matrix

The role of the EPS matrix, or "glycocalyx," was initially described in terms of functioning to provide intercellular adhesion and adherence to surfaces (Costerton et al. 1981). It is now understood that the EPS matrix has a broader range of functions that also includes aggregation of bacterial cells, water retention, protection, energy sink, ionic exchange, sorption of organic and inorganic compounds, nutrient source, and exchange of genetic information. The EPS matrix can account for more than 90 % of the dry mass of most biofilms and provides a "scaffold" for the biofilm, as well as a physical barrier to antimicrobial agents (Sutherland 2001; Flemming and Wingender 2010). The structure and specific components of the EPS matrix are highly heterogeneous and depend on species, strains, and the different environmental conditions. For example, components of the matrix can be cationic or anionic, hydrophilic or hydrophobic (Donlan and Costerton 2002; Flemming and Wingender 2010).

The major component of the EPS matrix is water which moves around, and not through, cell clusters (Stewart 2012). Water channels allow the inflow of nutrients and the outflow of waste materials (Lawrence et al. 1991). Fluid flow plays an important role in the movement of nutrients and metabolic products in and out of the biofilm. Slow moving fluid adjacent to the biofilm limits diffusive transport of solutes into and out of the biofilm and decreases opportunities for biofilm detachment and dispersion (Stewart 2012; McDougald et al. 2012). The percentage of water will depend on the specific system under examination but can be as high as 97 % in reactors (Zhang et al. 1998). In addition to water, major components of the highly hydrated EPS include exopolysaccharides, extracellular proteins, and extracellular DNA (eDNA) (Fig. 2).

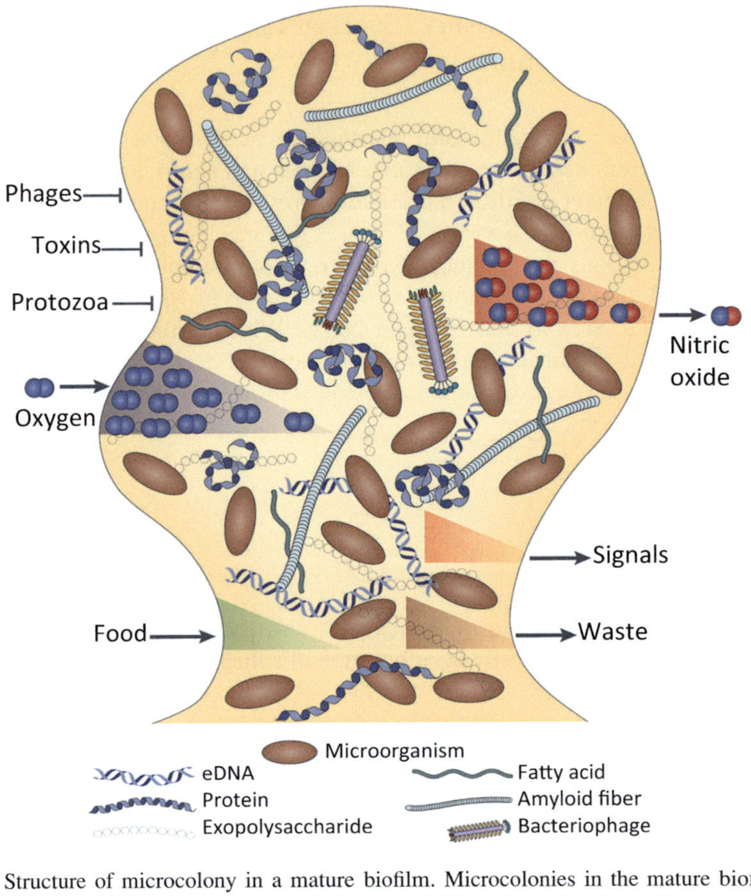

Fig. 2 Structure of microcolony in a mature biofilm. Microcolonies in the mature biofilm are characterized by an EPS matrix, composed of water, eDNA, polysaccharides, and proteins, among other components. The EPS matrix functions as a shield to protect the bacterial community or population from predators such as protozoa or lytic phages as well as from chemical toxins (e.g., biocides and antibiotics). The EPS matrix may help to sequester nutrients and, along with the underlying microorganisms, is also responsible for the establishment of gradients (e.g., oxygen and nutrients diffusing inward and waste products as well as signals such as nitric oxide diffusing outward) [Adapted by permission from Macmillan Publishers Ltd: Nature Reviews in Microbiology (McDougald et al. 2012), ©2012]

2.1.1 Exopolysaccharides

Exopolysaccharides provide structural stability (Sutherland 2001) and can act as a scaffold for proteins that mediate intercellular attachments and adhesion of the biofilm to a surface (Absalon et al. 2011). The importance of exopolysaccharides to biofilm formation has been confirmed in *P. aeruginosa* and *E. coli* monospecies biofilms by the inability of non-polysaccharide producing mutants to form mature biofilms (Danese et al. 2000; Ma et al. 2009). The production of exopolysaccharides can vary within genera. For example, the matrix of *S. aureus* and *Staphylococcus epidermidis* biofilms differs markedly in their surface polysaccharide poly-N-acetylglucosamine (PNAG), with biofilm integrity facilitated more by PNAG for *S. epidermis* than for *S. aureus* (Izano et al. 2008). In addition, in multispecies biofilms synergistic interactions might allow polysaccharide production by one species that allows the integration of nonproducing species into the biofilm (Sutherland 2001), as demonstrated in dual-species biofilms by the greater contribution of EPS to the matrix by *Enterobacter agglomerans* compared with *Klebsiella pneumoniae* (Skillman et al. 1998).

Exopolysaccharides can also play a role in antimicrobial resistance in biofilms. This was demonstrated by the sequestration of tobramycin by glucose polymers produced by *P. aeruginosa* in the periplasm, preventing the antibiotic from reaching its site of action (Mah et al. 2003). Another study has shown that the production of matrix polysaccharide by *S. epidermis* was enhanced by subinhibitory concentrations of tetracycline, quinupristin–dalfopristin, and erythromycin (Rachid et al. 2000). Exopolysaccharides can also play a role in protection of the biofilm from the host defenses; *S. epidermis* mutants defective in polysaccharide intercellular adhesin production had increased susceptibility to phagocytosis by neutrophils and other components of the innate host defense system (Vuong et al. 2004).

In contradistinction, exopolysaccharides in biofilms may have antimicrobial properties. Antibiofilm polysaccharides have been identified in gram-negative bacteria biofilms (Rendueles et al. 2011) and gram-positive bacteria supernatants (Kim et al. 2009) that appear to act on biofilms by means other than those that utilize bactericidal or bacteriostatic activity (Rendueles et al. 2012; Bandara et al. 2010; Benitez et al. 1997). For example, the presence of lipopolysaccharide (LPS), a well-established virulence factor in endodontic infections (Dahlen et al. 1981), inhibited biofilm formation of *Candida* species (Bandara et al. 2010) and partially inhibited the adherence of *Vibrio cholerae* to a human intestinal cell line (Benitez et al. 1997).

2.1.2 Extracellular Proteins

Extracellular proteins in the EPS matrix provide important structural and enzymatic functions. Matrix proteins are involved in maintaining biofilm architecture by linking bacteria and exopolysaccharides seen, for example, with the glucan-binding proteins in *Streptococcus mutans* (Lynch et al. 2007) and the protein Lec-B of *P. aeruginosa* (Tielker et al. 2005). Disruption of the biofilm-associated surface proteins (Bap) found in *S. aureus* (Cucarella et al. 2001) completely impaired biofilm formation (Trotonda et al. 2005). In *Bacillus subtilis* biofilms, an amphiphilic biofilm-surface layer protein, BslA, responsible for the hydrophobic surface layer, may play a role in antimicrobial resistance by facilitating liquid repellency (Kobayashi and Iwano 2012). Enzymes within the EPS matrix also digest polymers to provide carbon and energy sources for the biofilm (Flemming and Wingender 2010). In addition, enzymatic degradation of the EPS matrix allows dispersion of cells to form new biofilms (Sauer et al. 2004).

2.1.3 eDNA

eDNA functions in adhesion, aggregation, cohesion, and exchange of genetic information (Whitchurch et al. 2002; Flemming and Wingender 2010). It also plays a critical role in the initial establishment of biofilms as demonstrated by the dissolution of early biofilms of *P. aeruginosa* and *Enterococcus faecalis* treated with DNAse (Whitchurch et al. 2002; Thomas et al. 2008; Barnes et al. 2012). eDNA in the matrix can vary, even between species in the same genus. For example, the eDNA component was shown to be more important for the structural integrity of *S. aureus* compared to *S. epidermidis* biofilms (Izano et al. 2008). eDNA has also been shown to have antimicrobial activity in *P. aeruginosa* biofilms, causing cell lysis by chelating (Mg^{2+}) cations that stabilize lipopolysaccharide and the bacterial outer membrane (Mulcahy et al. 2008).

The primary source of eDNA appears to be autolyzed cells (Qin et al. 2007; Allesen-Holm et al. 2006; Thomas et al. 2009), controlled by a quorum-sensing mechanism in *P. aeruginosa* biofilms (Allesen-Holm et al. 2006) and via altruistic suicide and fractricidal release of DNA in *E. faecalis* (Thomas et al. 2009). Actively excreted eDNA appears to provide another source. A recent study showed the production of eDNA during early biofilm formation in *E. faecalis* in a manner that was independent of cell lysis (Barnes et al. 2012). Further, the well-defined structure of the eDNA suggested a novel form of DNA secretion by viable cells (Barnes et al. 2012). In the case of *E. faecalis*, the cumulative data suggests that active eDNA secretion may be one of the earliest events in biofilm development that occurs following initial surface attachment by planktonic cells. The eDNA component of the matrix plays an essential role in the development of microcolonies and ultimately mature biofilms with characteristic three-dimensional architecture (Fig. 3). As the population density of the adherent community increases, expression of the GelE and SpreE proteases in a subpopulation of cells serves to both immunize the producer cells from autolysis (SpreE) and render non-protease-secreting neighbors susceptible to autolysis (GelE) and eDNA release (Thomas et al. 2009). Thus, the first 8–16 h of biofilm development entail two different mechanisms of eDNA production. SEM analysis (Barnes et al. 2012) suggests that eDNA structure may be different for the material released by each mechanism. Most studies have focused on biofilms grown for at least 24 h.

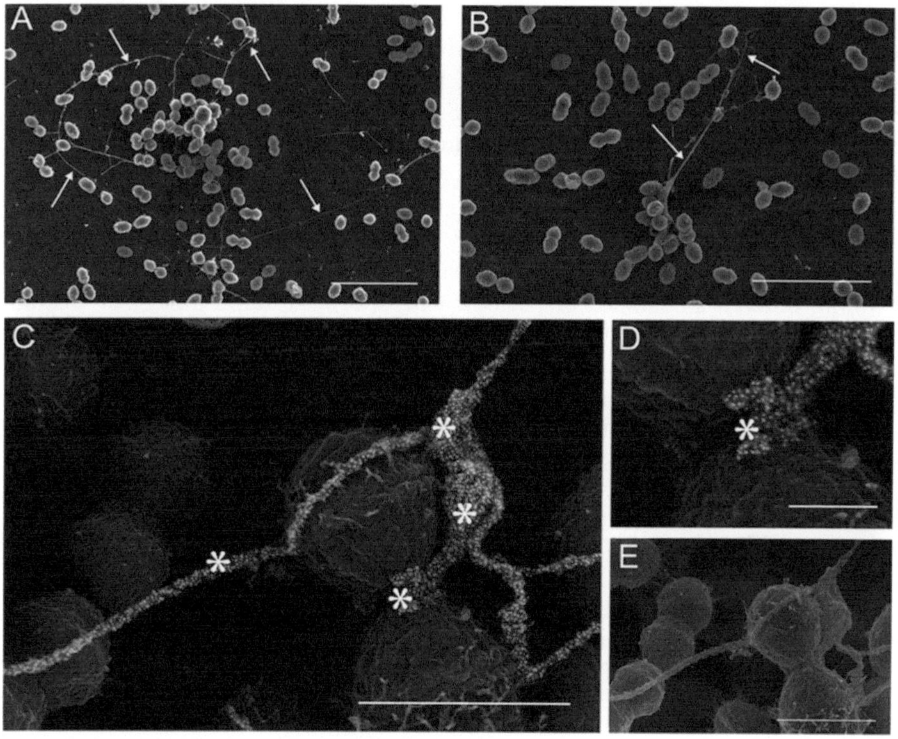

Fig. 3 Ultrastructural analysis of eDNA distribution in early biofilms. (**a** and **b**) Long intercellular strands (*yarn structures*) visible in early (4-h) *E. faecalis* OG1RF biofilms (*arrows*) (bar, 5 μm). (**c**) Strong punctate labeling of the strands visualized by using an anti-dsDNA monoclonal antibody conjugated to an immunogold particle (*asterisks*). The image shows the significant eDNA component (the central area is magnified in panel **d**). Bar, 250 nm. (**e**) Surface morphology can be appreciated in the corresponding secondary electron image. Bar, 1 μm [Reprinted from (Barnes et al. 2012), with permission]

As exemplified by these recent studies of eDNA production, it is clear that much remains to be learned about the critical events in early biofilm development, including how antibiotic sensitivities of biofilm cells change during development.

2.2 Cell–Cell Communication and Quorum Sensing

The fundamental processes in biofilm formation involve attachment to a surface (initial followed by irreversible), cell proliferation, adherence to other bacteria, production of an extracellular polymeric substance matrix, maturation, and dispersal (Costerton et al. 1999). Much has been learned about these processes from studies of the oral microflora and dental plaque (Kolenbrander et al. 2010). From these observations, it is apparent that spatial distribution of species within biofilms and cell–cell distance are critical factors for intermicrobial communication processes. Interactions are highly specific as demonstrated, for example, by the coinvasion of dentinal tubules by *Porphyromonas gingivalis* with *Streptococcus gordonii* that is facilitated by the recognition of the antigen I/II polypeptide (Love and Jenkinson 2002; Love et al. 2000). Similarly, in isolates recovered from root canal infections, both autoaggregation and coaggregation interactions have been observed, particularly in association with *Prevotella*, *Streptococcus*, and *Fusobacterium* species (Khemaleelakul et al. 2006) and between *E. faecalis* and *Fusobacterium nucleatum* and between *F. nucleatum* and *Streptococcus anginosus*, *Peptostreptococcus anaerobius*, and *Prevotella oralis* (Johnson et al. 2006). By forming ordered assemblies of bacteria with species-specific molecular interactions, bacteria are better able to adapt to fluctuating environmental conditions in biofilms (Fux et al. 2005).

Bacteria can communicate, cooperate, and alter their behavior in response to changes in their communal environment by individual and social processes (Elias and Banin 2012) (Fig. 4). Some bacteria communicate and coordinate behavior via

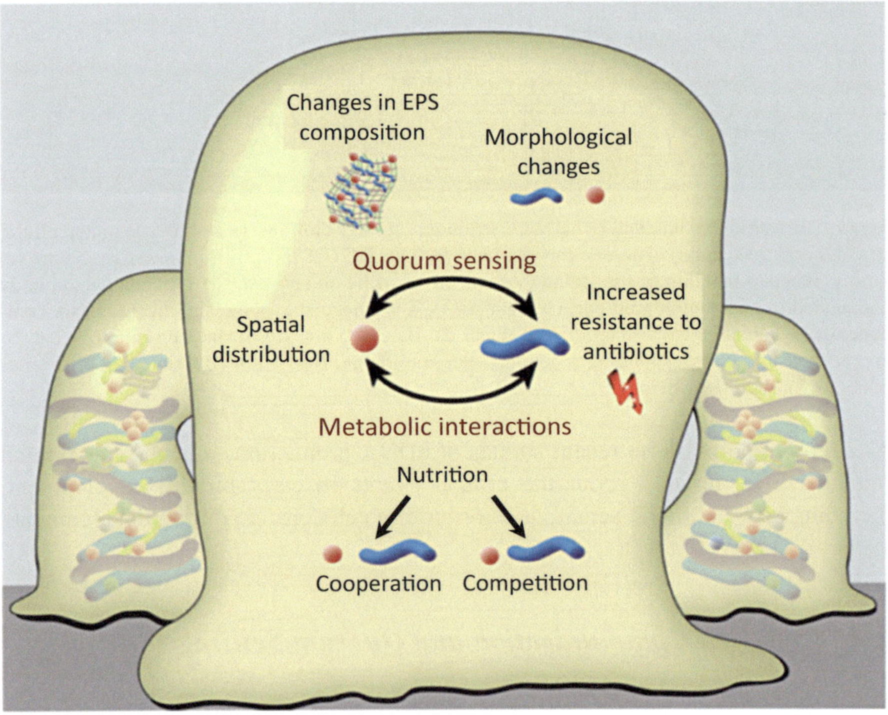

Fig. 4 Individual and social processes occurring within biofilm communities. Microorganisms within a mixed biofilm interact via quorum sensing and/or metabolically. Interactions can be synergistic or antagonistic and result in phenotypic changes, such as increased resistance to antimicrobial agents or to host defense systems, spatial distribution, or emergence of variants. Nutritional interactions can be either competitive or cooperative [Adapted by permission from John Wiley and Sons: FEMS Microbiology Reviews (Elias and Banin 2012) ©2012]

the constitutive synthesis of small signaling molecules using an intercellular signaling system called quorum sensing (Parsek and Greenberg 2005; Keller and Surette 2006), possibly in association with cyclic di-GMP (Srivastava and Waters 2012; Cotter and Stibitz 2007). When a high density population reaches a certain threshold ("quorum"), the concentration of normally low levels of certain signal molecules becomes high enough to act as autoinducers that trigger a synchronized response throughout the biofilm population. The increased expression of specific operons that enhance survival is important in bringing about the next stage of a colonization or virulence process (Parsek and Greenberg 2005). Binding of autoinducer to cognate receptors initiates transcription of genes involved in regulation of cell density. Gram-negative bacteria use *N*-acyl homoserine lactone-based signaling (Fuqua and Greenberg 2002). Small peptides are often involved in gram-positive bacteria, while autoinducer-2 (AI-2) signal is used by both gram-negative and gram-positive bacteria (Li and Nair 2012; Parsek and Greenberg 2005).

Quorum sensing allows the coordinated regulation of expression of key proteins in biofilms, with potential implications for pathogenicity. For example, AI-2 produced by the oropharyngeal flora in CF patients can modulate gene expression of *P. aeruginosa* to enhance pathogenicity (Duan et al. 2003), and *S. mutans* mutants with a defective quorum-sensing signaling system produced abnormal biofilms (Allegrucci and Sauer 2007). In *P. aeruginosa*, carbapenum antibiotic resistance and the *las* quorum-sensing system were shown to be under the control of CzcR, a key transcriptional regulator (Dieppois et al. 2012). Antimicrobials that disrupt quorum sensing appear to result in inhibition of pathogen virulence, but not necessarily lethality (Hentzer et al. 2003; Clatworthy et al. 2007). A key challenge remains to determine whether the induction of quorum sensing influences the antimicrobial tolerance of the biofilm (Parsek and Greenberg 2005).

2.3 Horizontal Gene Transfer

Horizontal gene transfer (HGT) allows the movement of genetic information both within and between species. The horizontal gene pool includes plasmids, bacteriophages, transposons, insertion sequences, and pathogenicity islands. These are particularly important from a clinical perspective because of their involvement in the dissemination of antibiotic resistance (Clewell and Francia 2004; Salyers et al. 1995).

HGT occurs by three basic methods: transformation, transduction, and conjugation. Extracellular DNA acquired by natural transformation includes fragments of DNA that can recombine with homologous regions of the recipient genome or with plasmids (Lorenz and Wackernagel 1994). The most efficient HGT process in bacteria is conjugation, with the requirement for cell-to-cell contact distinguishing conjugation from transduction and transformation. Elements similar to the conjugative transposon Tn*916* have been detected in tetracycline-resistant bacteria (*Streptococcus mitis* and *Neisseria* sp.) isolated from root canals (Rossi-Fedele et al. 2006). Some plasmids conjugatively transfer copies of themselves from one bacterial cell to another using small peptides called pheromones as essential signals in the process. Conjugative plasmids that make use of pheromones were first observed in an oral *E. faecalis* strain recovered from a patient with acute periodontitis (Dunny et al. 1978, 1979). Specifically, when conjugation functions are activated, there is a dramatic "clumping response" mediated by the appearance of a surface adhesin ("aggregation substance" or "AS") that facilitates the attachment of the donor cells to enterococcal binding substance (EBS) which is present on the surface of recipients as well as donors (Dunny et al. 1979). Even though the probability of random meetings of cells in suspension is greater than in biofilms, the relative spatial stability of bacteria in biofilms should favor conjugation (Hausner and Wuertz 1999). Recent data have shown that growth in biofilms alters the induction of conjugation by a sex pheromone in *E. faecalis*, resulting in increased plasmid copy number and heterogeneity under biofilm compared to planktonic growth conditions; however, it is likely that this process occurs only when donor cells are in extremely close proximity, or direct contact, with recipient cells (Cook et al. 2011).

Biofilm growth has also been shown to enhance genetic exchange via transformation, as exemplified by several streptococcal species naturally competent for transformation. In a series of seminal studies, Li et al. showed that peptide-mediated intercellular signaling actually co-regulates both biofilm formation and expression of competence in *S. mutans* (Li et al. 2001a, b, 2002). They also found that biofilm growth represents an efficient ecological niche for transformation; in fact their studies suggest that this may be where the process generally occurs in nature. A similar situation applies in the case of biofilm formation and competence in *S. pneumoniae* (Claverys and Havarstein 2007; Claverys et al. 2007; Havarstein et al. 2006). In this case, the fratricidal activities of the "early responder" cells to the competence stimulating peptide pheromone seem to cause lysis and eDNA release by neighboring cells. This likely contributes to the structural integrity of the biofilm matrix as well as providing a DNA substrate for uptake and genome incorporation by the highly competent cells in the population.

Biofilms are uniquely suited for DNA exchange since they sustain high bacterial density, and DNA can be trapped within the extracellular matrix (Sorensen et al. 2005; Kolenbrander et al. 2010). Open channels and pores may enable more frequent cell collisions, leading to rapid spread of plasmid-borne genes by conjugative gene transfer (Sorensen et al. 2005). It has been hypothesized that gene transfer takes place within biofilms with "bottlenecks" in the process due to the density of the biofilm structure increasing the likelihood of plasmid transfer (Molin and Tolker-Nielsen 2003). Nonetheless, when bacteria are dispersed from a "resistant" biofilm, they usually rapidly become susceptible to antibiotics (Spoering and Lewis 2001), which suggests that bacterial resistance in biofilms is not acquired via mutations or mobile genetic elements (Stewart and Costerton 2001).

The significance of these findings regarding gene transfer between microorganisms in root canal infections remains to be established. However, in enterococcal isolates recovered from dental patients in Sweden, phenotypic studies showed that

16 of the plasmid-positive strains exhibited a "clumping response" (characteristic of a response to pheromone) when exposed to a culture filtrate of a plasmid-free strain, suggesting the potential for conjugative transfer of genetic elements in these endodontic isolates (Sedgley et al. 2005). In addition, in an ex vivo model, antibiotic resistance gene transfer occurred between *S. gordonii* and *E. faecalis* in root canals of teeth, demonstrating the potential for the adoption of an optimal genetic profile for survival in the root canal (Sedgley et al. 2008).

2.4 Stress Responses

Environmental conditions in biofilms are constantly changing, and bacteria encounter a multitude of different types of stresses and challenges. Stress responses that positively impact the recruitment of resistance determinants, or promote physiological changes that compromise antimicrobial activity, include exposure to nutrient starvation/limitation (nutrient stress), reactive oxygen and nitrogen species (oxidative/nitrosative stress), membrane damage (envelope stress), elevated temperature (heat stress), and ribosome disruption (ribosomal stress) (Poole 2012).

Global response systems facilitate adaptation and survival by modulating intracellular metabolic processes in response to stress. Shared features of several stress responses include downregulation of error-correcting enzymes, upregulation and activation of error-prone DNA polymerases, and HGT of mobile genetic elements (Rice 1998; Foster 2007). Controlling these responses are a number of global regulators that include the alternative sigma factors RpoS (Hengge-Aronis 2002; Adnan et al. 2010) and RpoH (Guisbert et al. 2008), gene repressor LexA (Kelley 2006), and small molecule effectors such as (p)ppGpp induced by the stringent response (Potrykus and Cashel 2008). The role of *rpoS* may depend on strain-specific cofactors and specific growth conditions, as was demonstrated by impairment of biofilm growth by *rpoS* mutant *E. coli* (Schembri et al. 2003) and the higher antimicrobial tolerance shown by *rpoS* mutant *P. aeruginosa* (Whiteley et al. 2001). In addition, changes in (p)ppGpp pools can have profound effects on the ability of *E. faecalis* to form, develop, and maintain stable biofilms in vitro; strains lacking (p)ppGpp had a diminished capacity to sustain biofilm formation over an extended period of time and expressed abundant proteolytic activity (Chavez de Paz et al. 2012).

Toxin–antitoxin (TA) systems are also involved in responses to stress stimuli (Hayes and Van Melderen 2011). Within these two-component systems, toxins are stable proteins directed against specific intracellular targets, while antitoxins are degradable proteins or small RNAs that neutralize the toxin or inhibit toxin synthesis. TA systems may be important regulators of the switch from planktonic to biofilm lifestyles; in *E. coli* the antitoxin MqsA repressed *rpoS* which in turn reduced the concentration of the internal messenger cyclic di-GMP, resulting in increased motility and decreased biofilm formation (Wang et al. 2011). Among other functions, TA systems are also involved with persister cell formation

(Kim et al. 2010) and quorum sensing (Belitsky et al. 2011). The above diverse responses to stress have been described as "insurance effects" (Boles et al. 2004) and are likely to provide the biofilm with a greater chance of surviving environmental stresses such as exposure to antimicrobials.

2.5 Heterogeneity and Oxygen Gradients

The availability of nutrients and electron acceptors can vary throughout the biofilm. Subsequently, since the metabolic activity, growth status, and gene expression pattern of individual strains are heterogeneous, the growth of cells can range from rapidly growing to dormant (Stewart and Franklin 2008). This has been observed in *P. aeruginosa* biofilms using fluorescent reporter gene constructs; mature *P. aeruginosa* biofilms were shown to simultaneously harbor active, growing cells as well as large numbers of cells that were inactive and not growing (Werner et al. 2004). The high level of activity on the surface and limited or absent growth deeper in the biofilm reduce the susceptibility to antibiotics (Mah and O'Toole 2001) and are considered to contribute to the poor response of biofilm infections to antimicrobial chemotherapy (Borriello et al. 2004). For example, slow growth and dormancy by microbial cells provide protection against antibiotics such as the beta-lactams that are active on rapidly growing bacteria.

Similarly, cells located more deeply in the biofilm are exposed to decreased oxygen tension resulting in altered phenotypes in terms of growth rate and gene transcription. In *P. aeruginosa* biofilms, oxygen limitation and consequential low metabolic activity have been shown to contribute to antibiotic tolerance (Walters et al. 2003; Borriello et al. 2004). Oxygen depletion locally within the biofilm could cause bacteria in the anoxic zone to enter a stationary-phase state in which they are less susceptible to an antibiotic (Walters et al. 2003). Indeed, compared with cells at or near the surface, both oxygen levels and metabolic rates are reduced at the center of a microcolony (Sternberg et al. 1999; de Beer et al. 1994).

Oxygen dynamics in biofilms also depend on complex interactions between fluid flow, the distribution of the biomass and mass transfer resistance, or mass transport limitation occurring external to the biofilm subsequent to diffusion limitation occurring internally (Staal et al. 2011). An increase in external mass transfer resistance is thought to have multiple effects including exacerbation of oxygen or nutrient limitations in biofilms, an increase in the concentration of metabolic products in the biofilms, the development of tall, fingerlike biofilm clusters, and possible effects on quorum sensing (Stewart 2012). It is likely that these factors are also critical in the movement of antimicrobial agents throughout the biofilm matrix.

2.6 Low Penetration of Antimicrobials

The extent of exposure of cells in biofilms to antimicrobials will depend on many factors, including the thickness of the biofilm and the concentration of the antimicrobials and their ability to penetrate the matrix. While superficial cells can be rapidly exposed to high concentrations of antibiotics, in deeper parts of the biofilm, exposure to increasing concentrations of antibiotic will be more gradual and depend on the ability of the antibiotic to diffuse through the biofilm (Stewart and Costerton 2001). An inability of the antimicrobial agent to penetrate the entire depth of the biofilm is one mechanism of biofilm resistance (Costerton et al. 1999). Biofilms can reduce antibiotic penetration to a varying extent depending on the antibiotic (Suci et al. 1994; Vrany et al. 1997). However, while the biofilm matrix can slow diffusion of the agent (Konig et al. 2001), it does not necessarily prevent it (Spoering and Lewis 2001; Anderl et al. 2000). For example, rifampin penetrated *S. epidermidis* biofilms, but was ineffective in killing the bacteria (Zheng and Stewart 2002). Similarly ampicillin and ciprofloxacin penetrated through *K. pneumoniae* biofilms but failed to kill the bacteria (Zahller and Stewart 2002; Anderl et al. 2000). This response has been attributed to areas within the biofilms experiencing localized nutrient limitation that cause cells to enter stationary phase, thereby becoming less susceptible to killing by antibiotics that target dividing cells (Anderl et al. 2000). In addition, when bacteria are dispersed from a "resistant" biofilm, they usually rapidly become susceptible to antibiotics (Spoering and Lewis 2001), which suggests that bacterial resistance in biofilms is not acquired via mutations or mobile genetic elements (Stewart and Costerton 2001).

Interestingly, a recent study has identified "swimmer cells" propelled by flagella in *Bacillus thuringiensis* biofilms that form a subpopulation (0.1–1 %) that remain motile and move through the biofilm mass; their role might be to create pores in the biomass that facilitate the diffusion of nutrients and macromolecules (Houry et al. 2012). Alternatively, swimmers could express a bactericidal substance or provide the opportunity for penetration of environmental toxins such as antimicrobials into the biomass.

2.7 Upregulated Efflux Pumps

Efflux systems can actively pump toxic substances and antibiotics out of cells (Li and Nikaido 2009). For example, in *P. aeruginosa* grown under planktonic conditions, increased efflux pump activity resulting from mutations is a major resistance mechanism against aminoglycosides, fluoroquinolones, and tobramycin (Ciofu 2003; Jalal et al. 2000; Islam et al. 2009). Multidrug resistance efflux pumps may also contribute to antibiotic resistance in biofilms (Li and Nikaido 2009; Kvist et al. 2008; Zhang and Mah 2008). In addition, antibiotics at subinhibitory concentrations have been shown to induce gene expression of efflux pumps in *E. coli* that contributed to biofilm maturation (May et al. 2009).

From a therapeutic perspective, efflux pump inhibitors may be useful for decreasing biofilm formation (Kvist et al. 2008; Matsumura et al. 2011; Baugh et al. 2012). The inclusion of efflux pump inhibitors thioridazine, Phe-Arg β-naphthylamide (PAβN), and NMP resulted in decreased biofilm formation by *E. coli*, *Klebsiella*, *S. aureus*, and *Pseudomonas putida* strains (Kvist et al. 2008). Similarly, the addition of an efflux pump inhibitor (verapamil) to endodontic antimicrobial medicaments (calcium hydroxide, chitosan nanoparticles, and light-activated disinfection) enhanced their antibiofilm activity against *E. faecalis* biofilms (Upadya et al. 2011).

2.8 Persister Cells

Persister cells are nondividing dormant cells forming a small subpopulation of a susceptible and genetically homogeneous population of bacteria that survive antibiotic exposure (Lewis 2010). The cells are phenotypic variants genetically identical to the susceptible cells within a clonal population (Keren et al. 2004a; Wiuff et al. 2005) that demonstrate very slow or arrested growth and diminished protein synthesis (Shah et al. 2006; Lewis 2010). Persisters are thought to be formed through a combination of continuously occurring random events ("stochastic") or in response to an environmental stimulus ("deterministic") (Gefen and Balaban 2009; Lewis 2010). For the biofilm "community," the presence of these tolerant phenotypic variants ensures that the population survives. Following treatment of biofilms with antibiotics, the rapidly growing cells are killed, leaving the dormant cells to repopulate the biofilm.

As distinct to antimicrobial-resistant cells that prevent the interaction of an antibiotic with its intended target, persister cells appear to exhibit tolerance to bactericidal antibiotics independent of resistance mechanisms (Lewis 2010). This is supported by the analyses of high-level persistent (*hip*) mutants with *hip* gain-of-function loci (Moyed and Bertrand 1983). The *hip* locus shares the characteristics of toxin–antitoxin modules. Mutations of the *hipA7* gene increased tolerance in *E. coli*, likely by inducing (p)ppGpp synthesis and potentiating the transition to a dormant state upon application of stress (Korch et al. 2003). Furthermore, growth of *E. coli hipA7* mutants with ampicillin or ofloxacin resulted in a 10 to $10^{(5)}$-fold higher level of persister cells (Keren et al. 2004a). Mutant strains lacking persisters have not been isolated, indicating that dormancy mechanisms are redundant (Lewis 2010). In spite of this, investigators have shown that persistence results from the inhibition of translation of antibiotic targets by toxin/antitoxin modules (Keren et al. 2004b). In the case of *E. coli* persisters, there was increased expression of RelE, the toxin component of a toxin/antitoxin (TA) module (Keren et al. 2004b), and ciprofloxacin treatment promoted the formation of ciprofloxacin-resistant persisters; this action was dependent on the SOS response (Dorr et al. 2009) which activates expression of DNA repair enzymes. Knockout of the SOS response-inducible TA locus *tisAB/istR* markedly reduced the levels of ciprofloxacin-resistant persisters. Furthermore, ciprofloxacin was shown to induce expression of the *tisB* toxin gene, and TisB-expressing cells were tolerant to other antibiotics; the formation of dormant persisters was found to be associated with decreasing proton motive force and ATP levels in response to the toxin TisB (Dorr et al. 2010). These findings have been supported by other reports of the role of toxin–antitoxin (TA) gene pairs/module role in persister formation in *E. coli* biofilms (Wang and Wood 2011).

Persister cells have been associated with recalcitrant biofilm infections (Spoering and Lewis 2001; Lewis 2010; Fauvart et al. 2011). Overexpression of *hip* mutants over the course of antimicrobial treatment was shown in *P. aeruginosa* recovered from CF patients with chronic lung infections (Mulcahy et al. 2010) and *C. albicans* recovered from cancer patients with oral thrush (Lafleur et al. 2010). It is feasible that persister cells in immunocompetent patients are eventually targeted by immune cells (Lewis 2007). However, in chronic diseases, such as tuberculosis and CF lung infections where the immune system is compromised, this is less likely to occur, and persister cells could become a nidus for recurrent infectious episodes (Lewis 2010). A recent study has implicated oxidative stress in the formation of *E. coli* persister tolerance; and the intriguing scenario was raised that immune cells, by way of producing reactive oxygen species (ROS) and reactive nitrogen species (RNS) compounds, could activate multidrug tolerance in pathogens (Wu et al. 2012).

2.9 Biofilm Phenotype, Phase Variation, and Genetic Variation

Costerton and colleagues hypothesized that some cells in a biofilm adopt a distinct and protected biofilm phenotype that is a biologically programmed response to growth on a surface (Costerton et al. 1999). This hypothesis has been supported by observations of distinct protein expression patterns at different stages of biofilm growth shown by *P. aeruginosa* in proteomic studies (Sauer et al. 2002) and the existence of a regulatory protein (PvrR) that controls the conversion between antibiotic-susceptible and antibiotic-resistant forms (Drenkard and Ausubel 2002). It was proposed that antibiotic treatment and biofilm growth selected for the high frequency of antibiotic-resistant phenotypic variants of *P. aeruginosa* and contributed to chronic colonization of lungs in CF patients (Drenkard and Ausubel 2002). Further support was provided by DNA microarray analyses of biofilms and planktonic cultures grown under similar environmental conditions, where gene expression differed by only 6 % in *B. subtilis* (Stanley et al. 2003) and 1 % in *P. aeruginosa* (Whiteley et al. 2001). However, when *P. aeruginosa* biofilms were exposed to high levels of tobramycin, differential expression of 20 genes occurred, suggesting that existence in the biofilm mode induces moderate levels of resistance to antimicrobial treatments by inducing genes specific to that antibiotic (Whiteley et al. 2001).

Phase variation, or the random on–off switching of phenotypes identified in several bacterial species (Henderson et al. 1999), is thought to play an important role in the formation of diverse phenotypes in the biofilm community (Stewart and Franklin 2008; Tormo et al. 2007). Phase variation involves two general properties: reversible and inheritable gene mutations for inducing a phenotypic change and internal genome rearrangements that can utilize mobile genetic elements that have inserted into, or recombined with, the host genome (Chia et al. 2008). Pathogenic bacteria can utilize phase variation strategies to avoid detection by the host immune system by switching on the expression of proteins when they are needed and switching them off when they are likely to trigger an immune response. For example, in *S. aureus* biofilms a phase variation process controls expression of the *ica* genes involved in exopolysaccharide synthesis (Valle et al. 2007). Similarly, in *S. epidermidis*, biofilm-negative variants can be produced by the reversible inactivation of *ica* (Ziebuhr et al. 1999), staphylococcal accessory regulator *sarA*, and sigmaB regulator gene *rsbU* by the insertion sequence IS256 (Conlon et al. 2004).

Genetic variation occurs through mutations or recombinations that result in changes to the DNA sequence. Mutagenicity and efficient acquisition of resistance genes via HGT within biofilms (Molin and Tolker-Nielsen 2003) are both involved in the development of antibiotic resistance (Hoiby et al. 2010). Variants can comprise as much as 10 % of a mature biofilm population (Kirisits et al. 2005; Boles et al. 2004). Mutations in either the DNA oxidative lesion repair system (Oliver et al. 2002) or the mismatch repair system (Mandsberg et al. 2009) have contributed to the emergence of antibiotic-resistant strains of *P. aeruginosa*. For example, in CF patients administered repeated doses of antibiotics for lung infections, the hypermutability of *P. aeruginosa* in biofilms promotes the selection of mutants conferring resistance to multiple antibiotics such as beta-lactams because of mutations in the regulatory genes of beta-lactamase production (Ciofu 2003).

3 Antimicrobial Resistance in Mixed Species Biofilms

Multispecies biofilms have been shown to be significantly more resistant to antimicrobial treatment than single-species biofilms (Kara et al. 2006; Burmolle et al. 2006; Leriche et al. 2003). This has been attributed to the cooperative behavior of the biofilm community enabling survival upon exposure to antimicrobial agents (Elias and Banin 2012). Further, it has been shown that there are potential advantages to be gained by residing in multispecies biofilms compared to single-species biofilms (Burmolle et al. 2006). For example, when grown together as a multispecies biofilm, the biomass of four marine-derived isolates increased compared to single-species biofilms, and further, when exposed to hydrogen peroxide and tetracycline to evaluate oxidative stress and protein synthesis, multispecies biofilms demonstrated fitness advantages over single-species growth (Burmolle et al. 2006). Extracellular secreted factors and species-specific physical interactions between cells were associated with increased resistance to antimicrobial agents. Likewise, in dual-species biofilms of *C. albicans* and *S. epidermidis*, both species benefited by their association (Adam et al. 2002). Further, the presence of *Candida albicans* protected the slime-negative staphylococcus against vancomycin, and fluconazole penetration into the yeast was inhibited by the extracellular polymer produced by *S. epidermidis* (Adam et al. 2002).

Similarly, dual-species biofilms of oral plaque species *S. mutans* and *Veillonella parvula* were less susceptible to chlorhexidine, cetylpyridinium chloride, zinc chloride, erythromycin, hydrogen peroxide, and amine chloride than single-species biofilms of the same microorganisms (Kara et al. 2006; Luppens et al. 2008). Subsequent analyses of biofilm spatial arrangements showed differences between the single- and dual-species biofilms in microstructural alterations in response to chlorhexidine exposure. Dual-species biofilms, but not single-species biofilms, had formed distinct clusters that were considered to account for the increased resistance to chlorhexidine (Kara et al. 2007). Further, transcript levels in *S. mutans* growing in dual-species biofilms compared to single-species biofilm exhibited several differences, indicating that *V. parvula* induced gene expression changes in *S. mutans* (Luppens et al. 2008). Although not tested, the authors proposed that increased expression induced by *V. parvula* of *S. mutans* genes encoding ribosomal proteins could facilitate repair mechanisms after exposure of the dual-species biofilms to antimicrobials (Luppens et al. 2008). It is clear that much remains to be discovered about antibiotic sensitivities of multispecies biofilms.

4 Antimicrobial Resistance in the Root Canal Microflora

Intraradicular biofilms (Fig. 5) have been attributed as the primary cause of root canal infections and longstanding pathologic processes such as large periapical lesions and cysts (Ricucci et al. 2009; Ricucci and Siqueira 2010). Genera that have been frequently cultured from infected root canals and periapical abscesses include *Prevotella*, *Porphyromonas*, *Fusobacterium*, *Peptostreptococcus*, *Streptococcus*, *Lactobacillus*, *Enterococcus*, *Actinomyces*, *Propionibacterium*, and *Candida*. Analyses of endodontic infection samples using molecular techniques have shown a highly diverse flora composed of a large proportion of unidentifiable and unculturable species (Siqueira and Rocas 2005), and recent investigations using pyrosequencing have revealed a flora even more diverse than shown with traditional Sanger sequencing (Li et al. 2010; Ozok et al. 2012; Siqueira et al. 2011).

The antimicrobial activity of root canal irrigants, intracanal medicaments, and root filling materials have for the most part been evaluated using strains (predominantly *E. faecalis*) grown under planktonic conditions, although more recent studies have included biofilms. These studies will be further discussed in chapter

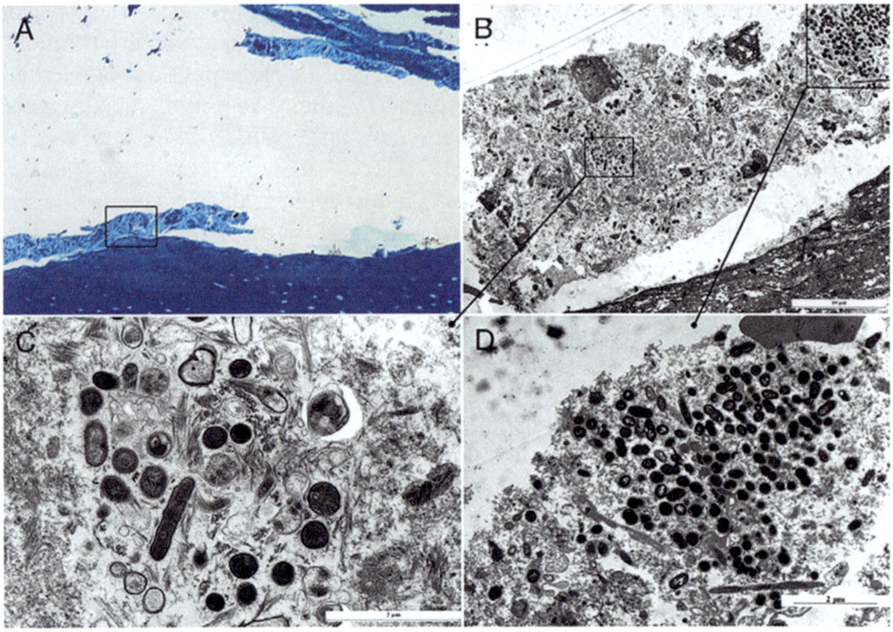

Fig. 5 Mesial root tip of an infected human lower molar tooth showing multispecies biofilm. (**a**) Semi-thin histologic section showing part of isthmus that communicates between the mesiobuccal and mesiolingual root canals. (**b**) Transmission electron microscopic examination of *boxed area* in (**a**) showing varied multispecies biofilm. (**c**) Higher magnification of *small insert area* from (**b**) showing multiple phenotypes within a complex biofilm. (**d**) Higher magnification of larger insert in (**b**) showing complex biofilm community [Reprinted from Journal of Endodontics, Vol 35(9), Carr GB, Schwartz RS, Schaudinn C, Gorur A, Costerton JW. Ultrastructural examination of failed molar retreatment with secondary apical periodontitis: an examination of endodontic biofilms in an endodontic retreatment failure, 1303–9, ©2009, with permission from Elsevier]

"Inter-appointment Medication with Calcium Hydroxide in Routine Cases of Root Canal Therapy." Information on antibiotic resistance characteristics of root canal isolates has been primarily acquired by using standardized MIC tests on a limited group of culturable species (Table 1).

The available data show that while antibiotic resistance is not uncommon, there is a low incidence of multidrug-resistant strains. A recent study using molecular techniques evaluated the effects of root canal treatment procedures (including interappointment medication with calcium hydroxide) on the eradication of microbial antibiotic resistance genes in 45 patients undergoing primary endodontic treatment or retreatment (Jungermann et al. 2011). Polymerase chain reaction was used to screen for genes associated with resistance to beta-lactams [bla_{TEM-1}, *cfxA*, and *blaZ*], tetracycline (*tetM*, *tetW*, *tetQ*), and vancomycin (*vanA*, *vanD*, and *vanE*). The authors reported that bla_{TEM-1} was more prevalent in primary than persistent root canal infections. After treatment procedures, the prevalence of *tetM* was unchanged. However, the prevalence of bla_{TEM-1} and *tetW* was significantly

Table 1 Studies on antibiotic resistance associated with root canal bacteria

Clinical features	Species (strains)	Antibiotics tested[a]	Resistance (% where provided)	Country	Reference
Root canals					
Primary treatment	Mixed anaerobes (38)	AMC, AMX, TET	None	France	Le Goff et al. (1997)
Primary treatment	*E. faecalis* (26), *E. faecium* (3)	AMP, CLI, ERY, MTZ, PEN, TET, VAN	AMP, CLI (96), MTZ (100), PEN, TET	Sweden	Dahlen et al. (2000)
Primary and acute apical abscess	*P. prevotii* (13), *F. necrophorum* (7)	AMC, AMX, AZM, CLI, ERY, MTZ, PEN	AZM, ERY	Brazil	de Sousa et al. (2003)
Primary and symptomatic	*F. necrophorum* (10), *F. nucleatum* (9), *P. micros* (10), *P. prevotii* (9)	AMC, AMX, AZM, CEC, CLI, ERY, MTZ, PEN	CLI, ERY, MTZ, PEN	Brazil	Jacinto et al. (2003)
	"Black-pigmented bacteria"				
Retreatment	*E. faecalis* (10), *Peptostreptococcus* (6)	AMC, AMX, AZM, ERY, PEN	AZM, ERY	Brazil	Pinheiro et al. (2003)
Retreatment	*E. faecalis* (21)	AMC, AMX, AZM, CHL, CIP, DOX,	AZM (38), DOX (14), ERY (10), TET (14)	Brazil	Pinheiro et al. (2004)
		ERY, MOX, PEN, TET, VAN			
Primary and retreatment	*E. faecalis* (31), *E. faecium* (2)	AMP, CHL, CLI, ERY, FUS, GEN, KAN, PEN, RIF, STR, TET, VAN	GEN (3), TET (16)	Sweden	Sedgley et al. (2005)
Primary and retreatment	*E. faecalis* (59)	AMP, CHL, CIP, CLI, CTX, ERY, GEN, LZD, PEN, Q-D, RIF, STR, TEC, TET, VAN	CTX (7), ERY (10), RIF (58), STR (7), TET (29)	Finland, Lithuania	Reynaud Af Geijersstam et al. (2007)
Primary and apical abscess	*P. gingivalis* (20)	AMC, AMX, AZM, CEC, CLI, ERY, MTZ, PEN, TET	AZM (40), ERY (5)	Brazil	Jacinto et al. (2006)
Primary	*F. nucleatum* (38)	AMC, AMX, AZM, CEC, CLI, ERY, MTZ, PEN, TET	CLI, ERY, MTZ, PEN	Brazil	Jacinto et al. (2008)
	F. necrophorum (20)				

(continued)

Table 1 (continued)

Clinical features	Species (strains)	Antibiotics tested[a]	Resistance (% where provided)	Country	Reference
Primary and symptomatic	*F. nucleatum* (44)	AMC, AMX, CLI, ERY, MTZ, PEN	CLI, ERY, PEN	Brazil	Gomes et al. (2011)
Collected 2000–2008	*P. intermedia*/*nigrescens* (35)		Increased CLI and PEN resistance over time		
	P. oralis (23), *P. micra* (34)				
Primary and retreatment	DNA from 45 samples	Resistance genes to beta-lactams, (*bla*$_{TEM-1}$, *cfxA*, and *blaZ*), TET (*tetM*, *tetQ*, *tetW*), VAN (*vanA*, *vanD*, *vanE*)	Preoperative: *bla*$_{TEM-1}$ (33), *blaZ* (2), *cfxA* (11), *tetM* (18), *tetW* (18), *tetQ* (9) Preobturation: *bla*$_{TEM-1}$ (9), *blaZ* (0), *cfxA* (0), *tetM* (22), *tetQ* (0), *tetW* (2)	USA	Jungermann et al. (2011)
Root canal and pus aspirates					
Primary and retreatment, apical abscess aspirates	Mixed (66)	AMC, AMP, AMX, CLI, ERY, MTZ, PEN, TET, VAN	MTZ (~50), TET (~40)	Lithuania	Skucaite et al. (2010)
Root canals and saliva					
Retreatment	*E. faecalis* (19)	AMP, CHL, ERY, GEN, MTZ, PEN, STR, TET, VAN	MTZ (100), STR (100)	China	Zhu et al. (2010)
Pus aspirates					
Acute apical abscess aspirates (17)	Mixed (118)	AMC, AMX, CLI, CLR, MTZ, PEN	AMX (15), CLI (11), MTZ (12), PEN (19)	Thailand	Khemaleelakul et al. (2002)
Apical abscess aspirates (12)	Mixed (98)	AMC, AMX, CLI, CLR, MTZ, PEN	AMX (9), CLI (4), MTZ (55), PEN (15)	USA	Baumgartner and Xia (2003)
From abscess (72), root canal (15)	*Prevotella* spp. (139)	*cfxA* and *cfxA2* (beta-lactamase genes) Production of lactamase	*cfxA* and *cfxA2* (31) Production of lactamase (31)	Japan	Iwahara et al. (2006)

Periradicular tissues					
22 periapical lesions	Mixed (53)	CLI, CTX, ERY, FOX, MTZ, PEN, TET	CLI (6), CTX (11), ERY (4), FOX (15), MTZ (25), PEN (19), TET (6)	USA	Vigil et al. (1997)

Method used was Etest® for all studies except Reynaud Af Geijersstam et al. (2007), agar dilution plus Etest®; Iwahara et al. (2006), real-time PCR; Jungermann et al. (2011), PCR

[a] *AMX* amoxicillin, *AMC* amoxicillin–clavulanic acid, *AMP* ampicillin, *AZM* azithromycin, *CEC* cefaclor, *CTX* cefotaxime, *FOX* cefoxitin, *CHL* chloramphenicol, *CIP* ciprofloxacin, *CLR* clarithromycin, *CLI* clindamycin, *DOX* doxycycline, *ERY* erythromycin, *FUS* fusidic acid, *GEN* gentamicin, *KAN* kanamycin, *LZD* linezolid, *MTZ* metronidazole, *MOX* moxyfloxacin, *PEN* penicillin, *Q-D* quinupristin–dalfopristin (Synercid), *RIF* rifampin, *STR* streptomycin, *TEC* teicoplanin, *TET* tetracycline, *VAN* vancomycin

reduced, while of *cfxA*, *blaZ*, and *tetQ* was eliminated. No specimens contained *vanA*, *vanD*, or *vanE* (Jungermann et al. 2011).

It is reasonable to expect that the close proximity of microorganisms in infected root canals (Fig. 5) would facilitate interspecies communication, although support is limited. In an ex vivo tooth model, the bidirectional transfer of a conjugative plasmid carrying antibiotic resistance determinants was demonstrated between two endodontic infection-associated species, *S. gordonii* and *E. faecalis* (Sedgley et al. 2008). While considerable advances have been made in the identification of the microflora associated with endodontic infections, presently there are no available data on antibiotic resistance in clinical root canal biofilms in situ. Future studies using metaproteomic, metagenomic, and metabolomic methods should provide greater insight. However, obstacles to overcome include the absence of standardized methods for determining the antibiotic sensitivity of cells in multispecies biofilms as well as challenges in recovering and analyzing intact root canal biofilms using nondestructive methods.

5 Concluding Remarks

A broad range of factors can contribute to the decreased susceptibility of biofilm microorganisms to antimicrobial agents. These include provision of a physical barrier to antimicrobial agents by the EPS matrix, facilitation of HGT of DNA trapped within the extracellular matrix, quorum sensing and stress responses resulting in the recruitment and expression of resistance determinants such as multidrug resistance efflux pumps, the presence of persister cells that survive antibiotic treatment, and metabolic heterogeneity throughout the biofilm resulting in slow growth and protection against antibiotics active on rapidly growing bacteria. While further work is needed to fully understand antimicrobial resistance in biofilm communities, the accumulative effects of various processes, rather than individual involvement, are likely to be important. Much remains to be learned about the critical events in the development of antimicrobial resistance in biofilm communities.

References

Absalon C, Van Dellen K, Watnick PI (2011) A communal bacterial adhesin anchors biofilm and bystander cells to surfaces. PLoS Pathog 7, e1002210

Adam B, Baillie GS, Douglas LJ (2002) Mixed species biofilms of *Candida albicans* and *Staphylococcus epidermidis*. J Med Microbiol 51:344–349

Adnan M, Morton G, Singh J et al (2010) Contribution of *rpoS* and *bolA* genes in biofilm formation in *Escherichia coli* K-12 MG1655. Mol Cell Biochem 342:207–213

Allegrucci M, Sauer K (2007) Characterization of colony morphology variants isolated from *Streptococcus pneumoniae* biofilms. J Bacteriol 189:2030–2038

Allesen-Holm M, Barken KB, Yang L et al (2006) A characterization of DNA release in *Pseudomonas aeruginosa* cultures and biofilms. Mol Microbiol 59:1114–1128

Anderl JN, Franklin MJ, Stewart PS (2000) Role of antibiotic penetration limitation in *Klebsiella pneumoniae* biofilm resistance to ampicillin and ciprofloxacin. Antimicrob Agents Chemother 44:1818–1824

Bandara HM, Lam OL, Watt RM et al (2010) Bacterial lipopolysaccharides variably modulate in vitro biofilm formation of *Candida* species. J Med Microbiol 59:1225–1234

Barnes AM, Ballering KS, Leibman RS et al (2012) *Enterococcus faecalis* produces abundant extracellular structures containing DNA in the absence of cell lysis during early biofilm formation. MBio 3:e00193–e00112

Baugh S, Ekanayaka AS, Piddock LJ et al (2012) Loss of or inhibition of all multidrug resistance efflux pumps of *Salmonella enterica* serovar Typhimurium results in impaired ability to form a biofilm. J Antimicrob Chemother 67:2409–2417

Baumgartner JC, Xia T (2003) Antibiotic susceptibility of bacteria associated with endodontic abscesses. J Endod 29:44–47

Belitsky M, Avshalom H, Erental A et al (2011) The *Escherichia coli* extracellular death factor EDF induces the endoribonucleolytic activities of the toxins MazF and ChpBK. Mol Cell 41:625–635

Benitez JA, Spelbrink RG, Silva A et al (1997) Adherence of *Vibrio cholerae* to cultured differentiated human intestinal cells: an in vitro colonization model. Infect Immun 65:3474–3477

Bhullar K, Waglechner N, Pawlowski A et al (2012) Antibiotic resistance is prevalent in an isolated cave microbiome. PLoS ONE 7, e34953

Boles BR, Thoendel M, Singh PK (2004) Self-generated diversity produces "insurance effects" in biofilm communities. Proc Natl Acad Sci USA 101:16630–16635

Borriello G, Werner E, Roe F et al (2004) Oxygen limitation contributes to antibiotic tolerance of *Pseudomonas aeruginosa* in biofilms. Antimicrob Agents Chemother 48:2659–2664

Burmolle M, Webb JS, Rao D et al (2006) Enhanced biofilm formation and increased resistance to antimicrobial agents and bacterial invasion are caused by synergistic interactions in multispecies biofilms. Appl Environ Microbiol 72:3916–3923

Carr GB, Schwartz RS, Schaudinn C et al (2009) Ultrastructural examination of failed molar retreatment with secondary apical periodontitis: an examination of endodontic biofilms in an endodontic retreatment failure. J Endod 35:1303–1309

Ceri H, Olson ME, Stremick C et al (1999) The calgary biofilm device: new technology for rapid determination of antibiotic susceptibilities of bacterial biofilms. J Clin Microbiol 37:1771–1776

Chavez de Paz LE, Lemos JA, Wickstrom C et al (2012) Role of (p)ppGpp in biofilm formation by *Enterococcus faecalis*. Appl Environ Microbiol 78:1627–1630

Chia N, Woese CR, Goldenfeld N (2008) A collective mechanism for phase variation in biofilms. Proc Natl Acad Sci USA 105:14597–14602

Ciofu O (2003) *Pseudomonas aeruginosa* chromosomal beta-lactamase in patients with cystic fibrosis and chronic lung infection. Mechanism of antibiotic resistance and target of the humoral immune response. APMIS Suppl 116:1–47

Clatworthy AE, Pierson E, Hung DT (2007) Targeting virulence: a new paradigm for antimicrobial therapy. Nat Chem Biol 3:541–548

Claverys JP, Havarstein LS (2007) Cannibalism and fratricide: mechanisms and raisons d'etre. Nat Rev Microbiol 5:219–229

Claverys JP, Martin B, Havarstein LS (2007) Competence-induced fratricide in streptococci. Mol Microbiol 64:1423–1433

Clewell DB, Francia MV (2004) Conjugation in gram-positive bacteria. In: Funnell BE, Phillips GJ (eds) Plasmid biology. ASM Press, Washington, D.C., pp 227–256

Conlon KM, Humphreys H, O'Gara JP (2004) Inactivations of *rsbU* and *sarA* by IS256 represent novel mechanisms of biofilm phenotypic variation in *Staphylococcus epidermidis*. J Bacteriol 186:6208–6219
Cook L, Chatterjee A, Barnes A et al (2011) Biofilm growth alters regulation of conjugation by a bacterial pheromone. Mol Microbiol 81:1499–1510
Costerton JW, Irvin RT, Cheng KJ (1981) The bacterial glycocalyx in nature and disease. Annu Rev Microbiol 35:299–324
Costerton JW, Stewart PS, Greenberg EP (1999) Bacterial biofilms: a common cause of persistent infections. Science 284:1318–1322
Cotter PA, Stibitz S (2007) c-di-GMP-mediated regulation of virulence and biofilm formation. Curr Opin Microbiol 10:17–23
Cucarella C, Solano C, Valle J et al (2001) Bap, a *Staphylococcus aureus* surface protein involved in biofilm formation. J Bacteriol 183:2888–2896
Dahlen G, Magnusson BC, Moller A (1981) Histological and histochemical study of the influence of lipopolysaccharide extracted from *Fusobacterium nucleatum* on the periapical tissues in the monkey *Macaca fascicularis*. Arch Oral Biol 26:591–598
Dahlen G, Samuelsson W, Molander A et al (2000) Identification and antimicrobial susceptibility of enterococci isolated from the root canal. Oral Microbiol Immunol 15:309–312
Danese PN, Pratt LA, Kolter R (2000) Exopolysaccharide production is required for development of *Escherichia coli* K-12 biofilm architecture. J Bacteriol 182:3593–3596
de Beer D, Stoodley P, Roe F et al (1994) Effects of biofilm structures on oxygen distribution and mass transport. Biotechnol Bioeng 43:1131–1138
de Sousa EL, Ferraz CC, Gomes BP et al (2003) Bacteriological study of root canals associated with periapical abscesses. Oral Surg Oral Med Oral Pathol Oral Radiol Endod 96:332–339
Dieppois G, Ducret V, Caille O et al (2012) The transcriptional regulator CzcR modulates antibiotic resistance and quorum sensing in *Pseudomonas aeruginosa*. PLoS ONE 7, e38148
Donlan RM, Costerton JW (2002) Biofilms: survival mechanisms of clinically relevant microorganisms. Clin Microbiol Rev 15:167–193
Dorr T, Lewis K, Vulic M (2009) SOS response induces persistence to fluoroquinolones in *Escherichia coli*. PLoS Genet 5, e1000760
Dorr T, Vulic M, Lewis K (2010) Ciprofloxacin causes persister formation by inducing the TisB toxin in *Escherichia coli*. PLoS Biol 8, e1000317
Drenkard E, Ausubel FM (2002) *Pseudomonas* biofilm formation and antibiotic resistance are linked to phenotypic variation. Nature 416:740–743
Duan K, Dammel C, Stein J et al (2003) Modulation of *Pseudomonas aeruginosa* gene expression by host microflora through interspecies communication. Mol Microbiol 50:1477–1491
Dunny GM, Brown BL, Clewell DB (1978) Induced cell aggregation and mating in *Streptococcus faecalis*: evidence for a bacterial sex pheromone. Proc Natl Acad Sci USA 75:3479–3483
Dunny GM, Craig RA, Carron RL et al (1979) Plasmid transfer in *Streptococcus faecalis*: production of multiple sex pheromones by recipients. Plasmid 2:454–465
Elias S, Banin E (2012) Multi-species biofilms: living with friendly neighbors. FEMS Microbiol Rev 36:990–1004
Fauvart M, De Groote VN, Michiels J (2011) Role of persister cells in chronic infections: clinical relevance and perspectives on anti-persister therapies. J Med Microbiol 60:699–709
Flemming HC, Wingender J (2010) The biofilm matrix. Nat Rev Microbiol 8:623–633
Foster PL (2007) Stress-induced mutagenesis in bacteria. Crit Rev Biochem Mol Biol 42:373–397
Fuqua C, Greenberg EP (2002) Listening in on bacteria: acyl-homoserine lactone signalling. Nat Rev Mol Cell Biol 3:685–695
Fux CA, Costerton JW, Stewart PS et al (2005) Survival strategies of infectious biofilms. Trends Microbiol 13:34–40
Gefen O, Balaban NQ (2009) The importance of being persistent: heterogeneity of bacterial populations under antibiotic stress. FEMS Microbiol Rev 33:704–717

Gomes BP, Jacinto RC, Montagner F et al (2011) Analysis of the antimicrobial susceptibility of anaerobic bacteria isolated from endodontic infections in Brazil during a period of nine years. J Endod 37:1058–1062

Guisbert E, Yura T, Rhodius VA et al (2008) Convergence of molecular, modeling, and systems approaches for an understanding of the *Escherichia coli* heat shock response. Microbiol Mol Biol Rev 72:545–554

Hausner M, Wuertz S (1999) High rates of conjugation in bacterial biofilms as determined by quantitative in situ analysis. Appl Environ Microbiol 65:3710–3713

Havarstein LS, Martin B, Johnsborg O et al (2006) New insights into the pneumococcal fratricide: relationship to clumping and identification of a novel immunity factor. Mol Microbiol 59:1297–1307

Hayes F, Van Melderen L (2011) Toxins-antitoxins: diversity, evolution and function. Crit Rev Biochem Mol Biol 46:386–408

Henderson IR, Owen P, Nataro JP (1999) Molecular switches–the ON and OFF of bacterial phase variation. Mol Microbiol 33:919–932

Hengge-Aronis R (2002) Signal transduction and regulatory mechanisms involved in control of the sigma(S) (RpoS) subunit of RNA polymerase. Microbiol Mol Biol Rev 66:373–395

Hentzer M, Wu H, Andersen JB et al (2003) Attenuation of *Pseudomonas aeruginosa* virulence by quorum sensing inhibitors. Embo J 22:3803–3815

Hoiby N, Bjarnsholt T, Givskov M et al (2010) Antibiotic resistance of bacterial biofilms. Int J Antimicrob Agents 35:322–332

Houry A, Gohar M, Deschamps J et al (2012) Bacterial swimmers that infiltrate and take over the biofilm matrix. Proc Natl Acad Sci USA 109:13088–13093

Islam S, Oh H, Jalal S et al (2009) Chromosomal mechanisms of aminoglycoside resistance in *Pseudomonas aeruginosa* isolates from cystic fibrosis patients. Clin Microbiol Infect 15:60–66

Iwahara K, Kuriyama T, Shimura S et al (2006) Detection of *cfxA* and *cfxA2*, the beta-lactamase genes of *Prevotella* spp., in clinical samples from dentoalveolar infection by real-time PCR. J Clin Microbiol 44:172–176

Izano EA, Amarante MA, Kher WB et al (2008) Differential roles of poly-N-acetylglucosamine surface polysaccharide and extracellular DNA in *Staphylococcus aureus* and *Staphylococcus epidermidis* biofilms. Appl Environ Microbiol 74:470–476

Jacinto RC, Gomes BP, Ferraz CC et al (2003) Microbiological analysis of infected root canals from symptomatic and asymptomatic teeth with periapical periodontitis and the antimicrobial susceptibility of some isolated anaerobic bacteria. Oral Microbiol Immunol 18:285–292

Jacinto RC, Gomes BP, Shah HN et al (2006) Incidence and antimicrobial susceptibility of *Porphyromonas gingivalis* isolated from mixed endodontic infections. Int Endod J 39:62–70

Jacinto RC, Montagner F, Signoretti FG et al (2008) Frequency, microbial interactions, and antimicrobial susceptibility of *Fusobacterium nucleatum* and *Fusobacterium necrophorum* isolated from primary endodontic infections. J Endod 34:1451–1456

Jalal S, Ciofu O, Hoiby N et al (2000) Molecular mechanisms of fluoroquinolone resistance in *Pseudomonas aeruginosa* isolates from cystic fibrosis patients. Antimicrob Agents Chemother 44:710–712

Johnson EM, Flannagan SE, Sedgley CM (2006) Coaggregation interactions between oral and endodontic *Enterococcus faecalis* and bacterial species isolated from persistent apical periodontitis. J Endod 32:946–950

Jungermann GB, Burns K, Nandakumar R et al (2011) Antibiotic resistance in primary and persistent endodontic infections. J Endod 37:1337–1344

Kara D, Luppens SB, Cate JM (2006) Differences between single- and dual-species biofilms of *Streptococcus mutans* and *Veillonella parvula* in growth, acidogenicity and susceptibility to chlorhexidine. Eur J Oral Sci 114:58–63

Kara D, Luppens SB, van Marle J et al (2007) Microstructural differences between single-species and dual-species biofilms of *Streptococcus mutans* and *Veillonella parvula*, before and after exposure to chlorhexidine. FEMS Microbiol Lett 271:90–97

Keller L, Surette MG (2006) Communication in bacteria: an ecological and evolutionary perspective. Nat Rev Microbiol 4:249–258

Kelley WL (2006) Lex marks the spot: the virulent side of SOS and a closer look at the LexA regulon. Mol Microbiol 62:1228–1238

Keren I, Kaldalu N, Spoering A et al (2004a) Persister cells and tolerance to antimicrobials. FEMS Microbiol Lett 230:13–18

Keren I, Shah D, Spoering A et al (2004b) Specialized persister cells and the mechanism of multidrug tolerance in *Escherichia coli*. J Bacteriol 186:8172–8180

Khemaleelakul S, Baumgartner JC, Pruksakorn S (2002) Identification of bacteria in acute endodontic infections and their antimicrobial susceptibility. Oral Surg Oral Med Oral Pathol Oral Radiol Endod 94:746–755

Khemaleelakul S, Baumgartner JC, Pruksakom S (2006) Autoaggregation and coaggregation of bacteria associated with acute endodontic infections. J Endod 32:312–318

Kim Y, Oh S, Kim SH (2009) Released exopolysaccharide (r-EPS) produced from probiotic bacteria reduce biofilm formation of enterohemorrhagic *Escherichia coli* O157:H7. Biochem Biophys Res Commun 379:324–329

Kim Y, Wang X, Zhang XS et al (2010) *Escherichia coli* toxin/antitoxin pair MqsR/MqsA regulate toxin CspD. Environ Microbiol 12:1105–1121

Kirisits MJ, Prost L, Starkey M et al (2005) Characterization of colony morphology variants isolated from *Pseudomonas aeruginosa* biofilms. Appl Environ Microbiol 71:4809–4821

Kobayashi K, Iwano M (2012) BslA(YuaB) forms a hydrophobic layer on the surface of *Bacillus subtilis* biofilms. Mol Microbiol 85:51–66

Kolenbrander PE, Palmer RJ Jr, Periasamy S et al (2010) Oral multispecies biofilm development and the key role of cell-cell distance. Nat Rev Microbiol 8:471–480

Konig C, Schwank S, Blaser J (2001) Factors compromising antibiotic activity against biofilms of *Staphylococcus epidermidis*. Eur J Clin Microbiol Infect Dis 20:20–26

Korch SB, Henderson TA, Hill TM (2003) Characterization of the *hipA7* allele of *Escherichia coli* and evidence that high persistence is governed by (p)ppGpp synthesis. Mol Microbiol 50:1199–1213

Kvist M, Hancock V, Klemm P (2008) Inactivation of efflux pumps abolishes bacterial biofilm formation. Appl Environ Microbiol 74:7376–7382

Lafleur MD, Qi Q, Lewis K (2010) Patients with long-term oral carriage harbor high-persister mutants of *Candida albicans*. Antimicrob Agents Chemother 54:39–44

Lawrence JR, Korber DR, Hoyle BD et al (1991) Optical sectioning of microbial biofilms. J Bacteriol 173:6558–6567

Le Goff A, Bunetel L, Mouton C et al (1997) Evaluation of root canal bacteria and their antimicrobial susceptibility in teeth with necrotic pulp. Oral Microbiol Immunol 12:318–322

Leriche V, Briandet R, Carpentier B (2003) Ecology of mixed biofilms subjected daily to a chlorinated alkaline solution: spatial distribution of bacterial species suggests a protective effect of one species to another. Environ Microbiol 5:64–71

Lewis K (2007) Persister cells, dormancy and infectious disease. Nat Rev Microbiol 5:48–56

Lewis K (2010) Persister cells. Annu Rev Microbiol 64:357–372

Li Z, Nair SK (2012) Quorum sensing: how bacteria can coordinate activity and synchronize their response to external signals. Protein Sci 21(10):1403–1417

Li XZ, Nikaido H (2009) Efflux-mediated drug resistance in bacteria: an update. Drugs 69:1555–1623

Li YH, Hanna MN, Svensater G et al (2001a) Cell density modulates acid adaptation in *Streptococcus mutans*: implications for survival in biofilms. J Bacteriol 183:6875–6884

Li YH, Lau PC, Lee JH et al (2001b) Natural genetic transformation of *Streptococcus mutans* growing in biofilms. J Bacteriol 183:897–908

Li YH, Tang N, Aspiras MB et al (2002) A quorum-sensing signaling system essential for genetic competence in *Streptococcus mutans* is involved in biofilm formation. J Bacteriol 184:2699–2708

Li L, Hsiao WW, Nandakumar R et al (2010) Analyzing endodontic infections by deep coverage pyrosequencing. J Dent Res 89:980–984
Lorenz MG, Wackernagel W (1994) Bacterial gene transfer by natural genetic transformation in the environment. Microbiol Rev 58:563–602
Love RM, Jenkinson HF (2002) Invasion of dentinal tubules by oral bacteria. Crit Rev Oral Biol Med 13:171–183
Love RM, McMillan MD, Park Y et al (2000) Coinvasion of dentinal tubules by *Porphyromonas gingivalis* and *Streptococcus gordonii* depends upon binding specificity of streptococcal antigen I/II adhesin. Infect Immun 68:1359–1365
Luppens SB, Kara D, Bandounas L et al (2008) Effect of *Veillonella parvula* on the antimicrobial resistance and gene expression of *Streptococcus mutans* grown in a dual-species biofilm. Oral Microbiol Immunol 23:183–189
Lynch DJ, Fountain TL, Mazurkiewicz JE et al (2007) Glucan-binding proteins are essential for shaping *Streptococcus mutans* biofilm architecture. FEMS Microbiol Lett 268:158–165
Ma L, Conover M, Lu H et al (2009) Assembly and development of the *Pseudomonas aeruginosa* biofilm matrix. PLoS Pathog 5, e1000354
Mah TF, O'Toole GA (2001) Mechanisms of biofilm resistance to antimicrobial agents. Trends Microbiol 9:34–39
Mah TF, Pitts B, Pellock B et al (2003) A genetic basis for *Pseudomonas aeruginosa* biofilm antibiotic resistance. Nature 426:306–310
Mandsberg LF, Ciofu O, Kirkby N et al (2009) Antibiotic resistance in *Pseudomonas aeruginosa* strains with increased mutation frequency due to inactivation of the DNA oxidative repair system. Antimicrob Agents Chemother 53:2483–2491
Matsumura K, Furukawa S, Ogihara H et al (2011) Roles of multidrug efflux pumps on the biofilm formation of *Escherichia coli* K-12. Biocontrol Sci 16:69–72
May T, Ito A, Okabe S (2009) Induction of multidrug resistance mechanism in *Escherichia coli* biofilms by interplay between tetracycline and ampicillin resistance genes. Antimicrob Agents Chemother 53:4628–4639
McDougald D, Rice SA, Barraud N et al (2012) Should we stay or should we go: mechanisms and ecological consequences for biofilm dispersal. Nat Rev Microbiol 10:39–50
Molin S, Tolker-Nielsen T (2003) Gene transfer occurs with enhanced efficiency in biofilms and induces enhanced stabilisation of the biofilm structure. Curr Opin Biotechnol 14:255–261
Moyed HS, Bertrand KP (1983) *hipA*, a newly recognized gene of *Escherichia coli* K-12 that affects frequency of persistence after inhibition of murein synthesis. J Bacteriol 155:768–775
Mulcahy H, Charron-Mazenod L, Lewenza S (2008) Extracellular DNA chelates cations and induces antibiotic resistance in *Pseudomonas aeruginosa* biofilms. PLoS Pathog 4, e1000213
Mulcahy LR, Burns JL, Lory S et al (2010) Emergence of *Pseudomonas aeruginosa* strains producing high levels of persister cells in patients with cystic fibrosis. J Bacteriol 192:6191–6199
Oliver A, Baquero F, Blazquez J (2002) The mismatch repair system (*mutS*, *mutL* and *uvrD* genes) in *Pseudomonas aeruginosa*: molecular characterization of naturally occurring mutants. Mol Microbiol 43:1641–1650
Ozok AR, Persoon IF, Huse SM et al (2012) Ecology of the microbiome of the infected root canal system: a comparison between apical and coronal root segments. Int Endod J 45:530–541
Parsek MR, Greenberg EP (2005) Sociomicrobiology: the connections between quorum sensing and biofilms. Trends Microbiol 13:27–33
Pinheiro ET, Gomes BP, Ferraz CC et al (2003) Evaluation of root canal microorganisms isolated from teeth with endodontic failure and their antimicrobial susceptibility. Oral Microbiol Immunol 18:100–103
Pinheiro ET, Gomes BP, Drucker DB et al (2004) Antimicrobial susceptibility of *Enterococcus faecalis* isolated from canals of root filled teeth with periapical lesions. Int Endod J 37:756–763
Poole K (2012) Bacterial stress responses as determinants of antimicrobial resistance. J Antimicrob Chemother 67(9):2069–2089

Potrykus K, Cashel M (2008) (p)ppGpp: still magical? Annu Rev Microbiol 62:35–51
Qin Z, Ou Y, Yang L et al (2007) Role of autolysin-mediated DNA release in biofilm formation of *Staphylococcus epidermidis*. Microbiology 153:2083–2092
Rachid S, Ohlsen K, Witte W et al (2000) Effect of subinhibitory antibiotic concentrations on polysaccharide intercellular adhesin expression in biofilm-forming *Staphylococcus epidermidis*. Antimicrob Agents Chemother 44:3357–3363
Rendueles O, Ghigo JM (2012) Multi-species biofilms: how to avoid unfriendly neighbors. FEMS Microbiol Rev 36(5):972–989
Rendueles O, Travier L, Latour-Lambert P et al (2011) Screening of *Escherichia coli* species biodiversity reveals new biofilm-associated antiadhesion polysaccharides. MBio 2:e00043–e00011
Rendueles O, Kaplan JB, Ghigo JM (2012) Antibiofilm polysaccharides. Environ Microbiol 15 (2):334–346
Reynaud Af Geijersstam A, Culak R, Molenaar L et al (2007) Comparative analysis of virulence determinants and mass spectral profiles of Finnish and Lithuanian endodontic *Enterococcus faecalis* isolates. Oral Microbiol Immunol 22:87–94
Rice LB (1998) Tn*916* family conjugative transposons and dissemination of antimicrobial resistance determinants. Antimicrob Agents Chemother 42:1871–1877
Ricucci D, Siqueira JF Jr (2010) Biofilms and apical periodontitis: study of prevalence and association with clinical and histopathologic findings. J Endod 36:1277–1288
Ricucci D, Siqueira JF Jr, Bate AL et al (2009) Histologic investigation of root canal-treated teeth with apical periodontitis: a retrospective study from twenty-four patients. J Endod 35:493–502
Roberts AP, Mullany P (2010) Oral biofilms: a reservoir of transferable, bacterial, antimicrobial resistance. Expert Rev Anti Infect Ther 8:1441–1450
Rossi-Fedele G, Scott W, Spratt D et al (2006) Incidence and behaviour of Tn*916*-like elements within tetracycline-resistant bacteria isolated from root canals. Oral Microbiol Immunol 21:218–222
Salyers AA, Shoemaker NB, Stevens AM et al (1995) Conjugative transposons: an unusual and diverse set of integrated gene transfer elements. Microbiol Rev 59:579–590
Sauer K, Camper AK, Ehrlich GD et al (2002) *Pseudomonas aeruginosa* displays multiple phenotypes during development as a biofilm. J Bacteriol 184:1140–1154
Sauer K, Cullen MC, Rickard AH et al (2004) Characterization of nutrient-induced dispersion in *Pseudomonas aeruginosa* PAO1 biofilm. J Bacteriol 186:7312–7326
Schembri MA, Kjaergaard K, Klemm P (2003) Global gene expression in *Escherichia coli* biofilms. Mol Microbiol 48:253–267
Sedgley CM, Molander A, Flannagan SE et al (2005) Virulence, phenotype and genotype characteristics of endodontic *Enterococcus* spp. Oral Microbiol Immunol 20:10–19
Sedgley CM, Lee EH, Martin MJ et al (2008) Antibiotic resistance gene transfer between *Streptococcus gordonii* and *Enterococcus faecalis* in root canals of teeth ex vivo. J Endod 34:570–574
Shah D, Zhang Z, Khodursky A et al (2006) Persisters: a distinct physiological state of *E. coli*. BMC Microbiol 6:53
Siqueira JF Jr, Rocas IN (2005) Uncultivated phylotypes and newly named species associated with primary and persistent endodontic infections. J Clin Microbiol 43:3314–3319
Siqueira JF Jr, Rocas IN, Ricucci D (2010) Biofilms in endodontic infections. Endod Topics 22:33–49
Siqueira JF Jr, Alves FR, Rocas IN (2011) Pyrosequencing analysis of the apical root canal microbiota. J Endod 37:1499–1503
Skillman LC, Sutherland IW, Jones MV (1998) The role of exopolysaccharides in dual species biofilm development. J Appl Microbiol 85(Suppl 1):13S–18S
Skucaite N, Peciuliene V, Vitkauskiene A et al (2010) Susceptibility of endodontic pathogens to antibiotics in patients with symptomatic apical periodontitis. J Endod 36:1611–1616

Sorensen SJ, Bailey M, Hansen LH et al (2005) Studying plasmid horizontal transfer in situ: a critical review. Nat Rev Microbiol 3:700–710

Spoering AL, Lewis K (2001) Biofilms and planktonic cells of *Pseudomonas aeruginosa* have similar resistance to killing by antimicrobials. J Bacteriol 183:6746–6751

Srivastava D, Waters CM (2012) A tangled web: regulatory connections between quorum sensing and cyclic Di-GMP. J Bacteriol 194:4485–4493

Staal M, Borisov SM, Rickelt LF et al (2011) Ultrabright planar optodes for luminescence life-time based microscopic imaging of O_2 dynamics in biofilms. J Microbiol Methods 85:67–74

Stanley NR, Britton RA, Grossman AD et al (2003) Identification of catabolite repression as a physiological regulator of biofilm formation by *Bacillus subtilis* by use of DNA microarrays. J Bacteriol 185:1951–1957

Sternberg C, Christensen BB, Johansen T et al (1999) Distribution of bacterial growth activity in flow-chamber biofilms. Appl Environ Microbiol 65:4108–4117

Stewart PS (2012) Mini-review: convection around biofilms. Biofouling 28:187–198

Stewart PS, Costerton JW (2001) Antibiotic resistance of bacteria in biofilms. Lancet 358:135–138

Stewart PS, Franklin MJ (2008) Physiological heterogeneity in biofilms. Nat Rev Microbiol 6:199–210

Suci PA, Mittelman MW, Yu FP et al (1994) Investigation of ciprofloxacin penetration into *Pseudomonas aeruginosa* biofilms. Antimicrob Agents Chemother 38:2125–2133

Sutherland I (2001) Biofilm exopolysaccharides: a strong and sticky framework. Microbiology 147:3–9

Thomas VC, Thurlow LR, Boyle D et al (2008) Regulation of autolysis-dependent extracellular DNA release by *Enterococcus faecalis* extracellular proteases influences biofilm development. J Bacteriol 190:5690–5698

Thomas VC, Hiromasa Y, Harms N et al (2009) A fratricidal mechanism is responsible for eDNA release and contributes to biofilm development of *Enterococcus faecalis*. Mol Microbiol 72:1022–1036

Tielker D, Hacker S, Loris R et al (2005) *Pseudomonas aeruginosa* lectin LecB is located in the outer membrane and is involved in biofilm formation. Microbiology 151:1313–1323

Tormo MA, Ubeda C, Marti M et al (2007) Phase-variable expression of the biofilm-associated protein (Bap) in *Staphylococcus aureus*. Microbiology 153:1702–1710

Trotonda MP, Manna AC, Cheung AL et al (2005) SarA positively controls *bap*-dependent biofilm formation in *Staphylococcus aureus*. J Bacteriol 187:5790–5798

Upadya M, Shrestha A, Kishen A (2011) Role of efflux pump inhibitors on the antibiofilm efficacy of calcium hydroxide, chitosan nanoparticles, and light-activated disinfection. J Endod 37:1422–1426

Valle J, Vergara-Irigaray M, Merino N et al (2007) sigma^B regulates IS256-mediated *Staphylococcus aureus* biofilm phenotypic variation. J Bacteriol 189:2886–2896

Vigil GV, Wayman BE, Dazey SE et al (1997) Identification and antibiotic sensitivity of bacteria isolated from periapical lesions. J Endod 23:110–114

Vrany JD, Stewart PS, Suci PA (1997) Comparison of recalcitrance to ciprofloxacin and levofloxacin exhibited by *Pseudomonas aeruginosa* bofilms displaying rapid-transport characteristics. Antimicrob Agents Chemother 41:1352–1358

Vuong C, Voyich JM, Fischer ER et al (2004) Polysaccharide intercellular adhesin (PIA) protects *Staphylococcus epidermidis* against major components of the human innate immune system. Cell Microbiol 6:269–275

Walters MC 3rd, Roe F, Bugnicourt A et al (2003) Contributions of antibiotic penetration, oxygen limitation, and low metabolic activity to tolerance of *Pseudomonas aeruginosa* biofilms to ciprofloxacin and tobramycin. Antimicrob Agents Chemother 47:317–323

Wang X, Wood TK (2011) Toxin-antitoxin systems influence biofilm and persister cell formation and the general stress response. Appl Environ Microbiol 77:5577–5583

Wang X, Kim Y, Hong SH et al (2011) Antitoxin MqsA helps mediate the bacterial general stress response. Nat Chem Biol 7:359–366

Werner E, Roe F, Bugnicourt A et al (2004) Stratified growth in *Pseudomonas aeruginosa* biofilms. Appl Environ Microbiol 70:6188–6196

Whitchurch CB, Tolker-Nielsen T, Ragas PC et al (2002) Extracellular DNA required for bacterial biofilm formation. Science 295:1487

Whiteley M, Bangera MG, Bumgarner RE et al (2001) Gene expression in *Pseudomonas aeruginosa* biofilms. Nature 413:860–864

Wiuff C, Zappala RM, Regoes RR et al (2005) Phenotypic tolerance: antibiotic enrichment of noninherited resistance in bacterial populations. Antimicrob Agents Chemother 49:1483–1494

Wu Y, Vulic M, Keren I et al (2012) Role of oxidative stress in persister tolerance. Antimicrob Agents Chemother 56(9):4922–4926

Zahller J, Stewart PS (2002) Transmission electron microscopic study of antibiotic action on *Klebsiella pneumoniae* biofilm. Antimicrob Agents Chemother 46:2679–2683

Zhang L, Mah TF (2008) Involvement of a novel efflux system in biofilm-specific resistance to antibiotics. J Bacteriol 190:4447–4452

Zhang XQ, Bishop PL, Kupferle MJ (1998) Measurement of polysaccharides and proteins in biofilm extracellular polymers. Water Sci Technol 37:345–348

Zheng Z, Stewart PS (2002) Penetration of rifampin through *Staphylococcus epidermidis* biofilms. Antimicrob Agents Chemother 46:900–903

Zhu X, Wang Q, Zhang C et al (2010) Prevalence, phenotype, and genotype of *Enterococcus faecalis* isolated from saliva and root canals in patients with persistent apical periodontitis. J Endod 36:1950–1955

Ziebuhr W, Krimmer V, Rachid S et al (1999) A novel mechanism of phase variation of virulence in *Staphylococcus epidermidis*: evidence for control of the polysaccharide intercellular adhesin synthesis by alternating insertion and excision of the insertion sequence element IS256. Mol Microbiol 32:345–356

Part II
Observational and Experimental Evidence

The Use of Scanning Electron Microscopy (SEM) in Visualizing the Root Canal Biofilm

Linda B. Peters, Brandon Peterson, David E. Jaramillo, and Luc van der Sluis

Abstract Apical periodontitis is caused by microorganisms in planktonic or biofilm state present in the root canal system and/or attached to the outer apical root surface. Knowledge about the microorganisms and biofilm structure within and external to the root canal system is important in order to effectively treat apical periodontitis. Scanning electron microscopy (SEM) has been used to visualize and morphologically describe the presence of biofilm and microorganisms associated with teeth with apical periodontitis. This chapter provides a short outline of the applications of SEM in endodontics with the intention to describe the benefits and shortcomings of this microscopic technique.

1 Introduction

The landmark study by Kakehashi et al. (1965) demonstrated the cause and effect relationship between infection of the root canal system and apical periodontitis. The aim of endodontic treatment is to eliminate infection in the root canal system and create an environment that allows healing of apical periodontitis. However, despite appropriate treatment procedures, healing of apical periodontitis is not always predictable (Ricucci and Langeland 1998; Ørstavik et al. 2004). Further, new treatment modalities have had only limited success in influencing the healing of apical periodontitis (Ng et al. 2008a, b).

L.B. Peters • L. van der Sluis (✉)
Center for Dentistry and Oral Hygiene, University Medical Center Groningen, Groningen, The Netherlands
e-mail: l.w.m.van.der.sluis@umcg.nl

B. Peterson
W. J. Kolff Institute, Department of Biomedical Engineering, University Medical Center Groningen, University of Groningen, Groningen, The Netherlands

D.E. Jaramillo
Department of Endodontics, University of Texas Health Science Center, Houston, TX, USA

L.E. Chávez de Paz et al. (eds.), *The Root Canal Biofilm*, Springer Series on Biofilms 9, DOI 10.1007/978-3-662-47415-0_4

The probable cause of reduced healing has been attributed to biofilm infections rather than an infection with planktonic microorganisms (Costerton et al. 1999). Persistent infection has also been strongly associated with the biofilm state (Parsek and Singh 2003; Costerton et al. 2005). It therefore follows that an enhanced understanding of root canal infections must include the study of endodontic biofilms. However, demonstration of biofilms in the root canal system presents challenges because few techniques are capable of simultaneously imaging both the extracellular matrix and microorganisms within biofilm communities (Schaudinn et al. 2009; Bridier et al. 2013).

In this chapter we focus on the use of scanning electron microscopy (SEM) to visualize and describe biofilms associated with infected root canals and dentinal tubules and extraradicular biofilms associated with persistent infections.

2 General Information on the SEM Technique

In the seventeenth century, Antonie van Leeuwenhoek used a simple microscope to visualize the poor penetration of vinegar into oral plaque. He wrote "the vinegar with which I wash my teeth, kill'd only those animale which were on the outside of the scurf., but dis not pass thro the whole substance of it" (van Leeuwenhoek 1684). Subsequent developments in technology provided the scanning electron microscope and, more recently, advanced microscopy techniques incorporating fluorescence and confocal laser scanning that can evaluate specific microbial targets. While the more advanced scanning microscopy techniques can provide valuable information about the viability and composition of oral biofilms, the resolution of these methods cannot compare to that of SEM.

Electron microscopy uses an electron-dense material, typically gold, to coat the surface of the object of interest. The electron-rich environment is more sensitive and thus able to magnify the object to larger magnifications by using just light sources and lenses. The most commonly used electron microscopy technique, SEM, scans the electron-dense material, providing a detailed, high-resolution image of the sputter-coated surfaces, including bacterial cell surfaces, and can identify cellular membrane damage (Appelbaum et al. 1979; Okte et al. 1999). However, the high vacuum conditions required during sample preparation for SEM distort hydrated local environments, such as the extracellular polymeric matrix. After vacuum, the extracellular matrix is reduced to a dark rim of condensed matter around bacterial cells (Stewart and Costerton 2001) and does not show its expansive connective network observed under hydrated conditions. Additionally, the electron-dense coating applied to the sample for detection "freezes" the sample, rendering a single sample unavailable for detection at multiple time points. The use of multiple microscopic techniques is currently required to fulfill the high-resolution and 3D reconstruction of hydrated biofilms (Bridier et al. 2013).

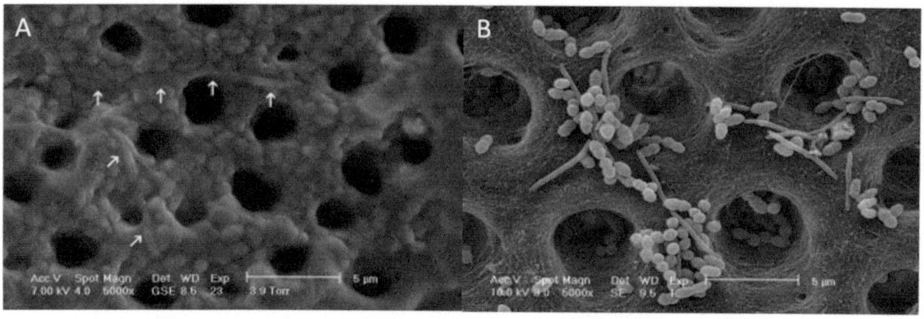

Fig. 1 Comparison of ESEM (**a**) and SEM (**b**) images demonstrating differences in resolution. (**a**) *Fusobacterium nucleatum* species (*arrows*) in mixed culture with *Enterococcus faecalis*; (**b**) Colonies of *F. nucleatum* and *E. faecalis*. Both species could clearly be distinguished (×5000, Fe-CSEM) [Reproduced with permission of John Wiley & Sons, Inc from Bergmans L, Moisiadis P, van Meerbeek B, Quirynen M, Lambrechts P. Microscopic observation of bacteria: review highlighting the use of environmental SEM. International Endodontic Journal 38(11):775–88 ©2005]

In an effort to overcome the difficulties of "traditional" SEM, many modified methods of electron microscopy have been developed that can target otherwise undetectable parts of the sample. These include environmental scanning electron microscopy (ESEM), transmission electron microscopy (TEM), and focused ion beam scanning electron microscopy (FIB-SEM). ESEM does not need vacuum conditions or a surface coating and consequently is capable of imaging a more "natural" sample (Danilatos and Postle 1982; Bergmans et al. 2005). TEM requires ultrathin samples and still uses vacuum conditions to prepare the samples; however, because the samples are thin, the interior components of the cell can be visualized (Reese and Guggenheim 2007). FIB-SEM uses a destructive milling process to access the layers of the sample in order to reconstruct a 3D representation (Wallace et al. 2011). Overall, SEM, TEM, and FIB-SEM are inadequate to visualize biofilms because the matrix structure is destroyed due to dehydration. In contrast, ESEM does not affect the matrix structure because dehydration of the biofilm is not necessary for the procedure. However, ESEM has a poor resolution and limited magnification (Alhede et al. 2012). In contrast, SEM can provide higher-resolution images that facilitate observations of microbial communities, such as those found in root canal biofilms (Fig. 1).

3 Visualization of Microorganisms/Biofilm in the Root Canal and Tubules by SEM

3.1 Root Canal

In light and electron microscopy studies, Nair described the dense plaque in the root canal that harbored large numbers of bacteria and polymorphonucleocytes that were

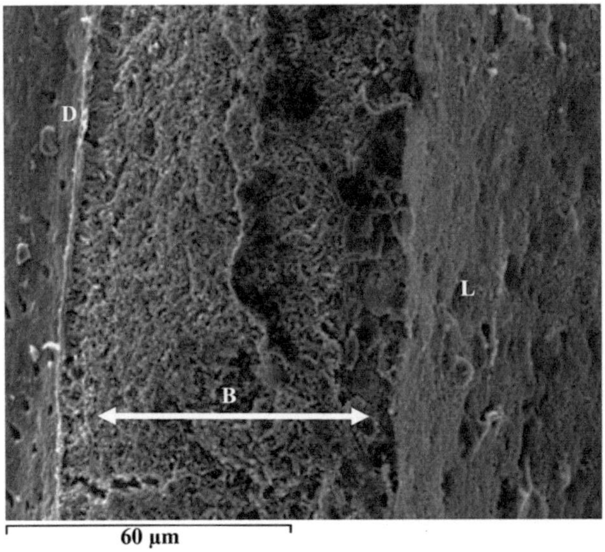

Fig. 2 Biofilm within the root canal lumen. SEM apical section through dentine (D) and biofilm (B) within the canal lumen (L) (μ bar represents 60 μm) [Reproduced with permission of John Wiley & Sons, Inc from Richardson N, Mordan NJ, Figueiredo JA, Ng YL, Gulabivala K. Microflora in teeth associated with apical periodontitis: a methodological observational study comparing two protocols and three microscopy techniques. International Endodontic Journal 42(10):908–921 ©2009]

invading the bacterial plaque on the root canal wall (Nair 1987). SEM studies have demonstrated biofilms within the canal lumen (Richardson et al. 2009) (Figs. 2 and 3) and adhering to the root canal wall surface in teeth with associated periapical pathology; these biofilms may have contained various cellular morphotypes such as cocci, rods, filaments, and spirochaetes (Molven et al. 1991; Richardson et al. 2009; Siqueira and Lopes 2001; Baldasso et al. 2012) (Figs. 4 and 5). These investigators also reported that more bacteria were found on the root canal wall when extensive carious lesions were present (Baldasso et al. 2012) (Fig. 6) and that the apical part of the root canal was more heavily infected than the coronal part (Molven et al. 1991; Richardson et al. 2009). Also, yeastlike structures and fungal hyphae have been observed under SEM to colonize infected root canals (Sen et al. 1995; Baldasso et al. 2012) (Fig. 7).

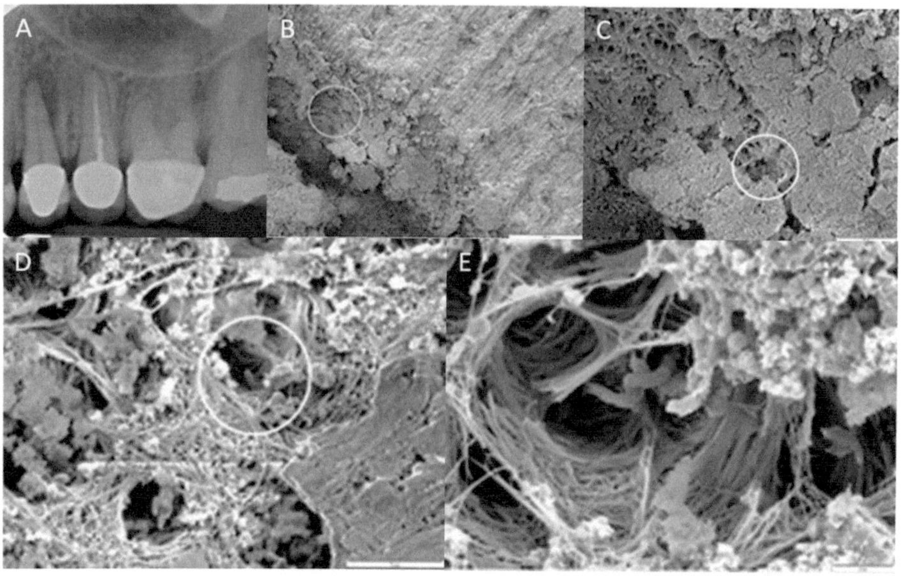

Fig. 3 Biofilm within the root canal. (**a**) Radiograph showing maxillary left first premolar. Crown delivered 3 years previously. Patient presented with intermittent pain, particularly upon chewing, and absence of deep probing. Diagnosis: necrotic pulp and symptomatic apical periodontitis. Patient elected extraction. (**b**) SEM image of middle section of the root canal wall showing necrotic pulp tissue and debris covering dentinal tubules; (**c**) Close-up of circled area in (**b**); (**d**) Close-up of circled area in (**c**) showing dentinal collagen fibers and bacteria inside the dentinal tubule; (**e**) Close-up of circled area in (**d**) showing bacteria inside dentinal tubule. (Images from David Jaramillo)

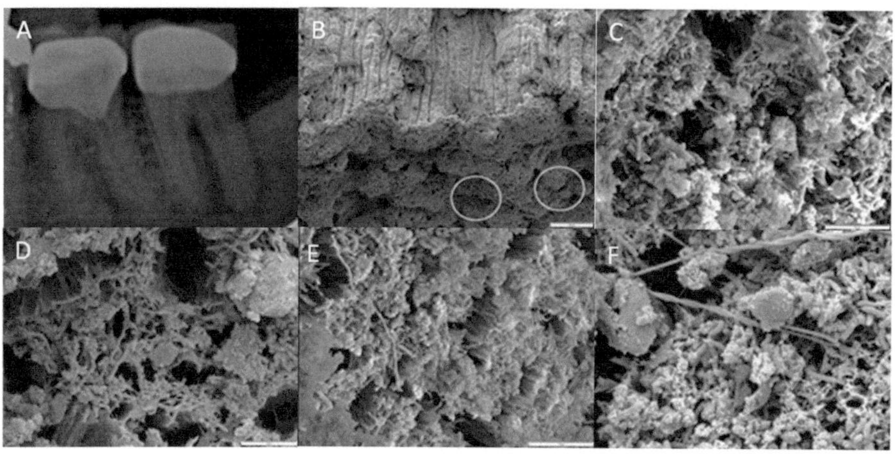

Fig. 4 Biofilm within the root canal. (**a**) Radiograph showing mandibular left second molar. Patient presented with no symptoms and absence of deep probing. Diagnosis: necrotic pulp and asymptomatic apical periodontitis. Patient elected extraction. (**b**) SEM image of middle section of the root canal wall showing dentinal wall as a niche for biofilms; (**c** and **d**) Close-up images of *left circled* area labeled in (**b**) showing dentinal collagen fibers and bacteria inside dentinal tubules; (**e** and **f**) Close-up images of *right circled* area in (**b**) showing multi-species biofilms with cocci, rods, and long filaments (Images from David Jaramillo)

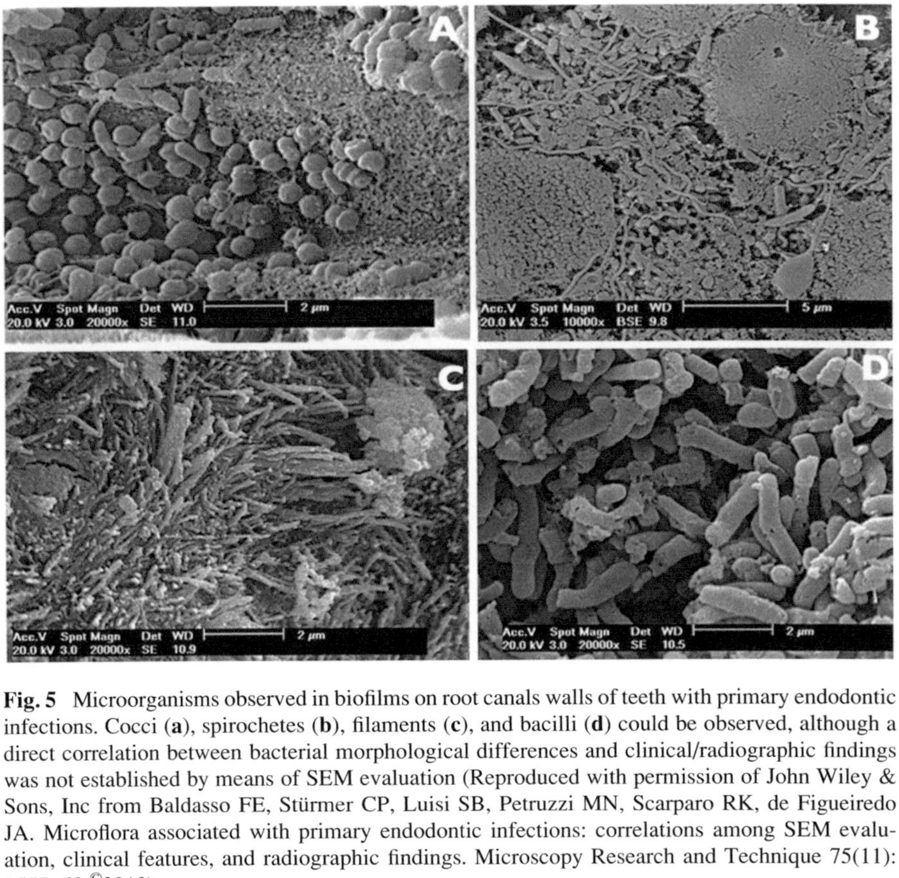

Fig. 5 Microorganisms observed in biofilms on root canals walls of teeth with primary endodontic infections. Cocci (**a**), spirochetes (**b**), filaments (**c**), and bacilli (**d**) could be observed, although a direct correlation between bacterial morphological differences and clinical/radiographic findings was not established by means of SEM evaluation (Reproduced with permission of John Wiley & Sons, Inc from Baldasso FE, Stürmer CP, Luisi SB, Petruzzi MN, Scarparo RK, de Figueiredo JA. Microflora associated with primary endodontic infections: correlations among SEM evaluation, clinical features, and radiographic findings. Microscopy Research and Technique 75(11): 1557–63 ©2012)

3.2 Dentinal Tubules

Using light microscopy and culture techniques, investigators have shown that bacteria enter the dentinal tubules (Peters et al. 2001; Nair et al. 2005; Vieira et al. 2012). SEM imaging has also demonstrated invasion of the dentinal tubules (Richardson et al. 2009) (Fig. 8) with cocci, rods, and yeasts invading to depths of up to 150 µm (Sen et al. 1995). Richardson et al. (2009) showed that the dentinal tubules in the coronal and middle section of the root were most heavily infected, while Sen et al. (1995) also observed heavily infected tubules in the apical area.

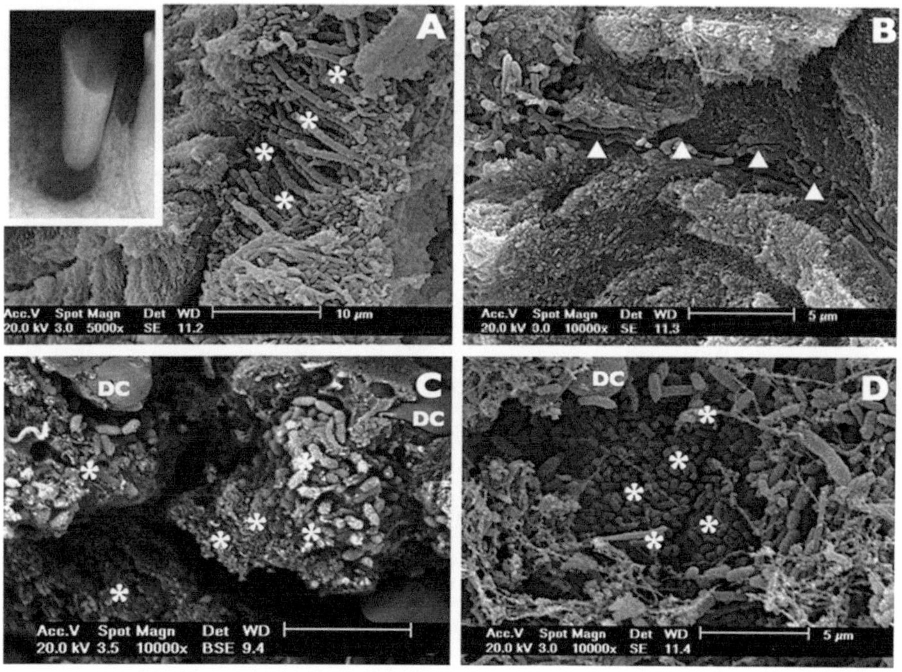

Fig. 6 Bacteria forming dense biofilms on the root canal walls of a lower canine with extensive caries and a periapical lesion. Biofilm, mainly comprised of cocci and rods (*), can be observed at cervical (**a**), middle (**b**), and apical (**c, d**) thirds of the canal. In some areas, bacteria could also be detected into dentinal tubules (~). (DC refers to defense cells.) (Reproduced with permission of John Wiley & Sons, Inc from Baldasso FE, Stürmer CP, Luisi SB, Petruzzi MN, Scarparo RK, de Figueiredo JA. Microflora associated with primary endodontic infections: correlations among SEM evaluation, clinical features, and radiographic findings. Microscopy Research and Technique 75(11):1557–63 ©2012)

Fig. 7 Fungal hyphae (FH) in root canal biofilm located at the cervical level of a maxillary first molar with extensive caries, periapical lesion, and clinically detectable periodontal pocket (Reproduced with permission of John Wiley & Sons, Inc from Baldasso FE, Stürmer CP, Luisi SB, Petruzzi MN, Scarparo RK, de Figueiredo JA. Microflora associated with primary endodontic infections: correlations among SEM evaluation, clinical features, and radiographic findings. Microscopy Research and Technique 75(11):1557–63 ©2012)

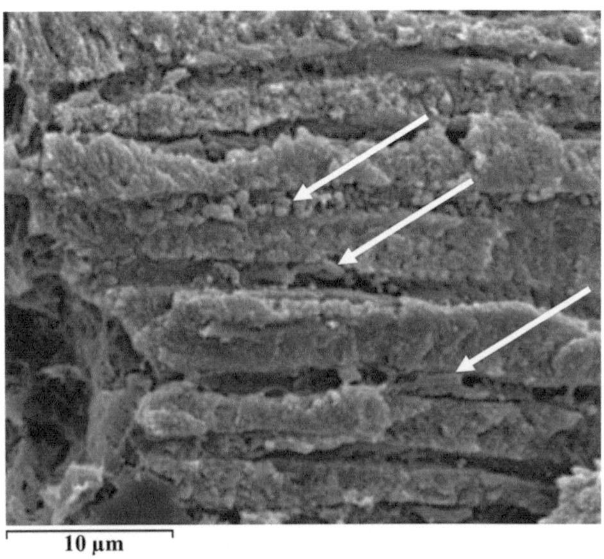

Fig. 8 Invasion by microorganisms into dentinal tubules of infected root canal. SEM of apical section showing bacteria (*arrows*) within the dentine tubules (μ bar represents 10 μm) (Reproduced with permission of John Wiley & Sons, Inc from Richardson N, Mordan NJ, Figueiredo JA, Ng YL, Gulabivala K. Microflora in teeth associated with apical periodontitis: a methodological observational study comparing two protocols and three microscopy techniques. International Endodontic Journal 42(10):908–921 ©2009)

4 Visualization of Extraradicular Microorganisms/Biofilm by SEM

Apical extraradicular biofilm is an important clinical phenomenon because they may be inherently resistant to antimicrobial agents and their location renders them difficult to remove by biomechanical preparation (Tronstad et al. 1990; Siqueira and Lopes 2001). This may cause failure of endodontic treatment as a consequence of persistent apical infection rather than intracanal persistent infection. Although it is not fully understood in which cases extraradicular biofilm infection can be expected, SEM studies have shown biofilm attached to the apical external root surface (Tronstad et al. 1990; Lomçali et al. 1996; Leonardo et al. 2002; Signoretti et al. 2011; Baldasso et al. 2012) (Fig. 9) and to apically extruded gutta-percha (Noiri et al. 2002) (Fig. 10).

Extraradicular biofilms have been proposed as etiological factors associated with failed endodontic treatment (Tronstad et al. 1990; Wang et al. 2013). In 1990, Tronstad et al. reported "apical plaque" on the root surface of teeth with apical periodontitis refractory to endodontic treatment visualized by SEM and described as "a continuous, smooth and structureless coating." At higher magnification predominantly cocci and rods were described as present in the "coating" and in

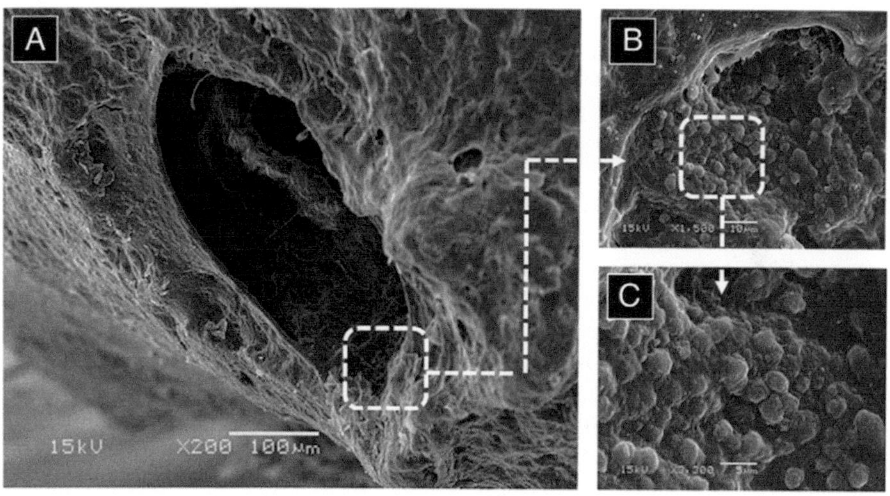

Fig. 9 Resected distal root apex of a failed endodontically retreated lower left first molar showing biofilm surrounding the apical foramen and external radicular surface. (**a**) Uninstrumented apical foramen (×200); (**b** and **c**) Bacterial colonies adhering to external radicular surface (×1500 and ×3300). (Reprinted from Journal of Endodontics, 37(12), Signoretti FG, Endo MS, Gomes BP, Montagner F, Tosello FB, Jacinto RC (2011) Persistent extraradicular infection in root-filled asymptomatic human tooth: scanning electron microscopic analysis and microbial investigation after apical microsurgery, 1696–700 ©2011, with permission from Elsevier)

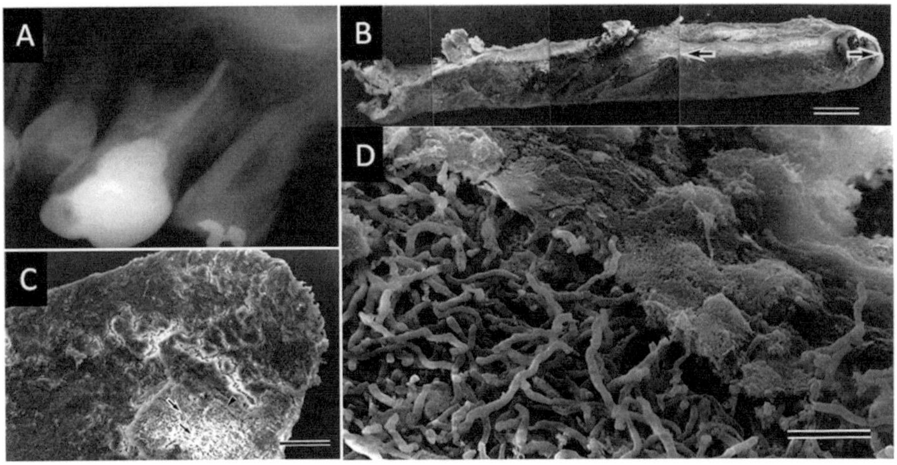

Fig. 10 Biofilm on extruded gutta-percha cone recovered from previously treated tooth with refractory periapical periodontitis. (**a**) Radiograph of tooth. Filling material (gutta-percha point) is observed outside the palatal root canal; (**b**) SEM image of gutta-percha specimen after removal. The apically extruded material (*arrows*) is approximately 2 mm in length (original magnification ×50; bar = 500 um); (**c**) SEM image of extruded gutta-percha (original magnification ×350; bar = 50 um); (**d**) High magnification of *arrowhead* area in (**c**). Glycocalyx structure is present in the upper right area but not in *lower left area*. Filamentous or spirochete-shaped bacteria are observed in *lower left* (original magnification ×3500; bar = 5 um) (Reprinted from Journal of Endodontics, 28(10), Noiri Y, Ehara A, Kawahara T, Takemura N, Ebisu S. Participation of bacterial biofilms in refractory and chronic periapical periodontitis,679–83, ©2011, with permission from Elsevier)

irregularities. Fibrillar forms were also seen mainly attached to cocci, and the bacteria were held together by an extracellular material. These findings were supported by Rocha et al. (2008) who observed biofilm on the external surfaces of the apical one third of roots of primary teeth with necrotic pulps and radiographically evident apical pathosis (Fig. 11); in contradistinction, teeth with vital pulps showed no evidence of apical infection, and normal collagen fibers were seen on the root surface. These findings are also supported by Wang et al. 2013 who reported the presence of extraradicular biofilm in all samples with persistent periapical periodontitis and in three samples with chronic periapical periodontitis. The cases with vital pulps were covered by fibers, with no extraradicular microorganisms present. Leonardo et al. (2002) also reported that teeth with pulp necrosis without radiographic signs of apical periodontitis harbored bacteria but the infection was confined to the lumen of the main root canal. When radiographic breakdown was clearly visible, the root surface was covered by different bacterial cells. The resorptive lacunae mainly harbored cocci and bacilli but spirochetes and filaments were also present on the outer surface of the root (Fig. 12). Lacunar resorptive zones

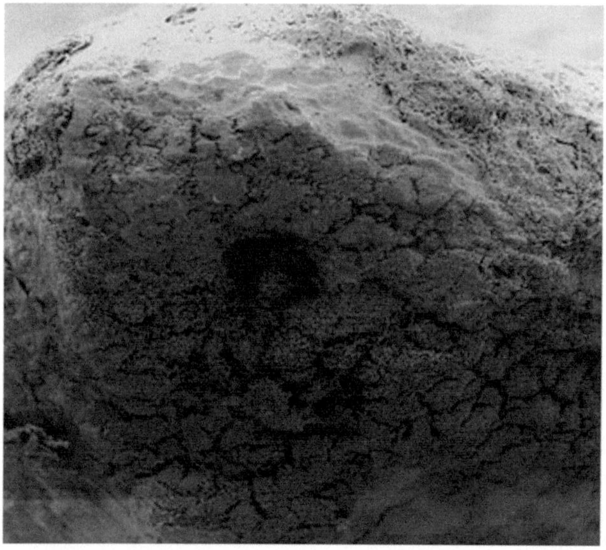

Fig. 11 Extraradicular biofilm associated with the apical one third of roots of primary teeth with necrotic pulps and radiographically evident apical pathosis (Reproduced with permission of John Wiley & Sons, Inc from Rocha CT, Rossi MA, Leonardo MR, Rocha LB, Nelson-Filho P, Silva LA. Biofilm on the apical region of roots in primary teeth with vital and necrotic pulps with or without radiographically evident apical pathosis. International Endodontic Journal 41(8):664–9 ©2008)

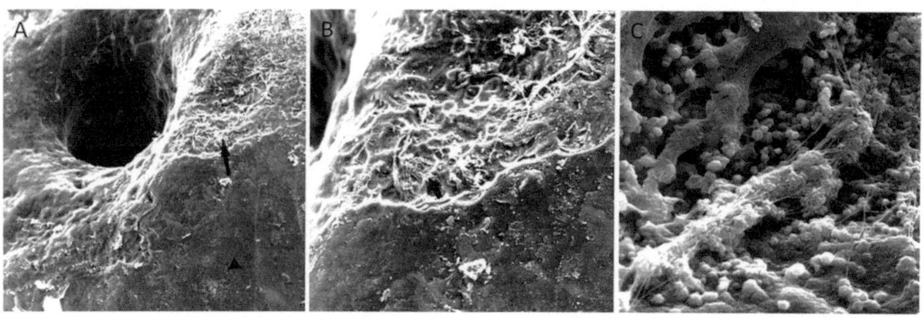

Fig. 12 External surface of the root apex of tooth with pulp necrosis and chronic periapical lesion. (**a**) Morphological changes in the radicular cementum close to the apical foramen showing areas of intact cementum (*arrowhead*) between areas of resorption (*arrow*) (×400). (**b**) Higher magnification of cementum resorption in (**a**) with areas of microorganisms (*arrow*) (×1650); (**c**) Higher magnification of (**b**) (*arrow*), showing the presence of cocci forming apical biofilm (×6000) (Reprinted from Journal of Endodontics, 28(12), Leonardo MR, Rossi MA, Silva LA, Ito IY, Bonifácio KC, EM evaluation of bacterial biofilm and microorganisms on the apical external root surface of human teeth, 815–818, ©2002, with permission from Elsevier)

close to the apical foramen on the outer surface of the root have also harbored yeasts according to another SEM study (Lomçali et al. 1996).

In contrast to the above studies, no evidence of extraradicular infection associated with periapical periodontitis in 26 out of 27 cases visualized by SEM was found by Siqueira and Lopes (2001). Although bacteria could be observed close to the apical foramen, they were confined to the canal lumen.

5 Concluding Remarks

The use of SEM to investigate endodontic infections has provided important information. However, it is difficult to show microorganisms embedded in the polysaccharide matrix (Richardson et al. 2009). The matrix is composed mainly of water and its three-dimensional structure changes when samples are prepared for SEM imaging by fixation and dehydration (Bridier et al. 2013). Further, while SEM can provide information on the size and morphology of microorganisms, definitive microbial identification is not feasible, nor is the ability to observe the changes in size and shape of microorganisms when embedded in a biofilm structure (Webster et al. 2004). It is apparent that a combination of different microscopic techniques is more likely to facilitate a deeper and more realistic analysis of biofilm architecture. For example, Schaudinn et al. (2009) demonstrated that by combining SEM with FISH/cLSM, the limitations of each technique could be overcome. In five teeth extracted due to failed endodontic treatment, the presence of bacteria and a matrix were shown by using SEM images overlaid with corresponding cLSM and FISH images of the same area (Fig. 13). Baldasso et al. (2012) also reported that by using the combination of clinical and radiographic findings with SEM imaging, a better understanding of the location and extent of the infection is possible.

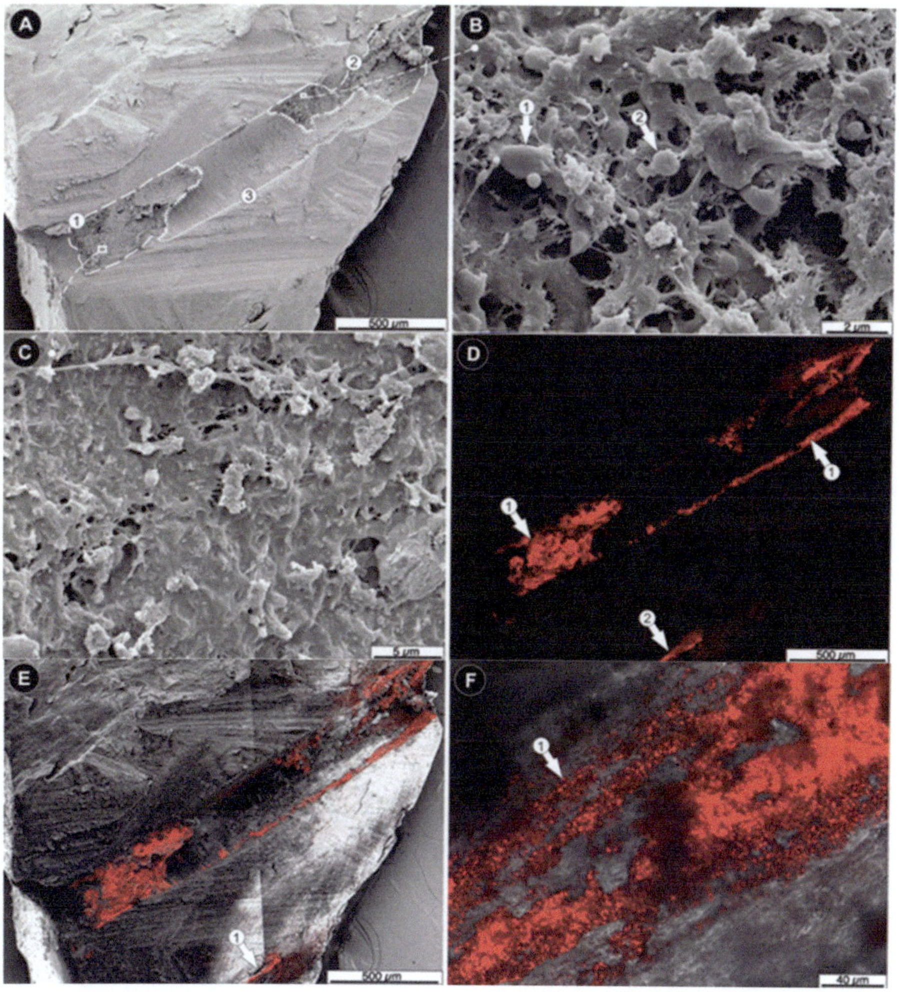

Fig. 13 Evaluation of root canal biofilm architecture using a combination of different microscopic techniques. (a) Overview of root canal of tooth extracted due to failed endodontic treatment. Certain parts of the canal surface were covered with a thick matrix layer (*encircled areas* 1, 2), whereas other regions showed only sparse and comparatively thin islands of matrix (*encircled area* 3). (**b**) The size and shape of some structures of the matrix suggested the presence of bacteria (*arrows* 1, 2), but without conclusive evidence. (**c**) Other areas of the matrix consisted of so densely composed material that no traces of bacterial presence could be found. (**d**) After labeling the split tooth with the EUB338 (Cy3) probe, distinct parts of the canal showed a strong red fluorescence signal in the cLSM (*arrow* 1). The red signal was also clearly present at a lateral spot at some distance from the main root canal (*arrow* 2). (**e**) The overlay of corresponding SEM and FISH/cLSM (Fig. 1a, d) images of the same regions revealed matching areas of the FISH signal and the amorphous matrix, suggesting the presence of bacterial biofilm. The laterally located FISH signal was within a lateral canal in the root (*arrow* 1). (**f**) Higher magnifications of the FISH-labeled root canal surface with cLSM indicated a biofilm, composed of bacteria with short rod or coccus-like morphology (*arrow* 1) (Reprinted by permission from John Wiley and Sons from Schaudinn C, Carr G, Gorur A, Jaramillo D, Costerton JW, Webster P. Imaging of endodontic biofilms by combined microscopy (FISH/cLSM—SEM). Journal of Microscopy 235(2):124–127 ©2009)

References

Alhede M, Qvortrup K, Liebrechts R, Høiby N, Givskov M, Bjarnsholt T (2012) Combination of microscopic techniques reveals a comprehensive visual impression of biofilm structure and composition. Fems Immunol Med Microbiol 65:335–342

Appelbaum B, Golub E, Holt SC, Rosan B (1979) In vitro studies of dental plaque formation: adsorption of oral streptococci to hydroxyapatite. Infect Immun 25:717–728

Baldasso FE, Stürmer CP, Luisi SB, Petruzzi MN, Scarparo RK, de Figueiredo JA (2012) Microflora associated with primary endodontic infections: correlations among sem evaluation, clinical features, and radiographic findings. Microsc Res Tech 75:1557–1563

Bergmans L, Moisiadis P, van Meerbeek B, Quirynen M, Lambrechts P (2005) Microscopic observation of bacteria: review highlighting the use of environmental SEM. Int Endod J 38: 775–788

Bridier A, Meylheuc T, Briandet R (2013) Realistic representation of bacillus subtilis biofilms architecture using combined microscopy (cLSM, ESEM and FESEM). Micron 48:65–69

Costerton JW, Stewart PS, Greenberg EP (1999) Bacterial biofilms: a common cause of persistent infections. Science 284(5418):1318–1322

Costerton JW, Montanaro L, Arciola CR (2005) Biofilm in implant infections: its production and regulation. Int J Artif Organs 28:1062–1068

Danilatos GD, Postle R (1982) The environmental scanning electron microscope and its applications. Scan Electron Microsc P-1:1–16

Kakehashi S, Stanley HR, Fitzgerald RJ (1965) The effects of surgical exposures of dental pulps in germ-free and conventional laboratory rats. Oral Surg Oral Med Oral Pathol 20:340–349

Leonardo MR, Rossi MA, Silva LA, Ito IY, Bonifácio KC (2002) EM evaluation of bacterial biofilm and microorganisms on the apical external root surface of human teeth. J Endod 28: 815–818

Lomçali G, Sen BH, Cankaya H (1996) Scanning electron microscopic observations of apical root surfaces of teeth with apical periodontitis. Endod Dent Traumatol 12:70–76

Molven O, Olsen I, Kerekes K (1991) Scanning electron microscopy of bacteria in the apical part of root canals in permanent teeth with periapical lesions. Endod Dent Traumatol 7:226–229

Nair PNR (1987) Light and electron microscopic studies of root canal flora and periapical lesions. J Endod 29–39

Nair PN, Henry S, Cano V, Vera J (2005) Microbial status of apical root canal system of human mandibular first molars with primary apical periodontitis after "one-visit" endodontic treatment. Oral Surg Oral Med Oral Pathol Oral Radiol Endod 99:231–252

Ng YL, Mann V, Rahbaran S, Lewsey J, Gulabivala K (2008a) Outcome of primary root canal treatment: systematic review of the literature – Part 2. Influence of clinical factors. Int Endod J 41:6–31

Ng YL, Mann V, Gulabivala K (2008b) Outcome of secondary root canal treatment: a systematic review of the literature. Int Endod J 41:1026–1046

Noiri Y, Ehara A, Kawahara T, Takemura N, Ebisu S (2002) Participation of bacterial biofilms in refractory and chronic periapical periodontitis. J Endod 28:679–683

Okte E, Sultan N, Dogan B, Asikainen S (1999) Bacterial adhesion of actinobacillus actinomycetemcomitans serotypes to titanium implants: SEM evaluation a preliminary report. J Periodontol 70:1376–1382

Ørstavik D, Qvist V, Stoltze K (2004) A multivariate analysis of the outcome of endodontic treatment. Eur J Oral Sci 112:224–230

Parsek MR, Singh PK (2003) Bacterial biofilms: an emerging link to disease pathogenesis. Annu Rev Microbiol 57:677–701

Peters LB, Wesselink PR, Buijs JF, van Winkelhoff AJ (2001) Viable bacteria in root dentinal tubules of teeth with apical periodontitis. J Endod 27:76–81

Reese S, Guggenheim B (2007) A novel TEM contrasting technique for extracellular polysaccharides in in vitro biofilms. Microsc Res Tech 70:816–822

Richardson N, Mordan NJ, Figueiredo JA, Ng YL, Gulabivala K (2009) Microflora in teeth associated with apical periodontitis: a methodological observational study comparing two protocols and three microscopy techniques. Int Endod J 42:908–921

Ricucci D, Langeland K (1998) Apical limit of root canal instrumentation and obturation, part 2. A histological study. Int Endod J 31:394–409

Rocha CT, Rossi MA, Leonardo MR, Rocha LB, Nelson-Filho P, Silva LA (2008) Biofilm on the apical region of roots in primary teeth with vital and necrotic pulps with or without radiographically evident apical pathosis. Int Endod J 41:664–669

Schaudinn C, Carr G, Gorur A, Jaramillo D, Costerton JW, Webster P (2009) Imaging of endodontic biofilms by combined microscopy (FISH/cLSM - SEM). J Microsc 235:124–127

Sen BH, Piskin B, Demirci T (1995) Observation of bacteria and fungi in infected root canals and dentinal tubules by sem. Endod Dent Traumatol 11:6–9

Signoretti FG, Endo MS, Gomes BP, Montagner F, Tosello FB, Jacinto RC (2011) Persistent extraradicular infection in root-filled asymptomatic human tooth: scanning electron microscopic analysis and microbial investigation after apical microsurgery. J Endod 37:1696–1700

Siqueira JF Jr, Lopes HP (2001) Bacteria on the apical root surfaces of untreated teeth with periradicular lesions: a scanning electron microscopy study. Int Endod J 34:216–220

Stewart PS, Costerton JW (2001) Antibiotic resistance of bacteria in biofilms. Lancet 358:135–138

Tronstad L, Barnett F, Cervone F (1990) Periapical bacterial plaque in teeth refractory to endodontic treatment. Endod Dent Traumatol 6:73–77

van Leeuwenhoek A (1684) Letter to the Royal Society of London, 17 Sept 1683

Vieira AR, Siqueira JF Jr, Ricucci D, Lopes WS (2012) Dentinal tubule infection as the cause of recurrent disease and late endodontic treatment failure: a case report. J Endod 38:250–254

Wallace PK, Arey B, Mahaffee WF (2011) Subsurface examination of a foliar biofilm using scanning electron- and focused-ion-beam microscopy. Micron 42:579–585

Wang J, Chen W, Jiang Y, Liang J (2013) Imaging of extraradicular biofilm using combined scanning electron microscopy and stereomicroscopy. Microsc Res Tech 76:979–983

Webster P, Wu S, Webster S, Rich KA, Mcdonald K (2004) Ultrastructural preservation of biofilms formed by non-typeable hemophilus influenza. Biofilms 1:165–182

Bacterial Biofilms and Endodontic Disease: Histobacteriological and Molecular Exploration

José F. Siqueira Jr., Domenico Ricucci, and Isabela N. Rôças

Abstract Recent evidence brought about by morphological studies has indicated that apical periodontitis is a disease caused by or at least highly associated with bacterial biofilms. Histobacteriological studies revealed that biofilm-like structures are the main form in which bacteria infecting the root canal system are organized. Bacterial biofilms can be found in virtually all areas of the root canal system, including the main canal, apical and lateral ramifications, isthmuses, and recesses. Biofilms are very frequent in the apical part of root canals of teeth with primary or posttreatment apical periodontitis. Morphology of endodontic biofilms can vary from case to case and a unique pattern has not been established. Bacterial biofilms are expected to be even more prevalent in the root canals of teeth associated with long-standing pathologic processes, including large apical radiolucencies and cysts. The bacterial diversity associated with endodontic biofilms is broader than previously anticipated, and several culture-difficult or as-yet-uncultivated bacteria can participate in these communities. The clinician should be aware that, when performing root canal treatment or retreatment, he or she is dealing with a biofilm infection, which may be very difficult to reach and eradicate and may require special strategies for successful management.

1 Introduction

Apical periodontitis is an inflammatory disease caused primarily by bacterial infection of the dental root canal. As long as the dental pulp is vital, it can protect itself from bacterial infection. Infection of the root canal can only occur in a tooth whose dental pulp became necrotic as a result of advanced caries, trauma, or

J.F. Siqueira Jr. (✉) • I.N. Rôças
Department of Endodontics and Molecular Microbiology Laboratory, Estácio de Sá University, Rio de Janeiro, RJ, Brazil
e-mail: jf_siqueira@yahoo.com

D. Ricucci
Private Practice, Cetraro, Italy

L.E. Chávez de Paz et al. (eds.), *The Root Canal Biofilm*, Springer Series on Biofilms 9, DOI 10.1007/978-3-662-47415-0_5

periodontal disease or was removed for previous root canal therapy. Primary endodontic infection is the one that develops in the necrotic root canal and represents the prime cause of primary apical periodontitis. Secondary or persistent infections occur in treated teeth with posttreatment apical periodontitis. Microorganisms (generally bacteria) colonizing the root canal space are usually members of the normal oral microbiota. Once the root canal infection advances and reaches the apical part of the root canal, bacteria elicit inflammatory changes in the tissues surrounding the root apex, giving rise to the development of apical periodontitis.

It has been shown by morphological studies that the root canal microbiota in untreated (primary infection) and treated (persistent/secondary infection) teeth is usually organized in structured communities resembling typical biofilms (Nair 1987; Molven et al. 1991; Siqueira et al. 2002a; Ricucci et al. 2009; Ricucci and Siqueira 2010a). These findings permitted to include apical periodontitis in the list of human diseases associated with biofilm infections. Actually, estimates indicate that about 80 % of the human infections in the developed world have been ascribed to biofilms (Costerton 2004). In the oral cavity, apical periodontitis joins caries, gingivitis, and marginal periodontitis as typical biofilm-associated diseases. In certain aspects, multispecies biofilm communities can behave as a multicellular organism, which can result in collective pathogenic effects on the host. This community-as-pathogen concept may well be applied to the etiology of apical periodontitis (Siqueira and Rôças 2009b). The recognition of apical periodontitis as a biofilm-associated disease has significant clinical implications, because biofilms are recognizably resistant to several antimicrobial agents and may require special strategies to be eliminated (Chávez de Paz et al. 2010; Kishen 2010; Kishen and Haapasalo 2010; Alves et al. 2013; Stojicic et al. 2013).

This chapter highlights the findings from histobacteriological and molecular biology identification studies evaluating diverse aspects of endodontic biofilms, including prevalence, morphology, association with clinical and histopathological conditions, and species diversity.

2 Biofilms and the Community-As-Pathogen Concept

Biofilm can be defined as a sessile multicellular microbial community characterized by cells that are firmly attached to a surface and enmeshed in a self-produced matrix of extracellular polymeric substances (EPS) (Donlan and Costerton 2002; Costerton 2007). In bacterial biofilms, individual cells grow and aggregate to form microcolonies (populations) that are embedded and nonrandomly distributed in the EPS matrix and separated by water channels (Costerton et al. 1999; Donlan and Costerton 2002; Socransky and Haffajee 2002; Stoodley et al. 2002). In most biofilms, the bacterial populations account for about 10–15 % and the EPS matrix can account for over 85–90 % of the biofilm dry mass (Lawrence et al. 1991; Flemming and Wingender 2010). Most of the matrix (about 95 %) is water (Lawrence et al. 1991). It has been reported that dental biofilms are usually multilayered and can be up to 300 (or even more) cell thick (Socransky and Haffajee 2002).

The EPS matrix that embeds bacterial populations in biofilms is hydrated biopolymers (generally polysaccharides but also DNA, proteins, and lipids) (Costerton 2007). They are secreted by the biofilm cells and are of vital importance to the community. EPS mediates adhesion to surfaces; traps and concentrates essential nutrients for the community members; keeps biofilm cells in close proximity favoring intercellular interactions such as quorum sensing, genetic exchanges, and pathogenic synergism; and also provides protection against phagocytosis and antimicrobial agents (Flemming and Wingender 2010).

Biofilm communities are characterized by a great genetic and phenotypic diversity. The bacterial transcriptional profile radically changes after transition from the planktonic to the sessile (biofilm) state. Genes expressed by cells in biofilms can differ by 20–70 % from those expressed by the same cells occurring in a planktonic state (Oosthuizen et al. 2002; Sauer et al. 2002; Beloin et al. 2004). This leads to a different biofilm phenotype that is usually more resistant to antimicrobial agents, stress, and host defenses when compared to counterparts living in planktonic state.

Of especial therapeutic interest is the gradient of metabolic activity that can be observed throughout the biofilm structure. Bacteria present in the bottom of the biofilm, and thereby close or directly attached to the host surface, are usually in a low metabolic state and are more resistant to antimicrobial agents (Lewis 2007; Rhoads et al. 2008). In contrast, those bacteria present in the most superficial areas have increased metabolic activity and are generally more vulnerable to the effects of antimicrobial agents (Rhoads et al. 2008). Therefore, a gradient is formed so that bacteria in diverse metabolic states are distributed throughout the biofilm structure. Biofilm communities have the ability to reconstitute themselves even after an event that resulted in dramatic ecological and structural changes by destroying a large proportion of the community (Wolcott et al. 2010; Wolcott and Dowd 2011). Such an event may consist of treatment procedures that eliminated a large portion of the community but did not succeed in completely eradicating it. The remaining biofilm left behind over a host surface may reconstitute itself, reactivating its metabolism and orchestrating the reconstruction of the community by means of quorum-sensing systems and other mechanisms (Wolcott and Dowd 2011), provided there is still space available for recolonization.

In summary, the most important features and strengths of the biofilm lifestyle are metabolic and genetic heterogeneity, interspecies cooperation, protection against exogenous threats, enhanced pathogenicity, and ability to reconstitute after drastic events (Costerton et al. 1987, 1995; Donlan and Costerton 2002; Socransky and Haffajee 2002; Stoodley et al. 2002; Marsh 2003, 2005; Hall-Stoodley et al. 2004; Percival et al. 2010).

Actually, most endogenous infections have been shown to be caused by mixed biofilm communities, which is in contrast to the "single-species etiology" concept established since Koch's classic studies. Therefore, it has been proposed that the microbial community as a whole is indeed the unit of pathogenicity for these endogenous diseases, including apical periodontitis (Jenkinson and Lamont 2005;

Kuramitsu et al. 2007; Siqueira and Rôças 2009b). The concept of the community as pathogen is based on the principle that "teamwork is what eventually counts." By this, the community behavior and the outcome of the host/bacterial community interaction will ultimately depend upon the community membership and the myriad of associations within the community. Virulence usually differs for a given bacterial species when it is living in pure culture, forming pairs with other species, or taking part of a large bacterial "society" coexisting with several other species (community) (Sundqvist et al. 1979; Baumgartner et al. 1992; Siqueira et al. 1998; Socransky et al. 1998; Kuramitsu et al. 2007; Siqueira and Rôças 2009b).

This concept holds that the pathogenesis of apical periodontitis is resultant of the concerted action of bacteria in a multispecies community. Bacterial virulence factors involved in the pathogenesis of apical periodontitis consist of a summation of structural cellular components, antigens, and secreted substances that accumulate in the biofilm (Siqueira and Rôças 2007). The concentration and virulence of this bacterial "soup" will depend upon the population density, species composition, and bacterial interactions in the community. Once the biofilm forms in the apical canal, this "soup" of antigens and virulence factors becomes in constant and direct contact with the periradicular tissues to cause damage and stimulate/modulate the host immune responses (Siqueira 2011).

3 Biofilm and Apical Periodontitis

The first report on the occurrence of biofilm-like structures in infected root canals is claimed to be from Nair (1987), who described those structures as "bacterial condensation on the surface of the dentin wall, forming thin- or thick-layered bacterial plaques." Similar observations were subsequently reported by in situ morphological investigations of teeth with primary or posttreatment apical periodontitis (Molven et al. 1991; Siqueira et al. 2002a; Carr et al. 2009; Ricucci et al. 2009; Schaudinn et al. 2009). In addition to the main root canal, bacterial biofilms have also been disclosed in anatomical variations of the root canal system, including apical ramifications, lateral canals, and isthmuses (Nair et al. 2005; Ricucci and Siqueira 2008, 2010b). Biofilms adhered to the apical root surface (extraradicular biofilms) have also been described in some teeth evincing posttreatment apical periodontitis (Tronstad et al. 1990; Ferreira et al. 2004; Ricucci et al. 2005).

3.1 *Histobacteriological Analysis of Endodontic Biofilm Communities*

The observations reported above contributed to the assumption of apical periodontitis as a disease caused by or at least associated with bacterial biofilms. However, the prevalence of biofilms and their association with diverse presentations of apical periodontitis were only determined after a recent histobacteriological and histopathological study by Ricucci and Siqueira (2010a). The study evaluated the prevalence of biofilms in the apical root canal of untreated teeth with primary apical periodontitis and treated teeth with posttreatment disease. Associations with clinical and histopathological findings were also determined. The main findings of this study were as follows:

1. Intraradicular biofilm arrangements were in general observed in the apical segment of 77 % of the root canals of teeth with apical periodontitis (80 % in untreated canals and 74 % in treated canals). There were cases in which the biofilm formed even onto the inflamed soft tissue near the apical foramen (Fig. 1).
2. Morphologically, intraradicular bacterial biofilms were usually thick and composed of several layers of bacterial cells. Different morphotypes were commonly seen per biofilm. The relative proportions between bacterial cells/populations and the extracellular matrix were highly variable. Therefore, endodontic biofilm morphology differed consistently from individual to individual (interindividual variability) and even when different areas of the same canal were examined (intraindividual variability) (Fig. 2).
3. Dentinal tubules underneath biofilms covering the walls of the main apical canal were very often invaded by bacteria from the bottom of the biofilm structure (Fig. 3). In addition, biofilms were also commonly seen covering the walls of apical ramifications, lateral canals, and isthmuses (Fig. 4).
4. Bacterial biofilms were visualized in 62 % and 82 % of the root canals of teeth with small and large apical periodontitis lesions, respectively. All root canals associated with very large lesions (>10 mm in radiographic diameter) were found to harbor intraradicular biofilms.
5. The prevalences of intraradicular biofilms in teeth associated with apical cysts, abscesses, and granulomas were 95 %, 83 %, and 69.5 %, respectively. Biofilms were significantly associated with epithelialized lesions (Fig. 5).
6. No correlation was found between biofilms and clinical symptoms or sinus tract presence.
7. Extraradicular biofilms were very infrequent, being observed in only 6 % of the cases (Fig. 6). Except for one case, they were always associated with intraradicular biofilms. All cases showing an extraradicular biofilm exhibited clinical symptoms. Thus, it seems that extraradicular infections in the form of biofilms or planktonic bacteria are not a common occurrence, are usually dependent on the intraradicular infection, and are more frequent in symptomatic teeth.

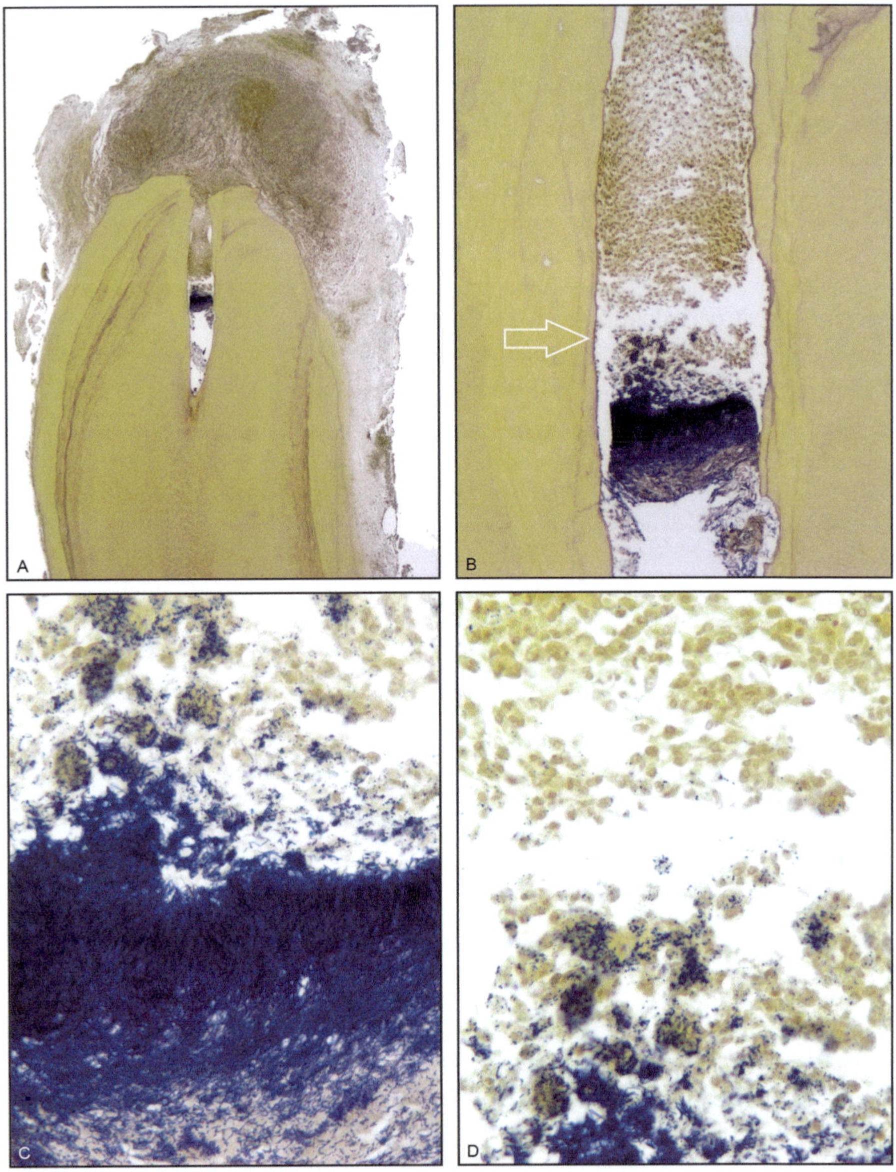

Fig. 1 (**a**) Distobuccal root of a maxillary first molar in a 55-year-old man. Repeated abscesses were reported. Overview of a section passing through the main foramen. No epithelium could be observed in this section and in the rest of the serial sections, leading to the diagnosis of non-epithelialized granuloma (Taylor's modified Brown & Brenn, original magnification ×16). (**b**) Detail of the apical canal. A thick biofilm fills completely the canal lumen just short of the foramen, while the most apical canal is occupied by granulation tissue (×100). (**c**) High power view of the biofilm showing high density of filamentous bacterial forms (×400). (**d**) Magnification of the area indicated by the *arrow* in (**b**). The biofilm structure is faced apically with a concentration of polymorphonuclear leukocytes (×400)

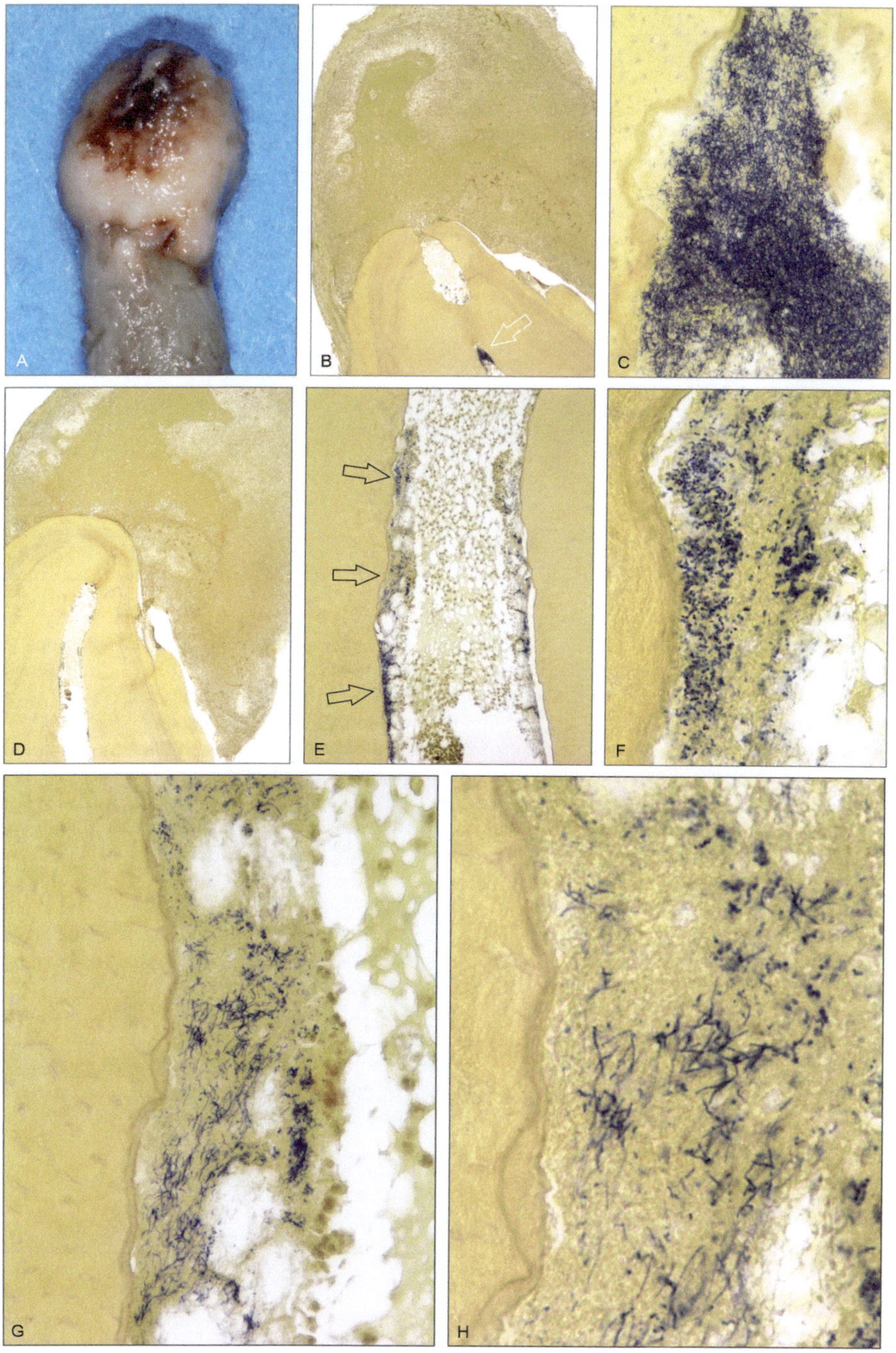

Fig. 2 (continued)

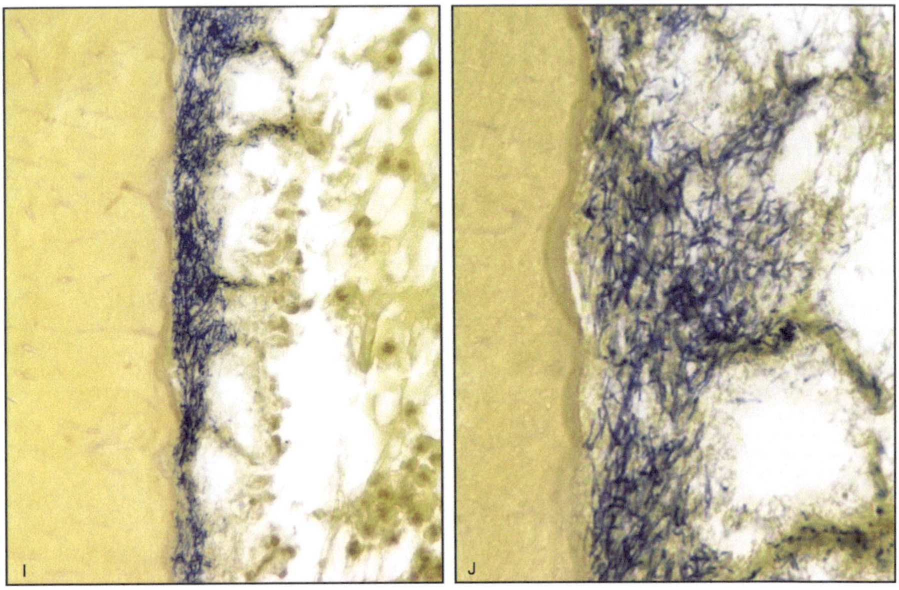

Fig. 2 (**a**) Palatal root of a maxillary first molar in a 48-year-old woman. The tooth was asymptomatic at the moment of extraction and no symptoms were reported previously. (**b**) Section passing through the main foramen. The lesion is a "pocket cyst" (Taylor's modified Brown & Brenn, original magnification ×16). (**c**) High magnification of the most coronal area indicated by the *arrow* in (**b**). The thick biofilm is composed mainly by filamentous forms (×400). (**d**) Section taken at a considerable distance from that shown in (**b**), passing through the canal but not encompassing the foramen (×16). (**e**) Detail of the root canal. A biofilm is present on the opposite root canal walls, with inflammatory tissue in between (×100). (**f**) High power view of the area of the left canal wall indicated by the *upper arrow* in (**e**). The biofilm is composed mainly by coccoid forms. Bacterial density is higher in the deepest part, while superficially the extracellular matrix is abundant (×1000). (**g, h**) Progressive magnifications of the area of the canal wall indicated by the *middle arrow* in (**e**). Few filamentous and coccoid forms in an abundant extracellular matrix (×400 and ×1000). (**i–j**) Progressive magnifications of the area of the canal wall indicated by the *lower arrow* in (**e**). The biofilm appears thinner at this level and composed by a high density of filamentous forms and reduced extracellular matrix (×400 and ×1000). *Considerations*. The morphology of the biofilm observed in different areas of the same canal may vary consistently. In this case, cocci were present in some areas, while filamentous forms dominated in others. Also the concentrations of bacterial cells may be different. In some areas, bacteria may appear more concentrated in the deepest parts, while in others they can be more numerous superficially, and an apparently bacteria-free extracellular matrix is attached to the canal wall

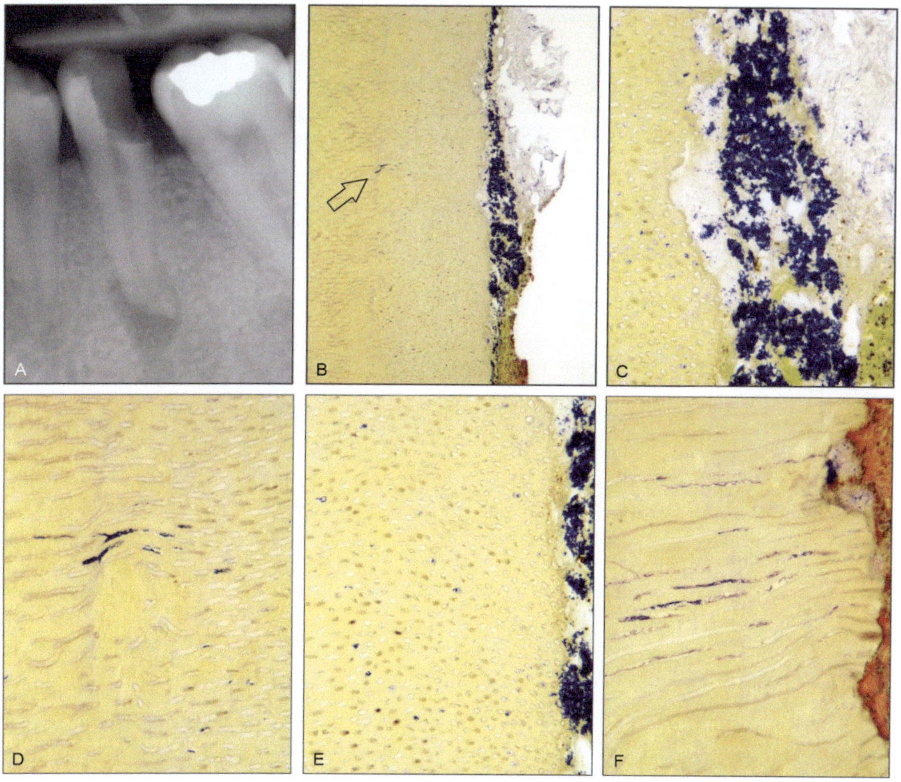

Fig. 3 (**a**) Mandibular premolar with long-standing pulp necrosis and apical radiolucency in a 32-year-old woman. An abscess with severe pain and swelling was present before extraction. (**b**) Transition between the middle and apical third. A biofilm is present on the root canal wall (Taylor's modified Brown & Brenn, original magnification ×100). (**c**) Detail from (**b**) (×400). (**d**) High power view of the area indicated by the *arrow* in (**b**). Some tubules are colonized by bacteria (×400). (**e**) Area of the root canal wall where dentin tubules were cut transversally. Some tubules are colonized by bacteria (×400). (**f**) Apical canal wall. Severe bacterial colonization of some dentinal tubules (×400)

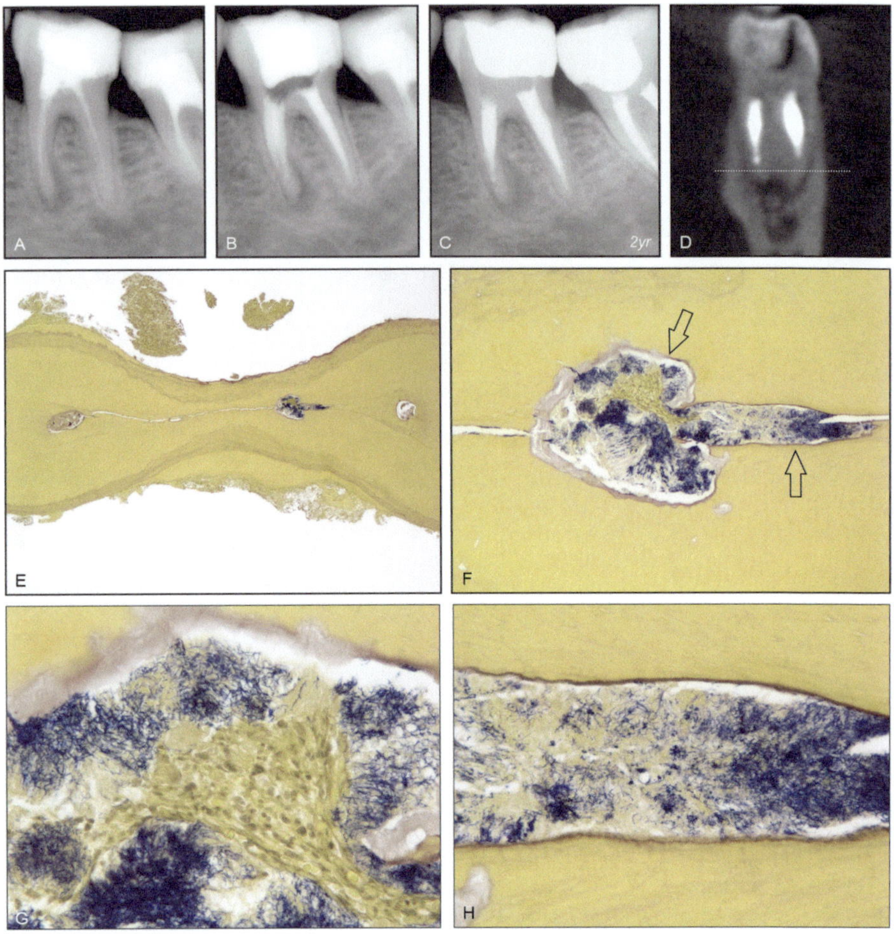

Fig. 4 (**a**) Mandibular first molar in a 39-year-old woman. The tooth had been root canal-treated 10 years before. A radiolucency is present on the mesial root apex. (**b**) Retreatment was performed. It was not possible to negotiate mesial canals for their full length. After more than 4 weeks of calcium hydroxide medication, the canals were filled. (**c**) 2 years later the patient presented with a flare-up. The radiolucency had remained the same size. (**d**) Apicoectomy was scheduled and a cone beam computed tomography scan was performed to ascertain the relationship of the root apex with the mandibular nerve. (**e**) Crosscut sections of the removed mesial root apex, taken approximately at the level of the *line* in (**d**). An isthmus connecting the two main canals is present, with an expansion (Taylor's modified Brown & Brenn, original magnification ×16). (**f**) The expansion is clogged with a thick biofilm (×100). (**g**) High power view from the area indicated by the *left arrow* in (**f**). Condensations of filamentous bacterial forms and accumulation of inflammatory cells (×400). (**h**) High power view from the area indicated by the *right arrow* in (**f**). The lumen is occupied by a thick bacterial biofilm at this level (×400)

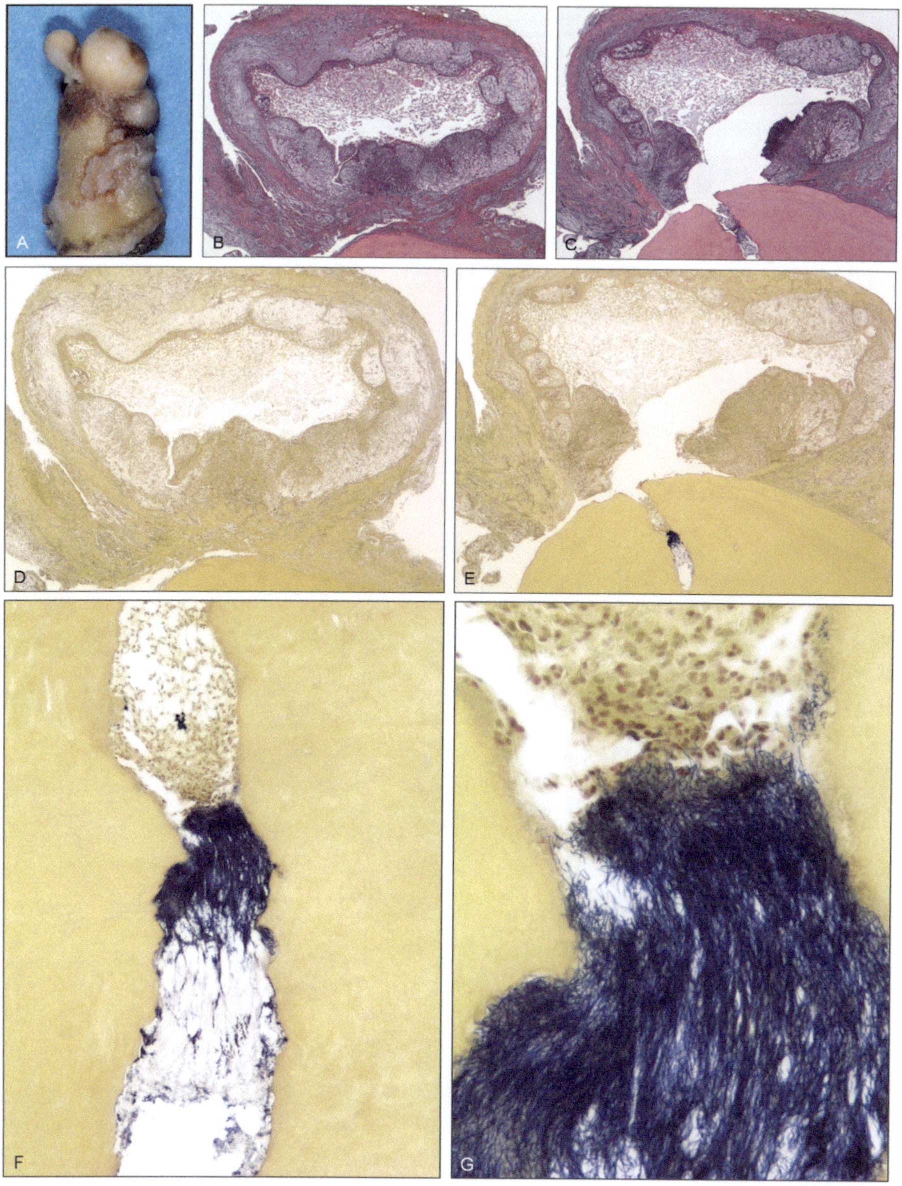

Fig. 5 (**a**) Maxillary third molar with a history of repeated abscess episodes in a 40-year-old man. Several "soft tissue lesions" can be seen attached to the root apex after extraction. (**b**) Section encompassing the middle and bigger soft tissue lesion. A cyst cavity completely lined by epithelium and containing necrotic tissue can be observed (H&E, original magnification ×16). (**c**) After approximately 120 sections, a communication can be observed between the cyst lumen and the root canal space, and the diagnosis of "pocket cyst" is made (×16). (**d, e**) Sections stained with the modified B&B technique confirm that the lesion is a pocket cyst (Taylor's modified Brown & Brenn, original magnification ×16). (**f, g**) Consecutive magnifications of the apical canal show the presence of a thick bacterial biofilm faced with a concentration of polymorphonuclear leukocytes (×100 and ×400)

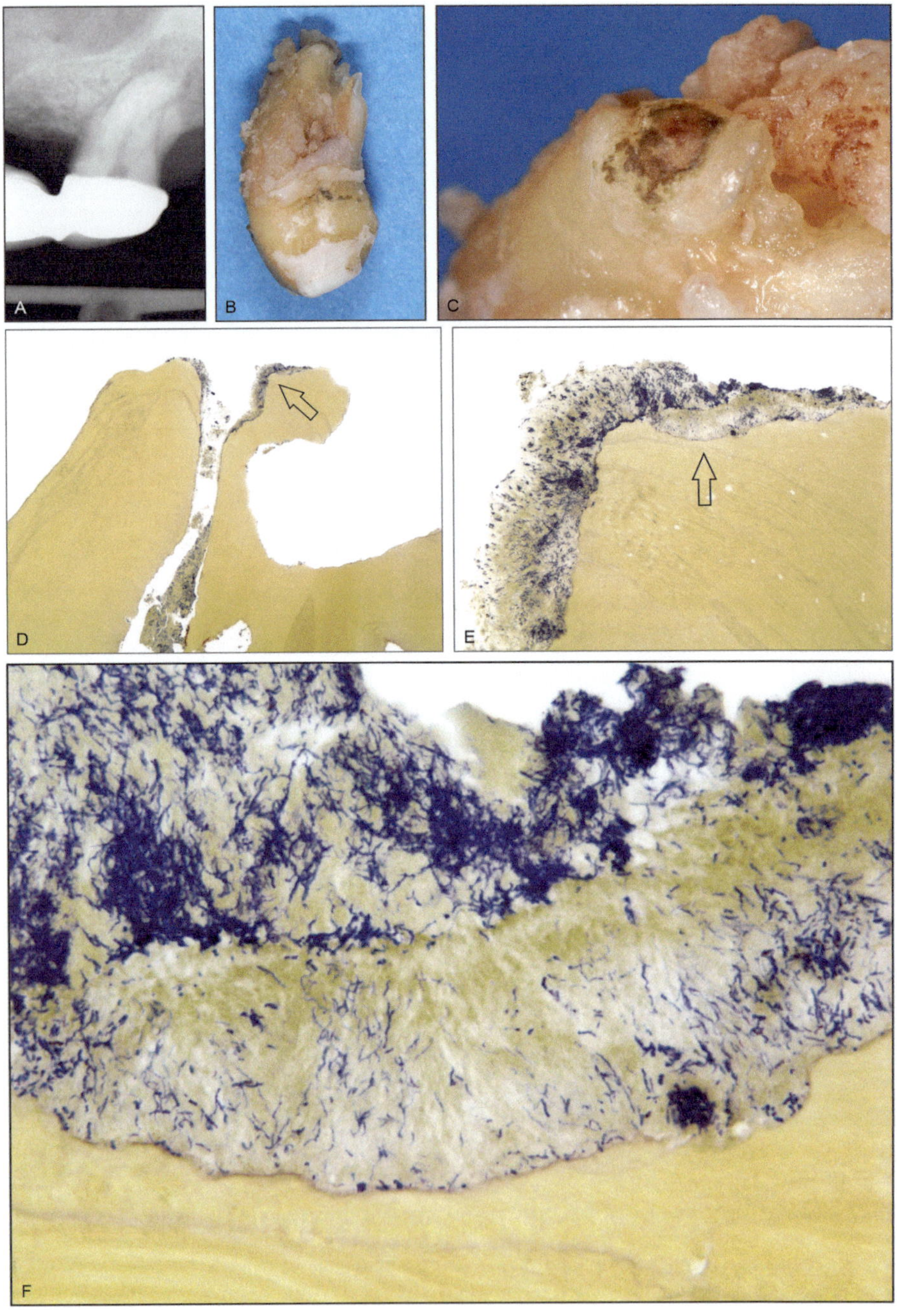

Fig. 6 (continued)

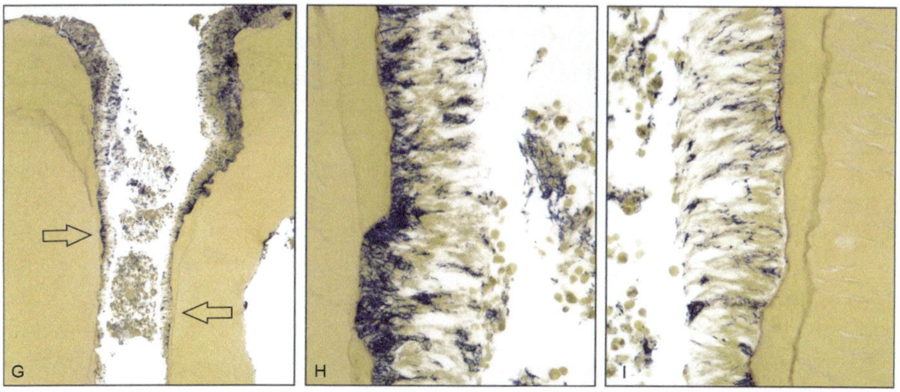

Fig. 6 (**a**) Symptomatic maxillary third molar in a 50-year-old man. At presentation, no sinus tract could be seen. The bridge had been decemented for long time (years). Endodontic treatment was recommended, but the patient opted for an alternative treatment plan including extraction of the tooth. (**b, c**) After extraction, calculus was noted at the tip of the mesiobuccal root, located concentrically around its foramen. (**d**) Sections passing through the mesiobuccal canal. Severe apical resorption (Taylor's modified Brown & Brenn, original magnification ×16). (**e**) Magnification of the area indicated by the *arrow* in (**d**). A dense biofilm extends uninterruptedly from the apical root canal wall to the external root surface (×100). (**f**) High power view of the external root surface indicated by the *arrow* in (**e**). The biofilm is composed of two layers (×400). (**g**) Detail of the apical canal in (**d**) (×50). (**h**) High magnification of the area of the left canal wall indicated by the *left arrow* in (**g**). Bacterial biofilm with a higher bacterial density in the deepest area (×400). (**i**) High magnification of the area of the right canal wall indicated by the *right arrow* in (**g**). The bacterial biofilm exhibits a reduced bacterial density at this level (×400)

8. Bacteria were also seen in the lumen of the main canal, ramifications, and isthmuses as flocs and planktonic cells, either intermixed with necrotic pulp tissue or possibly suspended in a fluid phase. Bacterial flocs may exhibit many of the same characteristics as biofilms and probably originate from growth of cell aggregates/coaggregates in a fluid, or they may have detached from biofilms (Hall-Stoodley et al. 2004; Hall-Stoodley and Stoodley 2009).

3.2 *Criteria to Classify Apical Periodontitis as a Biofilm-Induced Disease*

In 2003, Parsek and Singh (2003) proposed 4 criteria to determine whether a given infectious disease can be classified as a disease caused by biofilm communities. Later, a 5th criterion was suggested by Hall-Stoodley and Stoodley (2009), and a 6th criterion was added by Ricucci and Siqueira (2010a). The six criteria are as follows:

1. The infecting bacteria are adhered to or associated with a surface. By "associated with," the authors meant that bacterial aggregates/coaggregates do not need to be firmly attached to the surface.
2. Direct examination of infected tissue shows bacteria forming clusters or microcolonies encased in an extracellular matrix.
3. The infection is generally confined to a particular site, and although dissemination may occur, it is a secondary event.
4. The infection is difficult or impossible to eradicate with antibiotics despite the fact that the responsible microorganisms can be susceptible to killing in the planktonic state.
5. Ineffective host clearance. This may be evidenced by the location of microbial colonies in areas usually surrounded by host defense cells. Accumulation of polymorphonuclear neutrophils (PMNs) and macrophages near bacterial aggregates/coaggregates in situ considerably strengthens the point for biofilm involvement with disease causation.
6. Elimination or drastic disruption of the biofilm structure and ecology leads to remission of the disease process.

The findings from the Ricucci and Siqueira study (Ricucci and Siqueira 2010a) showing biofilm structures in the great majority of cases of primary and post-treatment apical periodontitis along with the observed morphological features of these biofilms were considered to fulfill 4 of the 6 criteria:

(a) Bacterial communities were observed adhered to or at least associated with the root canal dentin surface (criterion 1).
(b) Bacterial colonies were seen in the huge majority of the specimens encased in an amorphous extracellular matrix (criterion 2).
(c) Endodontic biofilms were often confined to the root canal system, in only a few cases extending to the external root surface, but dissemination through the lesion never occurred (criterion 3).
(d) In the great majority of cases, biofilms were directly faced by inflammatory cells (mostly PMNs) accumulated in the very apical part of the root canal system, including the main canal, apical ramifications, and isthmuses (criterion 5).

Criterion 4 was not assessed in that study, but it is well established that intraradicular endodontic infections cannot be effectively treated by systemic antibiotic therapy, even though most endodontic bacteria in the planktonic cell state are susceptible to currently used antibiotics (Khemaleelakul et al. 2002; Baumgartner and Xia 2003; Gomes et al. 2011). The lack of efficacy of systemic antibiotics against intraradicular infections is mainly due to the fact that the bacterial pathogens are not reached by the drug because they are located in an avascular necrotic space. The recognition of biofilms as the main mode of bacterial establishment in the root canal system further strengthens the explanations for the lack of antibiotic effectiveness against endodontic infections. As for criterion 6, the specific direct effects of treatment on endodontic biofilms have not been demonstrated, but culture studies demonstrate that a better treatment outcome can be achieved when the bacterial load in the root canal is reduced to levels that are undetectable by the culture approach (Engström et al. 1964; Sjögren et al. 1997; Sundqvist et al. 1998; Waltimo et al. 2005). Since biofilms are the main form in which endodontic infections are organized, then one may infer that biofilms were eliminated or substantially disrupted when culture yields negative results. In addition, the frequent observation of biofilms in treated canals with posttreatment disease (Ricucci and Siqueira 2008, 2010a; Ricucci et al. 2009) adds further potential for meeting this 6th criterion.

4 Molecular Analysis of Endodontic Biofilm Communities

The traditional method using paper points to sample infected root canals in the clinical setting does not allow specific sampling of bacteria present in biofilms. Bacteria also present in a planktonic state in the main root canal are conceivably the most accessible to sampling by this method. To improve detection of bacteria present in biofilms adhered to the root canal walls, it is important to use instruments to gently file the root canal walls circumferentially previously to sampling so as to dislodge bacteria from biofilms and suspend them in the root canal fluid. Even so, biofilms present in ramifications and isthmuses are highly likely to remain not sampled. Cryogenic pulverization of root fragments from extracted teeth improves detection of bacteria from the entire root canal system, but still it is not selective to biofilms as cells floating in the main canal can also be included in analysis. Thus far, no single study has selectively identified bacteria in endodontic biofilms. A possibility for further research is using fluorescence in situ hybridization for detection of target species in the endodontic biofilm or using a combination of paraffin-embedded samples and broad-range PCR for open-ended identification of bacteria in endodontic biofilms.

Even considering the limitations of the sampling approaches currently available to selectively detect bacteria from endodontic biofilms, one may consider that many of the floating bacteria suspended in the main canal can be conceivably cells that detached from the biofilms adhered to the root canal walls. However, the possibility also exists that they are bystanders, latecomers, or individuals that did not succeed in taking part of the community because of lack of competitivity. Anyway, all species present in the main canal allegedly assume importance as to the pathogenesis of apical periodontitis and whether or not they are or were members of the biofilm may have more implications in terms of resistance to treatment.

The diversity of the endodontic microbiota has been unveiled by numerous culture and molecular microbiology studies. As for the identification of microbial species, molecular methods have been directly used in clinical samples to detect the unexpected (open-ended analysis) or to target-specific taxa (closed-ended analysis). Examples of open-ended molecular techniques used for microbial identification in samples taken from infected canals include polymerase chain reaction (PCR) followed by cloning and Sanger sequencing and more recently pyrosequencing; closed-ended methods include species-specific PCR and checkerboard DNA–DNA hybridization. Application of these methods in endodontic microbiology research was reviewed elsewhere (Siqueira and Rôças 2009a).

Most culture and molecular studies have used the paper-point-sampling technique and only a few used cryopulverized root fragments. Collectively, nearly 500 different microbial species/phylotypes have been identified in endodontic samples from teeth with different forms of apical periodontitis (Siqueira and Rôças 2009c). These taxa are usually found in combinations involving many species/phylotypes in primary infections and fewer ones in secondary/persistent infections (Siqueira and Rôças 2005a).

Bacteria are by far the most common microorganisms found in endodontic infections and then implicated with the etiology of apical periodontitis. At high phylogenetic levels, endodontic bacteria fall into 18 phyla, with the most common representative species/phylotypes belonging to the phyla *Firmicutes*, *Bacteroidetes*, *Actinobacteria, Fusobacteria, Proteobacteria, Spirochaetes*, and *Synergistetes* (Munson et al. 2002; Siqueira and Rôças 2005b; Saito et al. 2006; Sakamoto et al. 2006, 2007; Rôças and Siqueira 2008; Li et al. 2010; Santos et al. 2011; Siqueira et al. 2011, 2012; Ozok et al. 2012; Hong et al. 2013). In spite of the relatively long list of bacterial taxa so far identified in samples from infected root canals, 20–30 bacterial species are invariably among the most prevalent in different studies and then considered to be major candidate endodontic pathogens (Siqueira and Rôças 2014). These species are shown in Table 1.

Microorganisms other than bacteria have also been found in endodontic infections, but usually in low prevalence. Archaea and fungi have been only occasionally found in intraradicular infections (Waltimo et al. 1997; Siqueira et al. 2002b; Vianna et al. 2006; Vickerman et al. 2007), though the latter can be more prevalent in treated teeth with posttreatment disease (Siqueira and Sen 2004).

Because of the polymicrobial nature of endodontic infections, with no single species being regarded as the main pathogen, molecular methods for bacterial community profiling have been conducted in an attempt to identify possible patterns related to clinical conditions and other factors (Siqueira and Rôças 2009b). Examples of these community profiling techniques include denaturing gradient gel electrophoresis, terminal-restriction fragment length polymorphism, and pyrosequencing. Community profiles are essentially determined by species richness (number of different species) and abundance (proportion of each member in the community). Many molecular microbiology techniques used to profile communities in different environments have been applied to the study of the microbial communities associated with human healthy and diseased sites (Siqueira et al. 2010). Application of these techniques to the study of endodontic communities has demonstrated that:

(a) The different types of endodontic infections are composed of multispecies bacterial communities (Siqueira et al. 2004; Machado de Oliveira et al. 2007; Chugal et al. 2011). This is also true for persistent/secondary infections

Table 1 Bacterial species/phylotypes commonly detected in endodontic infections

Species	Gram staining	Phylum	Type of infection
Bacteroidetes oral clone X083	Gram-negative	Bacteroidetes	Primary
Campylobacter rectus	Gram-negative	Proteobacteria	Primary
Dialister invisus	Gram-negative	Firmicutes	Primary; persistent/secondary
Fusobacterium nucleatum	Gram-negative	Fusobacteria	Primary; persistent/secondary; extraradicular
Porphyromonas endodontalis	Gram-negative	Bacteroidetes	Primary; extraradicular
Porphyromonas gingivalis	Gram-negative	Bacteroidetes	Primary; extraradicular
Prevotella intermedia	Gram-negative	Bacteroidetes	Primary; extraradicular
Prevotella nigrescens	Gram-negative	Bacteroidetes	Primary
Prevotella baroniae	Gram-negative	Bacteroidetes	Primary
Pyramidobacter piscolens	Gram-negative	Synergistetes	Primary; persistent/secondary
Tannerella forsythia	Gram-negative	Bacteroidetes	Primary
Treponema denticola	Gram-negative	Spirochaetes	Primary
Treponema socranskii	Gram-negative	Spirochaetes	Primary
Treponema maltophilum	Gram-negative	Spirochaetes	Primary
Veillonella parvula	Gram-negative	Firmicutes	Primary
Actinomyces israelii	Gram-positive	Actinobacteria	Primary; persistent/secondary; extraradicular
Enterococcus faecalis	Gram-positive	Firmicutes	Persistent/secondary
Filifactor alocis	Gram-positive	Firmicutes	Primary; persistent/secondary
Olsenella uli	Gram-positive	Actinobacteria	Primary; persistent/secondary
Parvimonas micra	Gram-positive	Firmicutes	Primary; persistent/secondary
Propionibacterium acnes	Gram-positive	Actinobacteria	Primary; persistent/secondary
Propionibacterium propionicum	Gram-positive	Actinobacteria	Primary; persistent/secondary; extraradicular

(continued)

Table 1 (continued)

Species	Gram staining	Phylum	Type of infection
Pseudoramibacter alactolyticus	Gram-positive	Firmicutes	Primary; persistent/secondary
Streptococcus anginosus	Gram-positive	Firmicutes	Primary; persistent/secondary
Streptococcus constellatus	Gram-positive	Firmicutes	Primary; persistent/secondary
Streptococcus intermedius	Gram-positive	Firmicutes	Primary; persistent/secondary
Streptococcus mitis	Gram-positive	Firmicutes	Primary; persistent/secondary

associated with treated teeth (Rôças et al. 2004, 2008; Siqueira and Rôças 2004; Blome et al. 2008; Sakamoto et al. 2008; Li et al. 2010; Chugal et al. 2011), which had been previously regarded as being composed of only one or two species on the basis of culture studies.

(b) There is a great interindividual variability in the composition of endodontic bacterial communities in teeth with the same clinical disease (Rôças et al. 2004; Siqueira et al. 2004; Machado de Oliveira et al. 2007; Li et al. 2010; Chugal et al. 2011; Santos et al. 2011; Hong et al. 2013). This means to say that every individual has its unique endodontic microbiota in terms of species diversity. This finding indicates that apical periodontitis has a heterogeneous etiology.

(c) Bacterial community structures follow specific patterns according to the clinical condition (asymptomatic apical periodontitis, acute apical abscesses, and posttreatment apical periodontitis) (Rôças et al. 2004; Siqueira et al. 2004; Sakamoto et al. 2006; Santos et al. 2011). Disease severity, as determined by the intensity of signs and symptoms, or response to endodontic treatment may be affected by the bacterial community composition. This finding indicated that some bacterial communities are more related to certain forms of the disease than others (Siqueira et al. 2004; Sakamoto et al. 2006; Santos et al. 2011);

(d) Community profiles also seem to follow patterns related to the geographic location of the individuals. Despite interindividual variability in the community structure, individuals from the same location share more similarities among them when compared to individuals from distant geographic locations (Rôças et al. 2004; Siqueira et al. 2004, 2008; Machado de Oliveira et al. 2007). This may have implications in terms of effectiveness of antimicrobial treatments in different countries.

(e) In addition to the high interindividual variability, there is also an intra-individual variability. Differences have been reported for the bacterial community profiles in different teeth from the same individual or even for communities living in different regions of the same tooth (apical versus cervical root canal) (Alves et al. 2009; Rôças et al. 2010).

5 Concluding Remarks

Apical periodontitis is a disease frequently associated with bacterial biofilms. Apparently, the older the disease process (as inferred by histopathological and radiographic features), the higher the possibility of bacterial biofilms being found in the apical root canal. The essential role of bacteria in causation of apical periodontitis is well established, and in most teeth with apical periodontitis, bacteria in the root canal are organized in biofilms. The concept of apical periodontitis is an infectious disease highly associated with bacterial biofilms should be understood by clinicians and researchers before planning the best therapeutic strategies. Thus, treatment of apical periodontitis in the huge majority of cases involves management of a biofilm infection present not only in the main apical root canal but also distributed throughout the entire root canal system.

References

Alves FR, Siqueira JF Jr, Carmo FL, Santos AL, Peixoto RS, Rôças IN, Rosado AS (2009) Bacterial community profiling of cryogenically ground samples from the apical and coronal root segments of teeth with apical periodontitis. J Endod 35:486–492

Alves FR, Neves MA, Silva MG, Rôças IN, Siqueira JF Jr (2013) Antibiofilm and antibacterial activities of farnesol and xylitol as potential endodontic irrigants. Braz Dent J 24:224–229

Baumgartner JC, Xia T (2003) Antibiotic susceptibility of bacteria associated with endodontic abscesses. J Endod 29:44–47

Baumgartner JC, Falkler WA Jr, Beckerman T (1992) Experimentally induced infection by oral anaerobic microorganisms in a mouse model. Oral Microbiol Immunol 7:253–256

Beloin C, Valle J, Latour-Lambert P, Faure P, Kzreminski M, Balestrino D, Haagensen JA, Molin S, Prensier G, Arbeille B, Ghigo JM (2004) Global impact of mature biofilm lifestyle on Escherichia coli K-12 gene expression. Mol Microbiol 51:659–674

Blome B, Braun A, Sobarzo V, Jepsen S (2008) Molecular identification and quantification of bacteria from endodontic infections using real-time polymerase chain reaction. Oral Microbiol Immunol 23:384–390

Carr GB, Schwartz RS, Schaudinn C, Gorur A, Costerton JW (2009) Ultrastructural examination of failed molar retreatment with secondary apical periodontitis: an examination of endodontic biofilms in an endodontic retreatment failure. J Endod 35:1303–1309

Chávez de Paz LE, Bergenholtz G, Svensater G (2010) The effects of antimicrobials on endodontic biofilm bacteria. J Endod 36:70–77

Chugal N, Wang JK, Wang R, He X, Kang M, Li J, Zhou X, Shi W, Lux R (2011) Molecular characterization of the microbial flora residing at the apical portion of infected root canals of human teeth. J Endod 37:1359–1364

Costerton B (2004) Microbial ecology comes of age and joins the general ecology community. Proc Natl Acad Sci USA 101:16983–16984

Costerton JW (2007) The biofilm primer. Springer, Berlin, Heidelberg

Costerton JW, Cheng KJ, Geesey GG, Ladd TI, Nickel JC, Dasgupta M, Marrie TJ (1987) Bacterial biofilms in nature and disease. Annu Rev Microbiol 41:435–464

Costerton JW, Lewandowski Z, Caldwell DE, Korber DR, Lappin-Scott HM (1995) Microbial biofilms. Annu Rev Microbiol 49:711–745

Costerton JW, Stewart PS, Greenberg EP (1999) Bacterial biofilms: a common cause of persistent infections. Science 284:1318–1322
Donlan RM, Costerton JW (2002) Biofilms: survival mechanisms of clinically relevant microorganisms. Clin Microbiol Rev 15:167–193
Engström B, Hard AF, Segerstad L, Ramström G, Frostell G (1964) Correlation of positive cultures with the prognosis for root canal treatment. Odontol Revy 15:257–270
Ferreira FB, Ferreira AL, Gomes BP, Souza-Filho FJ (2004) Resolution of persistent periapical infection by endodontic surgery. Int Endod J 37:61–69
Flemming HC, Wingender J (2010) The biofilm matrix. Nat Rev Microbiol 8:623–633
Gomes BP, Jacinto RC, Montagner F, Sousa EL, Ferraz CC (2011) Analysis of the antimicrobial susceptibility of anaerobic bacteria isolated from endodontic infections in Brazil during a period of nine years. J Endod 37:1058–1062
Hall-Stoodley L, Stoodley P (2009) Evolving concepts in biofilm infections. Cell Microbiol 11: 1034–1043
Hall-Stoodley L, Costerton JW, Stoodley P (2004) Bacterial biofilms: from the natural environment to infectious diseases. Nat Rev Microbiol 2:95–108
Hong BY, Lee TK, Lim SM, Chang SW, Park J, Han SH, Zhu Q, Safavi KE, Fouad AF, Kum KY (2013) Microbial analysis in primary and persistent endodontic infections by using pyrosequencing. J Endod 39:1136–1140
Jenkinson HF, Lamont RJ (2005) Oral microbial communities in sickness and in health. Trends Microbiol 13:589–595
Khemaleelakul S, Baumgartner JC, Pruksakorn S (2002) Identification of bacteria in acute endodontic infections and their antimicrobial susceptibility. Oral Surg Oral Med Oral Pathol Oral Radiol Endod 94:746–755
Kishen A (2010) Advanced therapeutic options for endodontic biofilms. Endod Topics 22:99–123
Kishen A, Haapasalo M (2010) Biofilm models and methods of biofilm assessment. Endod Topics 22:58–78
Kuramitsu HK, He X, Lux R, Anderson MH, Shi W (2007) Interspecies interactions within oral microbial communities. Microbiol Mol Biol Rev 71:653–670
Lawrence JR, Korber DR, Hoyle BD, Costerton JW, Caldwell DE (1991) Optical sectioning of microbial biofilms. J Bacteriol 173:6558–6567
Lewis K (2007) Persister cells, dormancy and infectious disease. Nat Rev Microbiol 5:48–56
Li L, Hsiao WW, Nandakumar R, Barbuto SM, Mongodin EF, Paster BJ, Fraser-Liggett CM, Fouad AF (2010) Analyzing endodontic infections by deep coverage pyrosequencing. J Dent Res 89:980–984
Machado de Oliveira JC, Siqueira JF Jr, Rôças IN, Baumgartner JC, Xia T, Peixoto RS, Rosado AS (2007) Bacterial community profiles of endodontic abscesses from Brazilian and USA subjects as compared by denaturing gradient gel electrophoresis analysis. Oral Microbiol Immunol 22: 14–18
Marsh PD (2003) Are dental diseases examples of ecological catastrophes? Microbiology 149: 279–294
Marsh PD (2005) Dental plaque: biological significance of a biofilm and community life-style. J Clin Periodontol 32(Suppl 6):7–15
Molven O, Olsen I, Kerekes K (1991) Scanning electron microscopy of bacteria in the apical part of root canals in permanent teeth with periapical lesions. Endod Dent Traumatol 7:226–229
Munson MA, Pitt-Ford T, Chong B, Weightman A, Wade WG (2002) Molecular and cultural analysis of the microflora associated with endodontic infections. J Dent Res 81: 761–766
Nair PNR (1987) Light and electron microscopic studies of root canal flora and periapical lesions. J Endod 13:29–39
Nair PN, Henry S, Cano V, Vera J (2005) Microbial status of apical root canal system of human mandibular first molars with primary apical periodontitis after "one-visit" endodontic treatment. Oral Surg Oral Med Oral Pathol Oral Radiol Endod 99:231–252

Oosthuizen MC, Steyn B, Theron J, Cosette P, Lindsay D, Von Holy A, Brozel VS (2002) Proteomic analysis reveals differential protein expression by Bacillus cereus during biofilm formation. Appl Environ Microbiol 68:2770–2780
Ozok AR, Persoon IF, Huse SM, Keijser BJ, Wesselink PR, Crielaard W, Zaura E (2012) Ecology of the microbiome of the infected root canal system: a comparison between apical and coronal root segments. Int Endod J 45:530–541
Parsek MR, Singh PK (2003) Bacterial biofilms: an emerging link to disease pathogenesis. Annu Rev Microbiol 57:677–701
Percival SL, Thomas JG, Williams DW (2010) Biofilms and bacterial imbalances in chronic wounds: anti-Koch. Int Wound J 7:169–175
Rhoads DD, Wolcott RD, Percival SL (2008) Biofilms in wounds: management strategies. J Wound Care 17:502–508
Ricucci D, Siqueira JF Jr (2008) Apical actinomycosis as a continuum of intraradicular and extraradicular infection: case report and critical review on its involvement with treatment failure. J Endod 34:1124–1129
Ricucci D, Siqueira JF Jr (2010a) Biofilms and apical periodontitis: study of prevalence and association with clinical and histopathologic findings. J Endod 36:1277–1288
Ricucci D, Siqueira JF Jr (2010b) Fate of the tissue in lateral canals and apical ramifications in response to pathologic conditions and treatment procedures. J Endod 36:1–15
Ricucci D, Martorano M, Bate AL, Pascon EA (2005) Calculus-like deposit on the apical external root surface of teeth with post-treatment apical periodontitis: report of two cases. Int Endod J 38:262–271
Ricucci D, Siqueira JF Jr, Bate AL, Pitt Ford TR (2009) Histologic investigation of root canal-treated teeth with apical periodontitis: a retrospective study from twenty-four patients. J Endod 35:493–502
Rôças IN, Siqueira JF Jr (2008) Root canal microbiota of teeth with chronic apical periodontitis. J Clin Microbiol 46:3599–3606
Rôças IN, Siqueira JF Jr, Aboim MC, Rosado AS (2004) Denaturing gradient gel electrophoresis analysis of bacterial communities associated with failed endodontic treatment. Oral Surg Oral Med Oral Pathol Oral Radiol Endod 98:741–749
Rôças IN, Hülsmann M, Siqueira JF Jr (2008) Microorganisms in root canal-treated teeth from a German population. J Endod 34:926–931
Rôças IN, Alves FR, Santos AL, Rosado AS, Siqueira JF Jr (2010) Apical root canal microbiota as determined by reverse-capture checkerboard analysis of cryogenically ground root samples from teeth with apical periodontitis. J Endod 36:1617–1621
Saito D, de Toledo LR, Rodrigues JLM, Tsai SM, Hofling JF, Gonçalves RB (2006) Identification of bacteria in endodontic infections by sequence analysis of 16S rDNA clone libraries. J Med Microbiol 55:101–107
Sakamoto M, Rôças IN, Siqueira JF Jr, Benno Y (2006) Molecular analysis of bacteria in asymptomatic and symptomatic endodontic infections. Oral Microbiol Immunol 21:112–122
Sakamoto M, Siqueira JF Jr, Rôças IN, Benno Y (2007) Bacterial reduction and persistence after endodontic treatment procedures. Oral Microbiol Immunol 22:19–23
Sakamoto M, Siqueira JF Jr, Rôças IN, Benno Y (2008) Molecular analysis of the root canal microbiota associated with endodontic treatment failures. Oral Microbiol Immunol 23: 275–281
Santos AL, Siqueira JF Jr, Rôças IN, Jesus EC, Rosado AS, Tiedje JM (2011) Comparing the bacterial diversity of acute and chronic dental root canal infections. PLoS ONE 6, e28088
Sauer K, Camper AK, Ehrlich GD, Costerton JW, Davies DG (2002) Pseudomonas aeruginosa displays multiple phenotypes during development as a biofilm. J Bacteriol 184:1140–1154
Schaudinn C, Carr G, Gorur A, Jaramillo D, Costerton JW, Webster P (2009) Imaging of endodontic biofilms by combined microscopy (FISH/cLSM - SEM). J Microsc 235:124–127
Siqueira JF Jr (2011) Treatment of endodontic infections. Quintessence, London

Siqueira JF Jr, Fouad AF, Rôças IN (2012) Pyrosequencing as a tool for better understanding of human microbiomes. J Oral Microbiol 4. doi:10.3402/jom.v3404i3400.10743
Siqueira JF Jr, Rôças IN (2004) Polymerase chain reaction-based analysis of microorganisms associated with failed endodontic treatment. Oral Surg Oral Med Oral Pathol Oral Radiol Endod 97:85–94
Siqueira JF Jr, Rôças IN (2005a) Exploiting molecular methods to explore endodontic infections: Part 2-redefining the endodontic microbiota. J Endod 31:488–498
Siqueira JF Jr, Rôças IN (2005b) Uncultivated phylotypes and newly named species associated with primary and persistent endodontic infections. J Clin Microbiol 43:3314–3319
Siqueira JF Jr, Rôças IN (2007) Bacterial pathogenesis and mediators in apical periodontitis. Braz Dent J 18:267–280
Siqueira JF Jr, Rôças IN (2009a) Molecular analysis of endodontic infections. In: Fouad AF (ed) Endodontic microbiology. Wiley-Blackwell, Ames, Iowa, pp 68–107
Siqueira JF Jr, Rôças IN (2009b) Community as the unit of pathogenicity: an emerging concept as to the microbial pathogenesis of apical periodontitis. Oral Surg Oral Med Oral Pathol Oral Radiol Endod 107:870–878
Siqueira JF Jr, Rôças IN (2009c) Diversity of endodontic microbiota revisited. J Dent Res 88: 969–981
Siqueira JF, Rôças IN (2014) Present status and future directions in endodontic microbiology. Endod Topics 30:3–22
Siqueira JF Jr, Sen BH (2004) Fungi in endodontic infections. Oral Surg Oral Med Oral Pathol Oral Radiol Endod 97:632–641
Siqueira JF Jr, Magalhaes FA, Lima KC, de Uzeda M (1998) Pathogenicity of facultative and obligate anaerobic bacteria in monoculture and combined with either *Prevotella intermedia* or *Prevotella nigrescens*. Oral Microbiol Immunol 13:368–372
Siqueira JF Jr, Rôças IN, Lopes HP (2002a) Patterns of microbial colonization in primary root canal infections. Oral Surg Oral Med Oral Pathol Oral Radiol Endod 93:174–178
Siqueira JF Jr, Rôças IN, Moraes SR, Santos KR (2002b) Direct amplification of rRNA gene sequences for identification of selected oral pathogens in root canal infections. Int Endod J 35:345–351
Siqueira JF Jr, Rôças IN, Rosado AS (2004) Investigation of bacterial communities associated with asymptomatic and symptomatic endodontic infections by denaturing gradient gel electrophoresis fingerprinting approach. Oral Microbiol Immunol 19:363–370
Siqueira JF Jr, Rôças IN, Debelian G, Carmo FL, Paiva SS, Alves FR, Rosado AS (2008) Profiling of root canal bacterial communities associated with chronic apical periodontitis from Brazilian and Norwegian subjects. J Endod 4:1457–1461
Siqueira JF Jr, Sakamoto M, Rosado AS (2010) Microbial community profiling using terminal restriction fragment length polymorphism (T-RFLP) and denaturing gradient gel electrophoresis (DGGE). Methods Mol Biol 666:71–85
Siqueira JF Jr, Alves FR, Rôças IN (2011) Pyrosequencing analysis of the apical root canal microbiota. J Endod 37:1499–1503
Sjögren U, Figdor D, Persson S, Sundqvist G (1997) Influence of infection at the time of root filling on the outcome of endodontic treatment of teeth with apical periodontitis. Int Endod J 30:297–306
Socransky SS, Haffajee AD (2002) Dental biofilms: difficult therapeutic targets. Periodontol 28: 12–55
Socransky SS, Haffajee AD, Cugini MA, Smith C, Kent RL Jr (1998) Microbial complexes in subgingival plaque. J Clin Periodontol 25:134–144
Stojicic S, Shen Y, Haapasalo M (2013) Effect of the source of biofilm bacteria, level of biofilm maturation, and type of disinfecting agent on the susceptibility of biofilm bacteria to antibacterial agents. J Endod 39:473–477
Stoodley P, Sauer K, Davies DG, Costerton JW (2002) Biofilms as complex differentiated communities. Annu Rev Microbiol 56:187–209

Sundqvist GK, Eckerbom MI, Larsson AP, Sjogren UT (1979) Capacity of anaerobic bacteria from necrotic dental pulps to induce purulent infections. Infect Immun 25:685–693
Sundqvist G, Figdor D, Persson S, Sjogren U (1998) Microbiologic analysis of teeth with failed endodontic treatment and the outcome of conservative re-treatment. Oral Surg Oral Med Oral Pathol Oral Radiol Endod 85:86–93
Tronstad L, Barnett F, Cervone F (1990) Periapical bacterial plaque in teeth refractory to endodontic treatment. Endod Dent Traumatol 6:73–77
Vianna ME, Conrads G, Gomes BPFA, Horz HP (2006) Identification and quantification of archaea involved in primary endodontic infections. J Clin Microbiol 44:1274–1282
Vickerman MM, Brossard KA, Funk DB, Jesionowski AM, Gill SR (2007) Phylogenetic analysis of bacterial and archaeal species in symptomatic and asymptomatic endodontic infections. J Med Microbiol 56:110–118
Waltimo TM, Siren EK, Torkko HL, Olsen I, Haapasalo MP (1997) Fungi in therapy-resistant apical periodontitis. Int Endod J 30:96–101
Waltimo T, Trope M, Haapasalo M, Ørstavik D (2005) Clinical efficacy of treatment procedures in endodontic infection control and one year follow-up of periapical healing. J Endod 31:863–866
Wolcott R, Dowd S (2011) The role of biofilms: are we hitting the right target? Plast Reconstr Surg 127(Suppl 1):28S–35S
Wolcott RD, Rumbaugh KP, James G, Schultz G, Phillips P, Yang Q, Watters C, Stewart PS, Dowd SE (2010) Biofilm maturity studies indicate sharp debridement opens a time- dependent therapeutic window. J Wound Care 19:320–328

Laboratory Models of Biofilms: Development and Assessment

Anil Kishen and Markus Haapasalo

Abstract Microbial biofilms are surface-adherent consortia formed by microbes and regulated by environmental changes. Currently, there are several reports on the existence of bacterial biofilms in infected root canal systems. Thus it is mandatory to simulate such biofilm modes in laboratory models for applied microbiological experiments in endodontics. This chapter covers different considerations for developing biofilm models in endodontics. In addition, different types of in vitro root canal biofilm models and the methods of assessment are described.

1 Introduction

Endodontic microbes dwell within the root canal as surface-adherent biofilm (*intraradicular biofilm*). These microbial presence and activities that are usually confined to the intracanal spaces may under certain conditions disseminated to locations beyond the apical foramen (*extraradicular biofilm*) (Costerton et al. 1994; Ramachandran Nair 1987; Nair 2006; Nair et al. 1990). The anatomical complexities in a root canal system tend to shelter the bacteria from disinfectants and instrumentation procedures (Nair et al. 1990). Further, the progression of bacterial infection within the root canals can alter the nutritional and environmental status within the root canal system, apparently rendering it more anaerobic and depleted of nutrients. These environmental changes result in a tough ecological niche for the surviving microorganisms within the root canal (Sundqvist and Figdor 2003). The biofilm mode of bacterial growth, in addition to higher resistance of antimicrobials, allow the resident bacteria to survive the unfavorable environmental and nutritional conditions (Baumgartner et al. 2008; Grenier and Mayrand 1986). Therefore, it is

A. Kishen (✉)
Department of Endodontics, University of Toronto, Toronto, ON, Canada
e-mail: Anil.Kishen@dentistry.utoronto.ca

M. Haapasalo
Division of Endodontics, The University of British Columbia, Vancouver, BC, Canada

L.E. Chávez de Paz et al. (eds.), *The Root Canal Biofilm*, Springer Series on Biofilms 9, DOI 10.1007/978-3-662-47415-0_6

crucial to consider bacterial biofilms as an indispensible in vitro model for microbiological investigations. These methods are used to assess different disinfectants and disinfection strategies in endodontics. They are also utilized to study microbial interactions within the root canals or interactions between bacteria and host immune cells (Al-Hashimi and Levine 1989). Currently they are more commonly employed to examine the antimicrobial efficacy of different irrigants/medicaments and irrigation procedures in endodontics (Handley et al. 1984).

2 Factors to Be Considered in Biofilm Model Systems

The antimicrobial resistance observed in biofilm bacteria is not largely due to the classic genetic mechanisms; instead this arises due to certain peculiarities associated with the biofilm growth. The types of resident bacterial species, nature of bacterial adherence to substrate, physicochemical characteristics of the substrate, thickness of the biofilm, bacterial cell density, amount of EPS, and phenotypical/genotypical modification of the resident bacteria are all features that could contribute to antimicrobial resistance in biofilm bacteria. Typically, different antimicrobial resistance mechanisms may act concurrently or synergistically in a biofilm. Understanding some of these mechanisms is key in developing biofilm model systems for different applications in endodontics (Baumgartner et al. 2008; Handley et al. 1985). In general, different factors associated with bacterial type, bacterial adherence, bacteria-substrate interaction, and biofilm ultrastructure should be standardized to develop useful biofilm models for in vitro experiments (Baumgartner et al. 2008). There is currently no universally accepted in vitro model that reproduces biofilm infection in endodontics.

2.1 Bacterial Type and Bacteria-Substrate Interaction

First stage in the formation of oral biofilms is the adsorption of macromolecules from tissue fluids such as saliva onto a biomaterial (natural or synthetic) surface, leading to the formation of a *conditioning layer*. The conditioning layer of adsorbed inorganic and organic molecules will alter the physical and chemical characteristics of the surface. The conditioning layer formed prior to the influx of microorganisms will selectively promote the adhesion of microbial cells to the surface. They may also serve as a source of nutrition for adherent bacteria. Bacteria can generally form biofilms on any surface that is coated with such conditioning fluids (Miron et al. 2001). This step is followed by the adhesion of microbial cells to the substrate. The adhesive potential of microbes is considered to be an important ecological and pathological determinant in biofilm-mediated infections.

The adherence of bacteria to a surface is influenced by (1) the *environmental conditions* such as pH, temperature, fluid flow rate, nutrient availability, etc.; (2) the *bacterial factors* such as the type of bacterial species/strain, growth phase of bacteria (log or stationary phase), type and charge of the surface molecules, etc.; and (3) the *substrate factors* such as the physical and chemical characteristics of the substrate. It is crucial to standardize these parameters to develop clinically realistic biofilm models that are ideal model systems for in vitro evaluations (Cowan et al. 1987; Costerton 1999).

The initial phase of bacteria-substrate interaction is determined by the physical and chemical properties of bacteria/substrate (e.g., surface energy and charge density) (*phase 1: bacterial adherence*). This phase is responsible for bringing microbial cells to the substrate surface. This reversible interaction is followed by the molecular-level nonspecific interactions between the bacterial surface and substrate. This phase is mediated by the bacterial surface structures such as fimbriae, pili, flagella, and EPS (*phase 2: initial nonspecific microbial-substrate adherence phase*). The bacterial surface structures form bridges between bacteria and conditioning film (Miron et al. 2001). *Porphyromonas gingivalis, Streptococcus mitis, Streptococcus salivarius, Prevotella intermedia, Prevotella nigrescens, Streptococcus mutans*, and *Actinomyces naeslundii* are some of the oral bacteria possessing surface structures (Miron et al. 2001; Cowan et al. 1987). The molecular bridges established between bacteria and substrate are a combination of electrostatic attraction and covalent/hydrogen bonding. Initially the bond between bacteria and substrate may not be strong. With time these bonds strengthen, making the bacterial attachment irreversible.

In the final stage, a more specific bacterial adhesion to the substrate is established via polysaccharide adhesin or ligand formation *(phase 3: specific microbial-substrate adherence phase)*. In this phase, adhesin or ligand molecules on the bacterial cell surface will bind to the specific receptors on the substrate. The specific bacterial adhesion is less affected by the environmental changes (Costerton 1999; Costerton and Lewandowski 1997). The phases involved in adherence of bacteria to a substrate are dynamic in nature and require time. The reversible and irreversible steps in the phase I of bacterial adherence occurs in a few seconds to minutes, while the phases 2 and 3 take a few hours to days to occur. The development and maturation of biofilm structure occurs subsequent to bacterial adherence. Therefore, it is important to provide sufficient bacteria-substrate interaction time and optimum environmental conditions while developing in vitro biofilms. The final biofilm model should represent all the important characteristics of a biofilm structure that would influence the requirement of the in vitro experiment.

2.2 *Biofilm Ultrastructure*

Two types of microbial interactions occur at the cellular level during the formation of biofilm. One is the process of recognition between a suspended cell and a cell already attached to substratum. This type of interaction is termed *coadhesion*. In the second type of interaction, genetically distinct cells in suspension recognize each other and clump together. This type of interaction is called *coaggregation*. This association is highly specific and occurs between coaggregating partners only. Interestingly, most oral bacteria recognize each other as coaggregating partners. *Fusobacterium nucleatum*, a gram-negative filamentous anaerobe, can coaggregate with all oral bacteria tested and can act as a bridging bacterium that binds together even non-aggregating bacteria (Baumgartner et al. 2008). The association of long-filamentous bacteria and surface-adsorbed spherical-shaped cocci produces the characteristic corncob structure of oral biofilms (Rosan et al. 1999). The attachment of cocci to filamentous bacteria is mediated via fimbriae of oral streptococci. Although the genetic makeup of bacteria is the chief determinant of coaggregation, the physicochemical characteristics of the environment also play a crucial role (Rosan et al. 1999). Bacteria in a coaggregated suspension are significantly more resistant to antimicrobials when compared to planktonic suspension, while bacteria in a biofilm mode showed the highest resistance to antimicrobials (Upadya and Kishen 2010). Hence antimicrobials selected on the basis of traditional susceptibility methods such as broth-based minimum inhibitory concentration (MIC) may not be very appropriate to eliminate coaggregated or biofilm bacteria.

During biofilm development, the resident bacteria proliferate with further accrual of EPS, leading to the expansion of biofilm structure. In this stage, the monolayer of microbes (primary colonizers) attracts the secondary colonizers forming the microcolonies, and the collection of microcolonies gives rise to the final structure of biofilm. A biofilm ultrastructure consists of bacterial cell populations attached irreversibly on a substrate and encased in a hydrated, polyanionic matrix of EPS, proteins, polysaccharides, and nucleic acids (del Pozo and Patel 2007; Vakulenko and Mobashery 2003). Typically, bacteria themselves account for 5–35 % of the total biofilm volume (Costerton and Stewart 2001). A mature biofilm will be a metabolically active community of microorganisms, where individuals share duties and benefits (Vakulenko and Mobashery 2003). For instance, some microorganism in the community helps in adhering to the solid surface, while some others create bridges between different species. This signifies the relevance of a polymicrobial biofilm over a monospecies biofilm. The physiological characteristics of the resident microorganisms in a biofilm offer an additional resistance to antimicrobial agents (Costerton and Stewart 2001; Sun et al. 2005). (Sun et al. 2005; Stoodley et al. 1994; Lim et al. 2009). The above facts further signify the relevance of developing multispecies biofilm models over monospecies biofilm models.

It is reported that biofilms formed in pure cultures of bacteria under laboratory conditions and the mixed-species biofilms formed in natural ecosystems showed a basic organization, in which the cells grow in a matrix-enclosed microcolonies separated by a network of open-water channels (Shen et al. 2010a). The thickness of EPS will influence the permeability of substances through the biofilm and provides a significant degree of protection or "barrier effect" against physical and chemical threats. Each step in the development of biofilm, starting from the adherence of bacteria to the final formation of a matured biofilm structure, as well as protein expression/slime production, is all modulated by a large number of variables. Some of these variables are the type of resident bacteria, growth/environmental conditions, and age of biofilm. Previous studies on endodontic biofilm models have shown that the mature biofilms with limited nutrient supply are more resistant to irrigants such as chlorhexidine and light-activated disinfection than early biofilms grown under normal nutrient conditions. These studies have emphasized relevance of the age or degree of maturation and nutritional condition as the major confounding factors in designing in vitro biofilm models that could be applied to test endodontic disinfectants (Kolenbrander et al. 1995; Jones 1972). Furthermore, it is important to use relevant bacterial species (primary colonizer), and provide ideal environmental conditions (substrate, fluid conditioning, nutritional conditions, and temperature) and optimum interaction time (matured biofilm) to achieve a standardized endodontic biofilm model for in vitro applications.

3 Bacterial Biofilm Models: In Vitro Development

Laboratory (in vitro) models are used to obtain standardized biofilms for different experimental purposes. Conventional biofilm models range from *monocultures in static growth conditions* to *diverse mixed cultures in dynamic growth conditions*. The static biofilm models used different substrates (e.g., glass, polycarbonate, silicon, hydroxyapatite, nitrocellulose, enamel, dentin) to grow biofilms, while the dynamic biofilm models used reactors or fermenting systems to grow biofilms on a substrate. Both aerobic and anaerobic environments can be employed for in vitro biofilm development. Figures 1 and 2 show different in vitro biofilms grown on different substrates. Given that the in vivo environmental conditions are commonly dynamic, studies evaluating biofilm formation under static conditions might be somewhat misleading, depending on the research question (McBain 2009). These in vitro bacterial biofilm models are routinely applied to (1) examine the adherence of specific bacterial species to any biomaterial surface (Kishen et al. 2008), (2) study the nature and pattern of early microbial biofilm formation on a particular substrate (George et al. 2010), (3) study the interaction between different biofilm bacteria and host immune cells (Mathew et al. 2010), and test the efficacy of antimicrobial agents or antimicrobial treatment strategies (Pratten and Ready 2010; Merritt et al. 2005).

Currently in endodontics most in vitro biofilms are utilized for testing antimicrobials and root canal irrigation strategies (Table 1). As a result, the activity of disinfectants showed noticeable discrepancies between in vitro and in vivo experiments. This discrepancy in antimicrobial efficacy may be attributed to the diversity of microbes, bacterial growth phase, type of biofilm models, and methods employed for analysis. So a number of parameters have to be considered in designing a biofilm model for applications in endodontics (Fig. 3).

If laboratory strains are used for bacterial adherence assays, it is imperative that they are representative of clinical isolates. In addition, assays that do not take into account the presence of saliva may be unsuitable to study adhesion and early

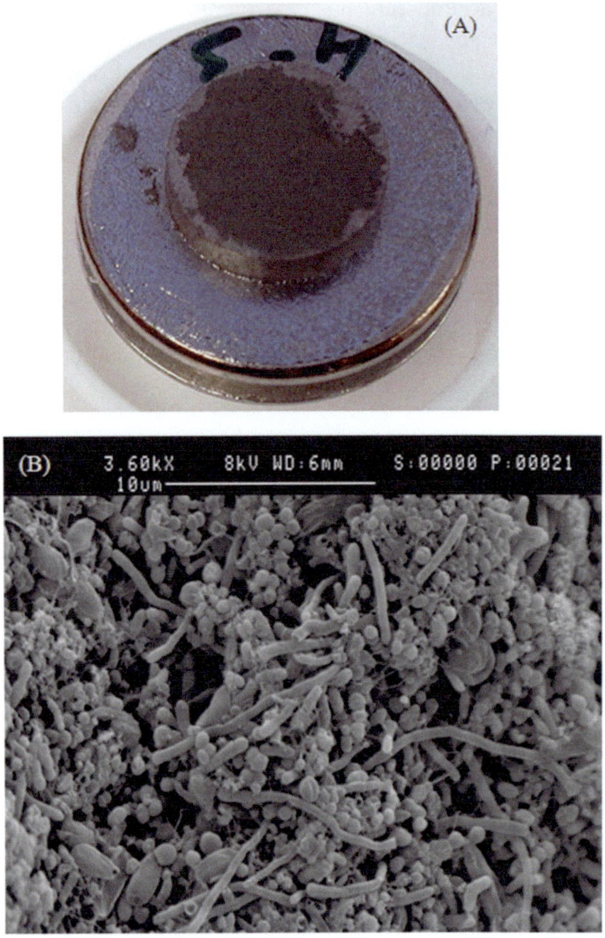

Fig. 1 (**a**) A photograph of a biofilm grown in vitro for 3 weeks on collagen-coated hydroxyapatite disk. The biofilm is coated with a palladium-gold mixture for SEM. (**b**) SEM image of a 6-month-old biofilm grown anaerobically in BHI broth on collagen-coated hydroxyapatite disk. Several bacterial morphotypes, including coiled spirochetes, can be seen (Kishen and Haapasalo 2010)

biofilm formation (George et al. 2010). Although there is little doubt that the results from planktonic killing studies must be interpreted with caution, a comparison of the planktonic and biofilm tests in a study indicated that planktonic killing tests may be useful for the preliminary screening of new disinfectants before proceeding into more complex biofilm models (Merritt et al. 2005).

4 Biofilm Devices: Flow Cells and Fermentors

Several in vitro devices are used to develop biofilms. Some of these devices irrigate fresh culture medium, which allows the growing biofilm to experience a continuous flow of medium supplemented with fresh nutrients. These in vitro devices are used to grow dynamic biofilm models. The flow cell system is one of the most utilized

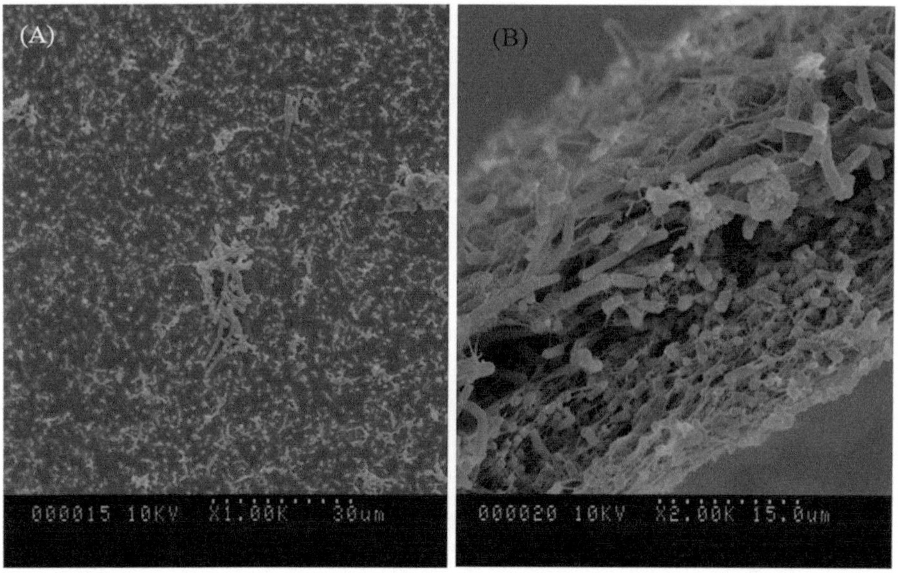

Fig. 2 (**a**) A multispecies biofilm grown anaerobically for 3 weeks on glass cover slip conditioned with media for 24 h. The biofilm was sparse and not uniform. (**b**) High-magnification image of the previous SEM picture showing mixed bacterial flora with abundant extracellular matrix (Kishen and Haapasalo 2010)

dynamic models. It consists of a transparent chamber of fixed depth through which the growth medium flows. The inlet tubing supplies growth medium and the outlet tubing drains the medium to the waste reservoir. The growth medium is passed through the cell with the aid of a peristaltic pump, which controls the flow rate of the medium. Prefabricated flow cell systems are available commercially, or they can be custom-made for any particular application. In conjunction with a microscope, charge-coupled device (CCD) camera, or confocal laser scanning microscopy (CLSM), these systems can be used to study early events in biofilm formation in real time (Pavarina et al. 2011).

Chemostats are also used to grow dynamic biofilms of bacteria on experimental substrates. One of the most important features of chemostats is that microbial biofilms can be grown at a constant rate and under constant culture conditions (temperature, pH). Similar to chemostats, there is another category of reactors in which biofilms are formed on thin filter membranes in a physiological steady state. These systems permit the evaluation of growth rate dependence and cell cycle specificity of antibacterial agents. Finally, there are constant-depth reactors in which surface growth is periodically removed to maintain a constant geometry of biofilms. In these reactors, microorganisms can be grown in a physiological steady state with all culture parameters constant. The systems can generate a large number of biofilms with comparable and reproducible data (Pavarina et al. 2011).

The static biofilm systems do not allow frequent refreshing of cutural medium. They generate biofilms that have exhausted important nutritional components at the

Table 1 List of literature using in vitro biofilm models for different endodontic applications

Authors	Type of model	Purpose	Preparation of biofilm
Torabinejad (2003)	In vitro *E. faecalis*	The antimicrobial effect of MTAD: an in vitro investigation	Extracted human teeth contaminated with *E. faecalis* for 4 weeks Dentinal shavings and CFU-based method were used for the analysis
Duggan and Sedgley (2007)	In vitro *E. faecalis* strains recovered from root canals, oral cavity, and non-oral sources	Biofilm formation of oral and endodontic *E. faecalis*	96-well plates for 24 h Crystal violet assay used for assessment (optical density at 570 nm)
George and Kishen (2007)	In vitro *E. faecalis* (gram-positive), *Actinobacillus actinomycetemcomitans* (gram-negative)	Methylene blue (MB) dissolved in different formulations: water, 70 % glycerol, and 70 % polyethylene glycol, and a mixture of glycerol-ethanol-water (30:20:50) was tested	Two-day-old biofilms in multiwell plates (polystyrene) Four-day-old biofilms in human teeth CFU-based method
George and Kishen (2008)	In vitro *E. faecalis*	This study aimed to investigate the effect of including an oxidizer and oxygen carrier in photosensitization formulation to disinfect a matured endodontic biofilm by light-activated disinfection	Human teeth (10-week-old biofilm) CFU-based method
McGill et al. (2008)	In vitro *E. faecalis*	The efficacy of dynamic irrigation using a commercially available system (RinsEndo®)	A collagen-based "biomolecular film" formed on extracted human teeth Digital image analysis of the canal surfaces (ipWin4)
Sainsbury et al. (2009)	Ex vivo	DIAGNOdent laser fluorescence assessment of endodontic infection	Extracted teeth with endodontic pathology Fluorescence emissions in the near-infrared range was measured

(continued)

Table 1 (continued)

Authors	Type of model	Purpose	Preparation of biofilm
Shen et al. (2009)	In vitro Multispecies	Evaluation of the effect of two chlorhexidine preparations on biofilm bacteria in vitro: a three-dimensional quantitative analysis	Collagen-coated hydroxyapatite (CHA) and uncoated hydroxyapatite (HA) disks Confocal laser scanning microscopy was used for the analysis of dead versus viable cells
Williamson et al. (2009)	In vitro *E. faecalis* (clinical isolate)	Antimicrobial susceptibility of monoculture biofilms to 6 % NaOCl, 2 % CHX, <6 % NaOCl with surface modifiers (Chlor-XTRA), and 2 % CHX with surface modifiers (CHX-Plus)	Glass substrate CFU-based method
Lim et al. (2009)	In vitro biofilms *E. faecalis*	The efficacy of an improved light-activated disinfection technique utilizing a specific photosensitizer formulation, liquid optical conduit, oxygen carrier, and light energy of appropriate wavelength was tested	Two different biofilms were tested: (1) four-day-old (immature) and (2) four-week-old (mature) human teeth CFU-based method
Shahriari et al. (2010)	In vitro *E. faecalis*	The study the effect of hydrogen peroxide on the antibacterial effect of chlorhexidine	Dentin tubes prepared from maxillary central and lateral incisors CFU-based method was used
Kishen et al. (2010)	In vitro *E. faecalis*	Efflux pump inhibitor potentiates antimicrobial photodynamic inactivation of *Enterococcus faecalis* biofilm	Microwell plates CFU-based method and confocal laser scanning microscopy
Hiraishi et al. (2010)	In vitro *E. faecalis*	Antimicrobial efficacy of 3.8 % silver diamine fluoride	Membrane filters CFU-based method

(continued)

Table 1 (continued)

Authors	Type of model	Purpose	Preparation of biofilm
Shrestha et al. (2010)	In vitro *E. faecalis*	Nanoparticulates for antibiofilm treatment and effect of aging on its antibacterial activity	Microwell plates/ saliva CFU-based method and confocal laser scanning microscopy
Liu et al. (2010)	In vitro *E. faecalis*	Biofilm formation capability of *Enterococcus faecalis* cells in starvation phase and its susceptibility to sodium hypochlorite	Human dentin and polystyrene blocks CFU-based method and SEM
Chávez de Paz et al. (2010)	In vitro (clinical isolates) *E. faecalis* *L. paracasei* *S. anginosus* *S. gordonii*	The effects of antimicrobials on endodontic biofilm bacteria	24-hour biofilm within a miniflow cell system Confocal microscopy and image analysis
Su et al. (2010)	In vivo	This study explored the effect of surgical endodontic treatment of refractory periapical periodontitis with extraradicular biofilm	Resected root-end samples
Soares et al. (2010)	In vitro *E. faecalis*	Effectiveness of chemomechanical preparation with alternating use of sodium hypochlorite and EDTA in eliminating intracanal *Enterococcus faecalis* biofilm	Human teeth (21-day-old biofilm) SEM and CFU-based method
Bhuva et al. (2010)	In vitro *E. faecalis*	The effectiveness of passive ultrasonic irrigation on intraradicular *Enterococcus faecalis* biofilms in extracted single-rooted human teeth	Human teeth SEM-based image analysis
Shen et al. (2010a)	In vitro Multispecies biofilm (subgingival plaque)	The synergistic antimicrobial effect by mechanical agitation and two chlorhexidine preparations	Collagen-coated hydroxyapatite (CHA) disks (3 weeks old) Confocal laser scanning microscopy

(continued)

Table 1 (continued)

Authors	Type of model	Purpose	Preparation of biofilm
Pappen et al. (2010)	In vitro Multispecies biofilm (subgingival plaque)	To investigate the antibacterial effect of Tetraclean, MTAD, and five experimental irrigants using both direct exposure test with planktonic cultures and mixed-species in vitro biofilm model	Collagen-coated hydroxyapatite (CHA) disks (2 weeks old) Confocal laser scanning microscopy
Shen et al. (2010b)	In vitro Multispecies biofilm (subgingival plaque)	The aim of this study was to enumerate viable bacteria at different growth stages of a multispecies oral biofilm and to compare results obtained with the LIVE/DEAD BacLight Kit with those from culturing and plate counting	Collagen-coated hydroxyapatite (CHA) disks Confocal laser scanning microscopy and CFU-based method
Lundstrom et al. (2010)	In vitro (multispecies) *Streptococcus sanguinis*, *Actinomyces viscosus*, *Fusobacterium nucleatum*, *Peptostreptococcus micros*, and *Prevotella nigrescens*	Bactericidal activity of stabilized chlorine dioxide as an endodontic irrigant in a polymicrobial biofilm tooth model system	Permanent bovine incisors coated with mucin and inoculated with standardized suspensions of bacteria (anaerobically for 14 days) CFU-based method
Hope et al. (2010)	In vitro *E. faecalis*	A direct comparison between extracted tooth and filter-membrane biofilm models of endodontic irrigation	Human teeth CFU-based method
Upadya and Kishen (2010)	In vitro *E. faecalis* and *P. aeruginosa*	To evaluate the efficacy of light-activated disinfection (LAD) using methylene blue (MB) and a non-coherent light source on gram-positive and gram-negative bacteria in different growth modes. The influence of different	Monospecies biofilms in 24-well polystyrene plates (4 days) CFU-based method Confocal laser scanning microscopy

(continued)

Table 1 (continued)

Authors	Type of model	Purpose	Preparation of biofilm
		photosensitizer (PS) formulations in the MB-mediated LAD of biofilms was also evaluated	
George et al. (2010)	In vitro *E. faecalis*	This study examined the biofilm-forming capacity of *E. faecalis* on gutta-percha points under different nutrient statuses and surface conditioning with saliva and serum	Gutta-percha conditioned with saliva or serum (2, 4, and 12 weeks) Biofilm growth for 2 weeks CFU-based method and SEM
Badr et al. (2011)	In vitro *E. faecalis*	A laboratory evaluation of the antibacterial and cytotoxic effect of licorice when used as root canal medicament	Grown on cellulose nitrate membrane filters CFU-based method
Shen et al. (2011)	In vitro Multispecies biofilm (subgingival plaque)	The aim of this study was to examine the susceptibility of multispecies biofilms at different phases of growth to root canal irrigants (2 % chlorhexidine (CHX) or CHX-Plus)	Collagen-coated hydroxyapatite (CHA) disks (2 days to months) Confocal laser scanning microscopy and CFU-based method

end of an overnight incubation. The key feature of this system is that numerous biofilms can be handled at any given time. It does not require time-consuming sterilization and setting up procedures, allowing it to be used as a high-throughput system for biofilm analysis (Merritt et al. 2005). This system provides a basis for the rapid screening of biofilm mutants (Kulasekara et al. 2005), biomass development, and biofilm-forming capacity (Watnick and Kolter 1999), as well as extracellular matrix composition (Friedman and Kolter 2004). However, this system is incompatible with CLSM, which is the preferred methodology to study the structure of biofilms. Structural evaluation of biofilm requires an irrigated or flow-through cell system, which allows the examination of biofilm development over a period of time. This system can be used to analyze the temporal and spatial expressions of selected genes and the complete life cycle of biofilm formation and dispersal.

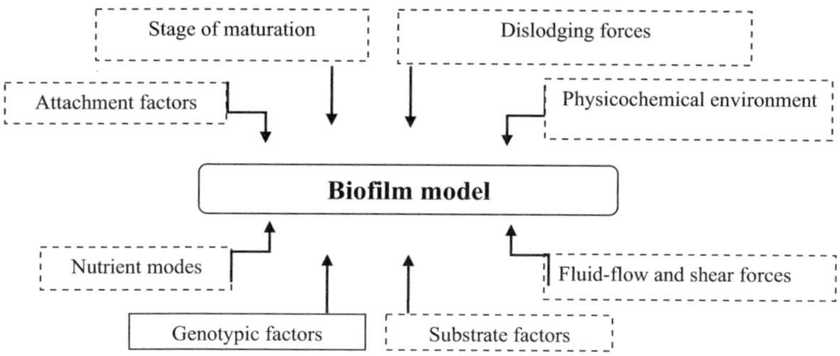

Fig. 3 Different features influencing the structure and development of in vitro biofilm

Several findings on the unique behavioral responses by biofilm cells that cannot be obtained using static systems are studied with the flow cell systems (Davey and O'Toole 2000).

5 Biofilm Assays

Biofilm assays are used to characterize (1) the number and type of microorganism, (2) the vitality (dead/living cells) of the resident microbial population, (3) the age, (4) the thickness (monolayered or multilayered), (5) the structure (homogeneous, irregular, dense, porous), and (6) the surface topography (peaks and valleys) of biofilms. Currently different techniques such as (1) microbiological culture method, (2) colorimetric techniques, (3) microscopy, (4) physical methods, (5) biochemical methods, and (6) molecular methods are used as biofilm assays. The basic steps required to assess the antibacterial efficacy of endodontic antimicrobials is shown in Fig. 4.

6 Microbiological Culture Techniques

Enumerating the colony-forming units (CFU) is an easy method for the quantification of biofilm bacteria. The CFU measurement provides information on the amount of viable bacteria present within a biofilm. However, the CFU-based method may detect only bacteria that are able to initiate cell division at a sufficient rate to form colonies, and their growth requirements are supported by the culture medium used. Several protocols recommend the removal of biofilm bacteria from the substrate by sonication or centrifugation process and use the supernatant to determine the CFU. The recovery of viable microbes is an important step in these experiments, and the bacteria can be sensitive to the procedures used for their removal from substrate. An in vitro study showed that in older, starved biofilms, bacteria are viable based on

Outline of methodologies to assess antibacterial efficacy of disinfectants

- Set the objective
- Bacterial strains: Choose appropriate bacterial strain(s): Gram positive, Gram negative
- Infection model: Planktonic suspension, coaggregated suspension, biofilm
- Inoculum: Adjust suspension of bacterial culture obtained under specific conditions
- Interaction time and temperature
- Eliminate test agent in the subcultures
- Method of assessments: Count of survivors (Colony Forming Units) (*quantitative analysis*); structural analysis (Scanning Electron Microscopy) (*Qualitative or quasi-quantitative*); Structural and viable/dead cell estimation (*quasi quantitative, quantitative*)
- Interpretation of the result

Fig. 4 The basic steps for assessing antimicrobial efficacy using in vitro biofilm model

green staining pattern as observed under CLSM, but over 99 % of these bacteria could not be grown when removed from the biofilm and grown using a culture method (Shen et al. 2010a). Ultrasonic vibrations and enzymes are suggested to remove bacterial biofilms before quantification. It is imperative in these cases to use the appropriate energy level of ultrasonic vibration and concentration of enzymes, since few studies have highlighted their possible lethal effect on bacterial cells (Johansen et al. 1997).

7 Colorimetric Techniques

The colorimetric assay is a semiquantitative method that applies the principles of dye uptake by bacterial cells to determine the biofilm biomass. In this assay, after the bacterial biofilm is stained with a dye (e.g., crystal violet), it is disrupted using a known quantity of alcohol or a surfactant (sodium dodecyl sulfate), and the intensity of the eluted dye is measured using a spectrophotometer (Fig. 5). This is an easy assay that allows the rapid quantification of biofilm bacteria. This test on the other hand may be difficult to interpret, since the optical density measured is an indication of the number of bacteria and is not a true indicator of EPS in the biofilm structure. The assay usually work well for strains, which are strong biofilm producers, while they may not be very useful to differentiate weak biofilm producers from biofilm-negative strains (McBain 2009).

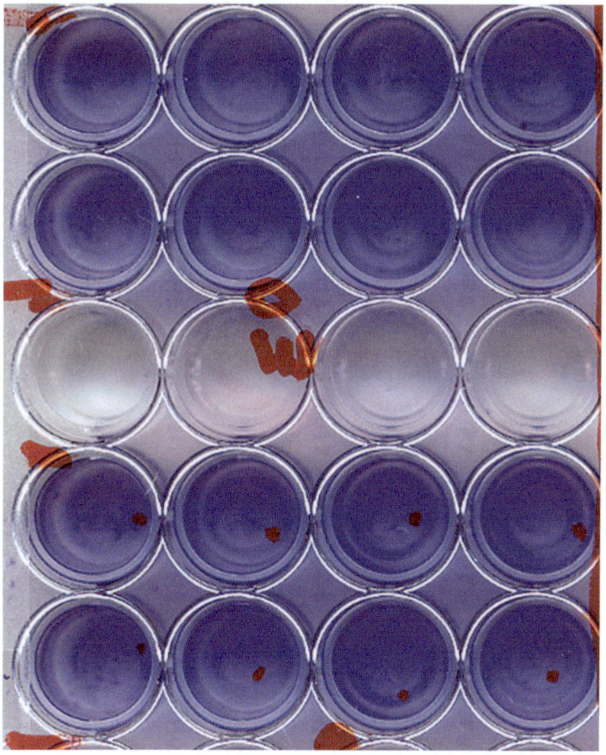

Fig. 5 A multiwell plate showing *E. faecalis* biofilms quantified using a calorimetric (*crystal violet*) assay (Kishen and Haapasalo 2010)

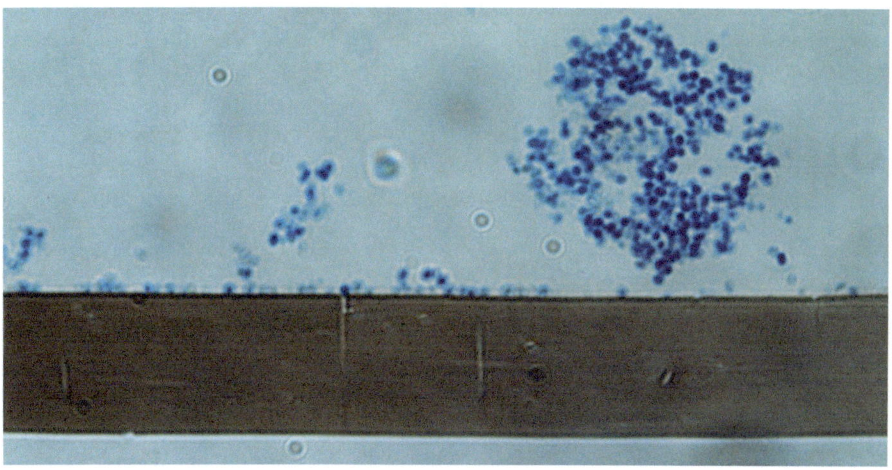

Fig. 6 Light microscopic image of a monospecies biofilm of *E. faecalis* grown on a polycarbonate membrane

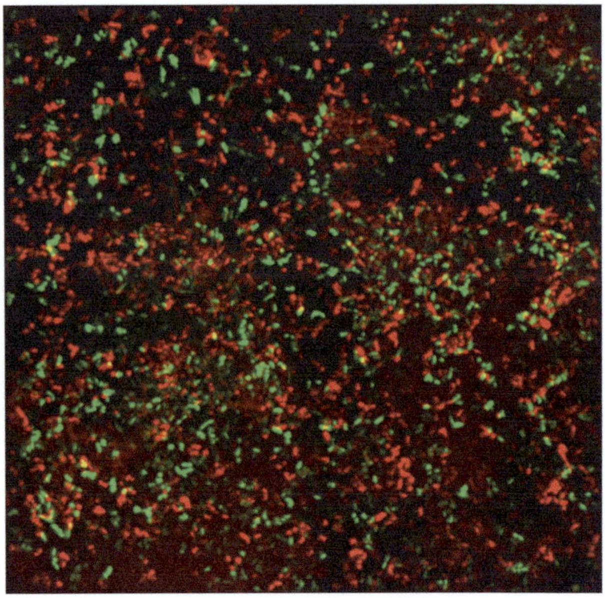

Fig. 7 A fluorescence microscopy image of bacterial colonies stained with LIVE/DEAD fluorescent stain

8 Microscopic Techniques

Light microscopy is the fundamental technique used for the examination of biofilms, either directly on in vitro samples or in vivo histopathological sections. It is a relatively inexpensive, rapid, and readily available method. Different microscopic methods have been used to assess the adherence of bacteria to substrates, structure of biofilms, and distribution/type/viability of bacteria in a biofilm structure (Figs. 1, 2, 5, 6, 7, 8, and 9). In a microscopic method, the bacterial biofilm is stained with a suitable dye that is fluorescent (e.g., propidium iodide) or non-fluorescent in nature (e.g., safranin). Most high-resolution light microscopy will enable counting of bacterial cells on a substrate surface. The biofilm slime may be stained with Alcian Blue, a phthalocyanine dye that stains acidic mucopolysaccharides and glycosaminoglycans in the EPS (Di Bonaventura et al. 2006). The stained portions will appear as a blue to bluish-green color. Bacterial cells may be visualized under fluorescence microscopes without using fluorescent probes by using plasmid-encoded green fluorescent protein (GFP). This transformed *E. coli* O157:H7 has been applied to study their attachment onto a surface (Burnett et al. 2000; Takeuchi and Frank 2001). The viability of these cells may be determined by staining the transformed cells with membrane-impermeable fluorescent dye (Takeuchi and Frank 2001).

Scanning electron microscopy (SEM) and transmission electron microscopy (Yavari et al. 2010) have been effective workhorses in biofilm analysis for many years. They have been employed for the morphological and structural

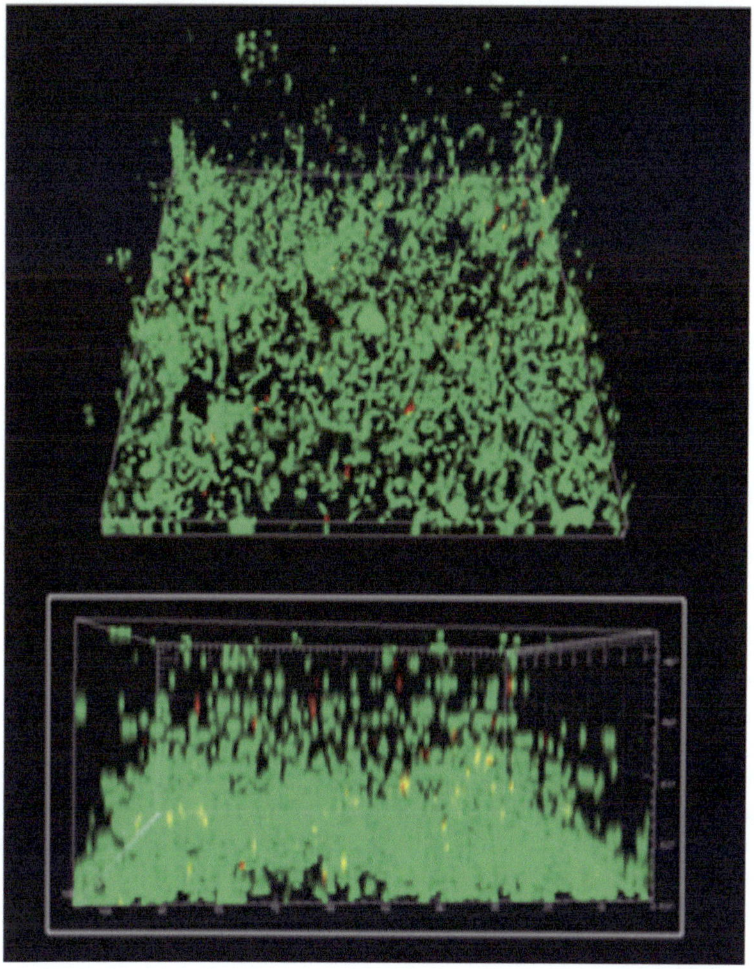

Fig. 8 A three-dimensional confocal laser scanning microscopy reconstruction of *E. faecalis* biofilm (×60)

characterization of biofilms. The main disadvantage with these techniques is the need for extensive sample preparation steps such as fixation, dehydration, freeze-drying or critical point drying, and sputtering. These steps can deeply affect the original biofilm morphology. Environmental SEM (ESEM) is a relatively new technique that represents a powerful alternative to conventional SEM (high vacuum), as it allows the imaging of biological samples in their original hydrated conditions at relatively high resolution (McKinlay et al. 2004). Structural modifications in microbial biofilm architecture, particularly an overall loss of matrix volume, were appreciable when comparing conventional high-vacuum SEM to ESEM images. Sutton et al. (1994) compared different dehydration techniques and showed that the freeze-dried samples presented significant detachment of

Fig. 9 Atomic force microscopy images showing the details of the bacterial cell surface in the nanometric range (Kishen and Haapasalo 2010)

microbial biofilm from the substrate, while more complex dehydration procedures such as critical point drying caused almost complete disappearance of the EPS matrix. Although proper fixation processes were applied, the collapse of the biofilm structure upon dehydration is mainly due to the lack of a self-sustaining scaffold in the EPS matrix. Yet, after fixation bacterial cells maintained their shape and size in vacuum and could be identified by SEM. In the ESEM mode, the semitransparent appearance of EPS and the low signal-to-noise ratio at high pressures result in a limited resolution of images. In brief, ESEM represents an effective technique for detecting highly hydrated bacterial biofilms by preserving the substantial EPS component. On the contrary, conventional high-vacuum SEM allowed detailed examination of cellular components and favored the detection of the three-dimensional hollow structures, but failed to show the actual biofilm architecture consisting of a large volume of EPS matrix surrounding the cells. The combined use of conventional SEM and ESEM techniques can therefore provide complementary information on different biofilm components, bacterial cells, and extracellular matrices (Sutton et al. 1994).

The epifluorescence microscope is used to study the microstructure of bacterial biofilm. Biofilms grown on biomaterial surfaces are usually stained with a fluorescent dye and viewed under an epifluorescence microscope. In a study, binary-species biofilms were stained using two different fluorescent probes for each microorganism and observed under an epifluorescence microscope using two excitation wavelengths. Two different images and background biofilms are captured with appropriate wavelengths. The images are then combined to construct a new image that simultaneously showed both microorganisms (Trachoo and Frank 2002). Epifluorescence microscopy is used to determine the viable cells, cell arrangement, microcolony formation, biofilm pH, and distribution of chemicals in a biofilm structure (Wolfaardt et al. 1993).

CLSM is a particularly important biofilm analysis technique that is restricted to 50 μm to 200 μm thick biofilm structures. CLSM has overcome some of the limitations exhibited by most of the earlier microscopic techniques such as epifluorescence, SEM, and TEM. Together with improvements in the molecular microbiological techniques, CLSM has become an important tool for studying biofilms. Green fluorescent protein (GFP) tagging of certain bacterial strains such as *Pseudomonas aeruginosa* is utilized to study biofilm formation. This method uses fluorescence imaging or CLSM to quantify biofilms. The preferred method of tagging has been to construct chromosomal insertions to ensure a stable gene dosage of the tag sequence (Harraghy et al. 2006; Sheppard and Shotton 1997). Recently a time-lapse CLSM together with the *gfp* reporter system has been applied to study the role of accessory gene regulation (*agr*) in biofilm formation. This approach has given an interesting insight into gene regulation during the course of biofilm development (Bagge et al. 2004). This technique is likely to be an important tool for future studies on the regulators and genes, involved in biofilm development.

CLSM generates a thin (~0.3 μm) plane of focus (optical sections) in which out-of-focus light is blocked by optical barriers or by applying the physics of light absorption as in multiphoton microscope (Bakke et al. 2001). These optical sections can then be stacked by software to generate a three-dimensionally reconstructed image of the entire biofilm. The CLSM images can be used to determine the thickness and distribution of cells in a biofilm. CLSM can also be used to determine the pH gradients in biofilms. The interior pH of biofilms can be measured by a fluorescence lifetime imaging technique using fluorescein as a pH indicator. The use of fluorescent dyes, in combination (LIVE/DEAD BacLight) with CLSM, has become a routine practice for biofilm analysis. The LIVE/DEAD Bacterial Viability Kit (Molecular Probes, Eugene, OR) contains separate vials of the two component dyes (SYTO 9 and propidium iodide). The dyes are used in 1:1 mixture for staining the biofilm bacteria following manufacturer's instructions. The dead cells emitted red light and the viable cells emitted green light under CLSM examination (George and Kishen 2008) (Fig. 8). In a recent in vitro study, it was shown that bacteria in the multispecies anaerobic biofilm grown under nutrient deprivation changed into a viable but nonculturable (VBNC) state but could be returned to the normal physiological state and cultured by reestablishing the supply of nutrients while it was still in the biofilm. The results from this study indicated that viability staining better reflected the "true viability" state of biofilm bacteria than culture method during starvation. This point has to be taken into account when employing culture-based methods for analyzing in vivo root canal biofilms (Shen et al. 2010b).

Fluorescence in situ hybridization (FISH) is a recognized tool for the specific and sensitive identification of target microbes within a complex microbial community. The FISH technique uses probes to target-specific 16S rRNA sequences in bacteria. It is applied for the simultaneous evaluation of the spatial distribution of both gram-positive and gram-negative bacteria in biofilms. Visualization of FISH-labeled cells in biofilms can be carried out by fluorescence microscopy and CLSM (Sheppard and Shotton 1997; Main et al. 1984). However, CLSM is preferred in a biofilm analysis, since it allows a three-dimensional noninvasive visualization of cells and computational reconstruction of the mature biofilms without distortion of their structure (Mattila-Sandholm and Wirtanen 1992).

9 Physical Methods: Thickness, Weight, Area, and Density Measurements

Physical parameters such as biofilm thickness, area, weight (wet and dry), and density estimates are used to quantify biofilm growth. Thickness measurement by light microscopy is usually effective in thin biofilms; however it may not work with thick biofilms. In this method, the biofilm is placed on the stage of a microscope, which has calibration scales on the fine control. The objective is lowered until the biofilm surface is in focus and the fine adjustment dial setting of the microscope recorded. The microscope objective is then focused on the substrate surface, preferably in an area with no biofilm (Stewart 1990). The difference in the fine adjustment setting can be used to calculate the thickness. A simple, manual-gauge needle method (Walker et al. 1994) and an electronic probe have also been used to measure biofilm thickness (Marshall et al. 1971). A properly prepared SEM sample or cryosection enables the estimation of biofilm thickness and also revealed layering of embedded bacterial cells (Marshall 1997). Biofilm wet weight is a useful measure of the biomass especially on tared substrates. This is a very simple and quick procedure. The substrate is weighed before biofilm growth and then weighed again in order to record the dry biofilm weight. In this case, assumption is made that no substrate solubilization has occurred during biofilm growth. If both wet and dry weight measurements on the same biofilm sample were measured, the approximate density may be determined. For comparative purposes, physical parameters such as film density and weight can be calculated per unit area of the substratum (Marshall et al. 1971; Marshall 1997).

10 Biochemical Methods: Biomass and Extracellular Matrix

The microbial biomass denotes the total number of microbes in a given area. The measurements of (1) microbial biomass are considered to be a rapid method and include measurements of wet or dry weight of the whole biofilm, (2) measurements of the cell contents (3) measurements of cellular activities or viable cells. In this line, adenosine triphosphate (ATP) bioluminescence is widely used to determine the metabolic activity of bacterial population. This technique requires a cell lysis step to release ATP, which is determined by a luciferin-luciferase reaction (de Beer et al. 1994). However, it should be noted that the rate of lysis and the ATP content vary depending on the microorganism. Hence, the ATP assay cannot be correlated with the initial number of microbial cells. Genetically modified strains containing genes for bioluminescence have also been developed (McAllister et al. 1994) and applied for some in vitro analysis. All these methods present advantages and disadvantages. Except for the microscopy-based techniques, most other methods require an optimum number of viable cells with the ability to multiply to a significant number in a brief period of time. Further research is required before employing rapid biochemical methods more routinely for the detection of viable bacteria in biofilm assays.

11 Molecular Biological Methods

Molecular biological techniques provide genetic information on the biofilm bacteria. Microarray analysis (Millsap et al. 1997) and the use of defined regulatory mutants (Lee et al. 1990; Bauer-Kreisel et al. 1996) have been important tools for studying biofilms. In addition, cloning and expression of bacterial virulence factors in less pathogenic organisms (Perez-Osorio and Franklin 2008; Kindaichi et al. 2006) are another important tool for assessing the role of bacteria in biofilm-mediated infections. It should be realized that when molecular-based analyses of bacterial adherence and biofilm formation are performed, the bacteria are mostly grown under ideal laboratory conditions. These conditions might not be a clinically realistic situation. Although such experiments can be used for a relative comparison between experimental groups, there is a possibility of disparity between laboratories and in vivo situations.

Enzyme-linked immunosorbent assay (ELISA) is a very sensitive method used to detect the presence of antigens or antibodies of interest in a sample. ELISA can be used for quantitative evaluation when used in conjunction with standard curves. ELISA is typically performed using one of the two detection methods: the direct or indirect assays. In direct ELISA, an enzyme-linked (labeled) antibody is used to directly detect the captured antigen or antibody of interest. In the more common indirect ELISA, a primary antibody is bound to the sample antigen/antibody and then a secondary labeled antibody (antiglobulin) is used to detect the primary antibody. For any ELISA procedure the sample antigens/antibodies of interest are concentrated and solubilized in an appropriate buffer. ELISA has been used as an alternative method to quantify biomass within biofilms and even protein expression in biofilms (Davey and O'Toole 2000; Heydorn et al. 2002). ELISA may be used to quantify the population of a particular bacterium in a mixed biofilm. An ELISA-based approach can circumvent errors due to cell clumping and EPS production, which can lead to significant errors in bacterial quantification. The disadvantages of ELISA are related to the cross-reactivity and nonspecificity of signals. This method is poorly suited for low concentrations of antigens (Davey and O'Toole 2000).

Detection of differential gene expression may serve as a novel high-resolution and specific assay to understand gene expressions in a biofilm community. However, current assays may only depict the average signal or response from all the cells in a biofilm. This measurement would not provide signals from the specific cell population in a biofilm. Further, there are several factors in an in vivo environment, which may influence biofilm formation, and questions are raised whether the assessment of gene expression in vitro is really an indicative of gene expression in vivo.

Polymerase chain reaction (PCR) is a method that allows exponential amplification of short DNA sequences. The method relies on thermal cycling and enzymatic replication of the DNA. Primers, which consist of short DNA fragments/ sequences complementary to the target region and a DNA polymerase are key components to enable selective and repeated amplification. As PCR progresses, the DNA generated is used as a template for replication, setting up a chain reaction in which the DNA template is exponentially amplified. This method of analysis is mostly used as a qualitative tool for detecting the presence or absence of a particular bacterial DNA. A real-time polymerase chain reaction, also called *quantitative real-time polymerase chain reaction* (qPCR), is based on PCR and is employed to amplify and simultaneously quantify a targeted DNA molecule. RT-PCR enables both detection and quantification of one or more specific sequences in a DNA sample. The key feature in RT-PCR is that the amplified DNA is detected as the reaction progresses in *real time* (Wimpenny et al. 2000). This is a new approach compared to the standard PCR, where the product of the reaction is detected at the end. Two common methods for detection of products in real-time PCR are: (1) nonspecific fluorescent dyes that intercalate with any double-stranded DNA and (2) sequence-specific DNA probes consisting of oligonucleotides that are labeled with a fluorescent reporter which permits detection only after hybridization of the probe with its complementary DNA target. RT-PCR can be used to estimate the number of copies of a target gene in a sample and is reported to be more sensitive than conventional qualitative PCR. This method has been used to detect and quantify bacterial population in a biofilm. Often, the RT-PCR is combined with reverse transcription to quantify messenger RNA and noncoding RNA in cells or tissues. Quantitative reverse transcriptase real-time PCR (qRT-PCR) can be used effectively to quantify the number of RNA transcripts of specific genes from bacteria growing in biofilms. qRT-PCR has a large dynamic range and may be used to verify gene expression data obtained from microarrays. In addition, qRT-PCR is sensitive and therefore may be used to quantify gene expression from biofilm samples where only a small amount of biological material is available (Wimpenny et al. 2000; Thurnheer et al. 2004).

12 Miscellaneous Advanced Techniques

Atomic force microscopy (AFM) has been applied to study the forces of interaction between bacterial cells and between bacterial cell and substrate (Postollec et al. 2006) (Fig. 9). In order to use AFM to determine bacteria-substrate interaction, the bacterial cell or substrate particle is attached to an AFM tip and the forces of interaction are studied. Briefly, as the AFM tip approaches the substrate and the gap between the two interacting bodies closes to nanometer range, the interacting forces developed are registered by the AFM tip (Razatos et al. 1998). The AFM force curves can be used to estimate the adhesion events and interaction forces in the interaction between the bacteria and substrate (Gaboriaud and Dufrene 2007). In an AFM analysis, positively charged polymers, such as polyethyleneimine and poly-L-lysine, are necessary to securely attach bacteria onto the cantilever tips. Based on this concept, an investigation was aimed to study the effects of endodontic irrigants on the adherence of *E. faecalis* to dentin (Sum et al. 2008). This study highlighted that chemicals that altered the physicochemical properties of dentin might influence the nature of bacterial adherence and adhesion force to dentin. The physical attachment of bacterial cells using positively charged polymers might promote structural rearrangements in bacterial cell surface structures (Vadillo-Rodriguez et al. 2004), which in turn may affect the value of forces measured. Recently micromanipulators are employed to study individual cells or biofilm compartments. Laser-based optical tweezers are a noninvasive and noncontact tool that can probe interaction between microscopic objects such as bacteria and collagen with sub-pN sensitivity. The optical tweezers technique gives more quantitative information about the forces of interaction between bacteria and substrate (Sum et al. 2008).

Fourier transform infrared (FTIR) spectroscopy is used to characterize the chemical composition of biofilm structures. In an FTIR spectroscopic analysis, infrared radiation is interacted with a test sample. During this interaction, some of the infrared radiation is absorbed by the sample. The resulting spectrum represents the molecular-level absorption/transmission, which is a molecular fingerprint of the sample. The FTIR spectroscopy can be used for the qualitative and quantitative analyses of the chemical constituents on a biofilm structure (Kishen et al. 2006). Similarly, biophysical techniques such as solid-state nuclear magnetic resonance (NMR) are a powerful analytical tool to study the chemical constituents of a biofilm structure (Grivet et al. 2003). NMR spectroscopy is useful to obtain metabolic information in planktonic cells, adherent bacterial cells, and in situ biofilm bacteria (Majors et al. 2005).

13 Concluding Remarks

A variety of biofilm models are used for different experimental purposes in endodontics. Making a rational choice regarding the best model to be used for a particular research remains to be a problem. Generally, biofilm model that closely reproduce in vivo condition or standardized model for repeated assessments should be selected. There is, however, no single, ideal biofilm model for all applications. Direct, nondestructive visualization of biofilms is advantageous in monitoring changes in biofilm bacteria and structures. In spite of all the recent advances, the quantification of bacterial biofilm and evaluation of the efficacy of disinfectants remain to be a major challenge in endodontics. Future efforts are warranted to standardize the type of in vitro biofilm models and in vivo test methods for assessing antimicrobials.

References

Al-Hashimi I, Levine MJ (1989) Characterization of in vivo salivary-derived enamel pellicle. Arch Oral Biol 34(4):289–295

Badr AE, Omar N, Badria FA (2011) A laboratory evaluation of the antibacterial and cytotoxic effect of Liquorice when used as root canal medicament. Int Endod J 44(1):51–58

Bagge N et al (2004) Dynamics and spatial distribution of beta-lactamase expression in Pseudomonas aeruginosa biofilms. Antimicrob Agents Chemother 48(4):1168–1174

Bakke R, Kommedal R, Kalvenes S (2001) Quantification of biofilm accumulation by an optical approach. J Microbiol Methods 44(1):13–26

Bauer-Kreisel P, Eisenbeis M, Scholz-Muramatsu H (1996) Quantification of dehalospirillum multivorans in mixed-culture biofilms with an enzyme-linked immunosorbent assay. Appl Environ Microbiol 62(8):3050–3052

Baumgartner JC, Siqueira J Jr, Sedgley C, Kishen A (2008) Microbiology of endodontic disease. In: Ingle JI, Bakland LK, Baumgartner JC (eds) Ingle's endodontics, BC Decker, Ontario

Bhuva B, Patel S, Wilson R, Niazi S, Beighton D, Mannocci F (2010) The effectiveness of passive ultrasonic irrigation on intraradicular Enterococcus faecalis biofilms in extracted single-rooted human teeth. Int Endod J 43(3):241–250

Burnett SL, Chen J, Beuchat LR (2000) Attachment of Escherichia coli O157:H7 to the surfaces and internal structures of apples as detected by confocal scanning laser microscopy. Appl Environ Microbiol 66(11):4679–4687

Chávez de Paz LE, Bergenholtz G, Svensäter G (2010) The effects of antimicrobials on endodontic biofilm bacteria. J Endod 36(1):70–7

Costerton JW (1999) Introduction to biofilm. Int J Antimicrob Agents 11(3–4):217–221, discussion 237–9

Costerton JWL, Lewandowski Z (1997) The biofilm lifestyle. Adv Dent Res 11:192–195

Costerton JW, Stewart PS (2001) Battling biofilms. Sci Am 285(1):74–81

Costerton JW et al (1994) Biofilms, the customized microniche. J Bacteriol 176(8):2137–2142

Cowan MM, Taylor KG, Doyle RJ (1987) Energetics of the initial phase of adhesion of Streptococcus sanguis to hydroxylapatite. J Bacteriol 169(7):2995–3000

Davey ME, O'Toole GA (2000) Microbial biofilms: from ecology to molecular genetics. Microbiol Mol Biol Rev 64(4):847–867

de Beer D et al (1994) Effects of biofilm structures on oxygen distribution and mass transport. Biotechnol Bioeng 43(11):1131–1138

del Pozo JL, Patel R (2007) The challenge of treating biofilm-associated bacterial infections. Clin Pharmacol Ther 82(2):204–209

Di Bonaventura G et al (2006) Biofilm formation by the emerging fungal pathogen Trichosporon asahii: development, architecture, and antifungal resistance. Antimicrob Agents Chemother 50 (10):3269–3276

Duggan JM, Sedgley CM (2007) Biofilm formation of oral and endodontic Enterococcus faecalis. J Endod 33(7):815–818

Friedman L, Kolter R (2004) Genes involved in matrix formation in Pseudomonas aeruginosa PA14 biofilms. Mol Microbiol 51(3):675–690

Gaboriaud F, Dufrene YF (2007) Atomic force microscopy of microbial cells: application to nanomechanical properties, surface forces and molecular recognition forces. Colloids Surf B: Biointerfaces 54(1):10–19

George S, Kishen A (2007) Photophysical, photochemical, and photobiological characterization of methylene blue formulations for light-activated root canal disinfection. J Biomed Opt 12 (3):034029

George S, Kishen A (2008) Augmenting the antibiofilm efficacy of advanced noninvasive light activated disinfection with emulsified oxidizer and oxygen carrier. J Endod 34(9):1119–1123

George S, Basrani B, Kishen A (2010) Possibilities of gutta-percha-centered infection in endodontically treated teeth: an in vitro study. J Endod 36(7):1241–1244

Grenier D, Mayrand D (1986) Nutritional relationships between oral bacteria. Infect Immun 53 (3):616–620

Grivet JP, Delort AM, Portais JC (2003) NMR and microbiology: from physiology to metabolomics. Biochimie 85(9):823–840

Handley PS, Carter PL, Fielding J (1984) Streptococcus salivarius strains carry either fibrils or fimbriae on the cell surface. J Bacteriol 157(1):64–72

Handley PS et al (1985) Surface structures (peritrichous fibrils and tufts of fibrils) found on Streptococcus sanguis strains may be related to their ability to coaggregate with other oral genera. Infect Immun 47(1):217–227

Harraghy N et al (2006) Advances in in vitro and in vivo models for studying the staphylococcal factors involved in implant infections. Int J Artif Organs 29(4):368–378

Heydorn A et al (2002) Statistical analysis of Pseudomonas aeruginosa biofilm development: impact of mutations in genes involved in twitching motility, cell-to-cell signaling, and stationary-phase sigma factor expression. Appl Environ Microbiol 68(4):2008–2017

Hiraishi N, Yiu CK, King NM, Tagami J, Tay FR (2010) Antimicrobial efficacy of 3.8% silver diamine fluoride and its effect on root dentin. J Endod 36(6):1026–1029

Hope CK, Garton SG, Wang Q, Burnside G, Farrelly PJ (2010) A direct comparison between extracted tooth and filtermembrane biofilm models of endodontic irrigation using Enterococcus faecalis. Arch Microbiol 192(9):775–781

Johansen C, Falholt P, Gram L (1997) Enzymatic removal and disinfection of bacterial biofilms. Appl Environ Microbiol 63(9):3724–3728

Jones SJ (1972) A special relationship between spherical and filamentous microorganisms in mature human dental plaque. Arch Oral Biol 17(3):613–616

Kindaichi T et al (2006) Population dynamics and in situ kinetics of nitrifying bacteria in autotrophic nitrifying biofilms as determined by real-time quantitative PCR. Biotechnol Bioeng 94(6):1111–1121

Kishen A, George S, Kumar R (2006) Enterococcus faecalis-mediated biomineralized biofilm formation on root canal dentine in vitro. J Biomed Mater Res A 77(2):406–415

Kishen A, Haapasalo M (2010) Biofilm models in endodontics. Endod Top 22(1):58–78

Kishen A et al (2008) Influence of irrigation regimens on the adherence of Enterococcus faecalis to root canal dentin. J Endod 34(7):850–854

Kishen A, Upadya M, Tegos GP, Hamblin MR (2010) Efflux pump inhibitor potentiates antimicrobial photodynamic inactivation of Enterococcus faecalis biofilm. Photochem Photobiol 86 (6):1343–1349

Kolenbrander PE et al (1995) Intergeneric coaggregation of oral Treponema spp. with Fusobacterium spp. and intrageneric coaggregation among Fusobacterium spp. Infect Immun 63(12):4584–4588

Kulasekara HD et al (2005) A novel two-component system controls the expression of Pseudomonas aeruginosa fimbrial cup genes. Mol Microbiol 55(2):368–380

Lee HA et al (1990) Enzyme-linked immunosorbent assay for Salmonella typhimurium in food: feasibility of 1-day Salmonella detection. Appl Environ Microbiol 56(6):1541–1546

Lim Z et al (2009) Light activated disinfection: an alternative endodontic disinfection strategy. Aust Dent J 54(2):108–114

Liu H, Wei X, Ling J, Wang W, Huang X (2010) Biofilm formation capability of Enterococcus faecalis cells in starvation phase and its susceptibility to sodium hypochlorite. J Endod 36 (4):630–635

Lundstrom JR, Williamson AE, Villhauer AL, Dawson DV, Drake DR (2010) Bactericidal activity of stabilized chlorine dioxide as an endodontic irrigant in a polymicrobial biofilm tooth model system. J Endod 36(11):1874–1878
Main C et al (1984) Instrumentation for measurement of dental plaque thickness in situ. J Biomed Eng 6(2):151–154
Majors PD et al (2005) NMR methods for in situ biofilm metabolism studies. J Microbiol Methods 62(3):337–344
Marshall KC (1997) Colonization, adhesion and biofilms. In: Hurst CJK, Knudsen GR, McInerney MJ (eds) Manual of environmental microbiology. American Society for Microbiology Press, Washington, DC, pp 358–365
Marshall KC, Stout R, Mitchell R (1971) Selective sorption of bacteria from seawater. Can J Microbiol 17(11):1413–1416
Mathew S, Yaw-Chyn L, Kishen A (2010) Immunogenic potential of Enterococcus faecalis biofilm under simulated growth conditions. J Endod 36(5):832–836
Mattila-Sandholm TW, Wirtanen G (1992) Biofilm formation in the industry: a review. Food Rev Int 8:573–603
McAllister TA et al (1994) Microbial attachment and feed digestion in the rumen. J Anim Sci 72 (11):3004–3018
McBain AJ (2009) Chapter 4: in vitro biofilm models: an overview. Adv Appl Microbiol 69:99–132
McGill S, Gulabivala K, Mordan N, Ng YL (2008) The efficacy of dynamic irrigation using a commercially available system (RinsEndo) determined by removal of a collagen 'bio-molecular film' from an ex vivo model. Int Endod J 41(7):602–608
McKinlay KJ et al (2004) Comparison of environmental scanning electron microscopy with high vacuum scanning electron microscopy as applied to the assessment of cell morphology. J Biomed Mater Res A 69(2):359–366
Merritt JH, Kadouri DE, O'Toole GA (2005) Growing and analyzing static biofilms. Curr Protoc Microbiol (Chapter 1: p. Unit 1B 1)
Millsap KW et al (1997) Adhesion of Lactobacillus species in urine and phosphate buffer to silicone rubber and glass under flow. Biomaterials 18(1):87–91
Miron J, Ben-Ghedalia D, Morrison M (2001) Invited review: adhesion mechanisms of rumen cellulolytic bacteria. J Dairy Sci 84(6):1294–1309
Nair PN (2006) On the causes of persistent apical periodontitis: a review. Int Endod J 39 (4):249–281
Nair PN et al (1990) Intraradicular bacteria and fungi in root-filled, asymptomatic human teeth with therapy-resistant periapical lesions: a long-term light and electron microscopic follow-up study. J Endod 16(12):580–588
Pavarina ACD, Dovigo LN, Sanitá PV, Machado AL, Giampaolo ET, Vergani CE (2011) Dynamic models for in vitro biofilm formation. In: Bailey WC (ed) Biofilms: formation, development and properties. Nova Science, New York
Pappen FG, Shen Y, Qian W, Leonardo MR, Giardino L, Haapasalo M (2010) In vitro antibacterial action of Tetraclean, MTAD and five experimental irrigation solutions. Int Endod J 43(6):528–535
Perez-Osorio AC, Franklin MJ (2008) qRT-PCR of Microbial Biofilms. CSH Protoc 2008:p. pdb prot5066
Postollec F et al (2006) Interactive forces between co-aggregating and non-co-aggregating oral bacterial pairs. J Dent Res 85(3):231–234
Pratten J, Ready D (2010) Use of biofilm model systems to study antimicrobial susceptibility. Methods Mol Biol 642:203–215
Ramachandran Nair PN (1987) Light and electron microscopic studies of root canal flora and periapical lesions. J Endod 13(1):29–39
Razatos A et al (1998) Molecular determinants of bacterial adhesion monitored by atomic force microscopy. Proc Natl Acad Sci 95(19):11059–11064

Rosan B, Correeia FF, DiRienzo JM (1999) Corncobs: a model for oral microbial biofilms. In: Busscher HJ , Evans LV (eds) Oral biofilms and plaque control: concepts in dental plaque formation. Harwood Academic Publishers, India, pp 145–162

Sainsbury AL, Bird PS, Walsh LJ (2009) DIAGNOdent laser fluorescence assessment of endodontic infection. J Endod 35(10):1404–7

Shahriari S, Mohammadi Z, Mokhtari MM, Yousefi R (2010) Effect of hydrogen peroxide on the antibacterial substantivity of chlorhexidine. Int J Dent 2010:946384

Shen Y, Qian W, Chung C, Olsen I, Haapasalo M (2009) Evaluation of the effect of two chlorhexidine preparations on biofilm bacteria in vitro: a three-dimensional quantitative analysis. J Endod 35(7):981–985

Shen Y, Stojicic S, Haapasalo M (2010a) Bacterial viability in starved and revitalized biofilms: comparison of viability staining and direct culture. J Endod 36(11):1820–1823

Shen Y, Stojicic S, Qian W, Olsen I, Haapasalo M (2010b) The synergistic antimicrobial effect by mechanical agitation and two chlorhexidine preparations on biofilm bacteria. J Endod 36(1):100–104

Shen Y, Stojicic S, Haapasalo M (2011) Antimicrobial efficacy of chlorhexidine against bacteria in biofilms at different stages of development. J Endod 37(5):657–661

Sheppard CRJS, Shotton DM (1997) Confocal laser scanning microscopy. BIOS Scientific, Oxford

Shrestha A, Shi Z, Neoh KG, Kishen A (2010) Nanoparticulates for antibiofilm treatment and effect of aging on its antibacterial activity. J Endod 36(6):1030–1035

Soares JA, de Carvalho MA R, Cunha Santos SM, Mendonça RM, Ribeiro-Sobrinho AP, Brito-Júnior M, Magalhães PP, Santos MH, de Macêdo Farias L (2010) Effectiveness of chemomechanical preparation with alternating use of sodium hypochlorite and EDTA in eliminating intracanal Enterococcus faecalis biofilm. J Endod 36(5):894–898

Stewart GS (1990) In vivo bioluminescence: new potentials for microbiology. Lett Appl Microbiol 10(1):1–8

Stoodley P, Debeer D, Lewandowski Z (1994) Liquid flow in biofilm systems. Appl Environ Microbiol 60(8):2711–2716

Su L, Gao Y, Yu C, Wang H, Yu Q (2010) Surgical endodontic treatment of refractory periapical periodontitis with extraradicular biofilm. Oral Surg Oral Med Oral Pathol Oral Radiol Endod 110(1):e40–4

Sum C et al (2008) Influence of endodontic chemical treatment on Enterococcus faecalis adherence to collagen studied with laser scanning confocal microscopy and optical tweezers: a preliminary study. J Biomed Opt 13(4):044017

Sun D, Accavitti MA, Bryers JD (2005) Inhibition of biofilm formation by monoclonal antibodies against Staphylococcus epidermidis RP62A accumulation-associated protein. Clin Diagn Lab Immunol 12(1):93–100

Sundqvist G, Figdor D (2003) Life as an endodontic pathogen: ecological differences between the untreated and root-filled root canals. Endod Topics 6(1):3–28

Sutton NA, Hughes N, Handley PS (1994) A comparison of conventional SEM techniques, low temperature SEM and the electroscan wet scanning electron microscope to study the structure of a biofilm of Streptococcus crista CR3. J Appl Bacteriol 76(5):448–454

Takeuchi K, Frank JF (2001) Expression of red-shifted green fluorescent protein by Escherichia coli O157:H7 as a marker for the detection of cells on fresh produce. J Food Prot 64 (3):298–304

Thurnheer T, Gmur R, Guggenheim B (2004) Multiplex FISH analysis of a six-species bacterial biofilm. J Microbiol Methods 56(1):37–47

Torabinejad M, Shabahang S, Aprecio RM, Kettering JD (2003) The antimicrobial effect of MTAD: an in vitro investigation. J Endod 29(6):400–403

Trachoo N, Frank JF (2002) Effectiveness of chemical sanitizers against Campylobacter jejuni-containing biofilms. J Food Prot 65(7):1117–1121

Upadya MH, Kishen A (2010) Influence of bacterial growth modes on the susceptibility to light-activated disinfection. Int Endod J 43(11):978–987

Vadillo-Rodriguez V et al (2004) Comparison of atomic force microscopy interaction forces between bacteria and silicon nitride substrata for three commonly used immobilization methods. Appl Environ Microbiol 70(9):5441–5446

Vakulenko SB, Mobashery S (2003) Versatility of aminoglycosides and prospects for their future. Clin Microbiol Rev 16(3):430–450

Walker AJ, Stewart GSAB, Sheppard F, Bloomfield SF, Holah JT (1994) Bioluminescence imaging as a tool for studying biocide challenge upon plantonic and surface attached bacteria. Bin Comp Microbiol 6:16–17

Watnick PI, Kolter R (1999) Steps in the development of a Vibrio cholerae El Tor biofilm. Mol Microbiol 34(3):586–595

Williamson AE, Marker KL, Drake DR, Dawson DV, Walton RE (2009) Resin-based versus gutta-percha-based root canal obturation: influence on bacterial leakage in an in vitro model system. Oral Surg Oral Med Oral Pathol Oral Radiol Endod 108(2):292–296

Wimpenny J, Manz W, Szewzyk U (2000) Heterogeneity in biofilms. FEMS Microbiol Rev 24 (5):661–671

Wolfaardt GM et al (1993) Development of steady-state diffusion gradients for the cultivation of degradative microbial consortia. Appl Environ Microbiol 59(8):2388–2396

Yavari HR et al (2010) Effect of Er, Cr: YSGG laser irradiation on Enterococcus faecalis in infected root canals. Photomed Laser Surg 28(Suppl 1):S91–S96

Root Canal Anatomy: Implications in Biofilm Disinfection

Marco A. Versiani and Ronald Ordinola-Zapata

Abstract The primary goals of endodontic treatment are to disinfect and to seal the root canal system as effectively as possible, aiming to establish or maintain healthy periapical tissues. Treating complex and anomalous anatomy requires knowledge of the internal anatomy of all types of the teeth before undertaking endodontic therapy. Recently, three-dimensional imaging using micro-computed tomography has been used to reveal the internal anatomy of the teeth to the clinician, bringing new perspectives on the overall quality of the endodontic treatment and confirming the inability of shaping tools in acting within the anatomical complexity of the root canal. Furthermore, the disinfecting effects of instruments and irrigants may be additionally hampered in the presence of accessory canals, ramifications, intercanal connections, fins, isthmuses, and apical deltas, which can be ideal sites for microbial biofilm formation and cannot be properly assessed by conventional techniques. Apart from biofilms, these hard-to-reach areas may also be packed with dentin debris generated and pushed therein by endodontic instruments, interfering with disinfection by both preventing the irrigant flow into them and by neutralizing its efficacy. This chapter provides an overview of recent information on the complexity of root canal anatomy and discusses its relationship to the understanding of the principles and problems of microbial biofilm elimination.

1 Introduction

The goal of endodontic therapy is the removal of all vital or necrotic tissue, microorganisms, and microbial by-products from the root canal space (Siqueira 2005; Hargreaves and Cohen 2011). Among the treatment steps, chemomechanical procedures play a pivotal role in eliminating or reducing bacterial populations from

M.A. Versiani (✉)
Department of Endodontics, Dental School of Ribeirão Preto, University of São Paulo, São Paulo, Brazil
e-mail: marcoversiani@yahoo.com

R. Ordinola-Zapata
Dental School of Bauru, University of São Paulo, São Paulo, Brazil

L.E. Chávez de Paz et al. (eds.), *The Root Canal Biofilm*, Springer Series on Biofilms 9, DOI 10.1007/978-3-662-47415-0_7

the main root canal, but the disinfecting effects of instruments and irrigants may be somewhat hampered in cases with complex anatomy and the resistance of microbial biofilms (Siqueira 2005; Shen et al. 2012; Siqueira et al. 2013). Thus, a thorough understanding of the canal morphology and its variations in all groups of teeth is a basic requirement for the success of the endodontic therapy.

The concept of a "single" root canal with a "single" apical foramen is mistaken (Siqueira 2005; Hargreaves and Cohen 2011). The root canal space is often complex, with canals that divide and rejoin, isthmuses, fins, anastomosis, accessory canals, and apical deltas. For this reason, it has been referred as a system. The complexity of the root canal system has been demonstrated by several authors since nineteenth century. Carabelli (1842) was probably the first author to provide a comprehensive description of the number and location of root canals. Some years later, the first systematic study on the root canal anatomy using sectioned teeth was published by Mühlreiter (1870). After a few decades, Black (1890) systematized the dental terminology and detailed the internal and external anatomy of the teeth, and in 1894, Gysi and Röse published pictures of histological sections of human teeth, demonstrating the complexity of the root canal. In the early twentieth century, Preiswerk (1903) introduced the "modeling technique" and obtained metal replicas of the internal anatomy of all groups of teeth by injection of molten metal. Fischer (1908) and Dewey (1916) improved this method and obtained full replicas of the canal ramifications and lateral branches by injection of collodion solution or paraffin into the root canal system. Hess (1917) and Hess and Zurcher (1925) used the demineralizing method to study the canal morphology of approximately 3000 teeth, after injection of natural rubber within the root canal space. At this time, diaphanization (clearing technique) was also used for several authors to the study of the root canal anatomy (Adloff 1913; Fasoli 1913; Moral 1915; Okumura 1927). In this method, hard tissues of the teeth were rendered transparent through demineralization after injecting fluid materials, such as molten metal (Adloff 1913), gelatin-containing cinnabar (Fasoli 1913), and different types of ink (Moral 1915; Okumura 1927), into the root canal system.

In the following decades, the morphology of the root canal system was described by using several in vivo and ex vivo methods such as three-dimensional wax models (Meyer and Scheele 1955), conventional (Mueller 1933; Barker et al. 1969; Sykaras 1971; Pineda and Kuttler 1972; Kaffe et al. 1985; Fabra-Campos 1989; Pécora et al. 1991; Pécora et al. 1993a; Estrela et al. 1995; Sousa Neto et al. 1998; Ferraz et al. 2001; Baratto-Filho et al. 2002) and digital radiography (Nattress and Martin 1991; Burger et al. 1999; Schäfer et al. 2002), resin injection (Barker et al. 1973, 1974a, b), macroscopic (Green 1955, 1973; Baratto-Filho et al. 2002) and microscopic evaluation (Green 1956; Burch and Hulen 1972; Kerekes and Tronstad 1977a, b, c), tooth sectioning on different planes (Green 1955; Green 1958; Weine et al. 1969; Kerekes and Tronstad 1977a, b, c; Mauger et al. 1998), clearing techniques (Robertson et al. 1980; Vertucci 1984; Pécora et al. 1992a, b; Pécora et al. 1993b, c; Rocha et al. 1996; Guerisoli et al. 1998; Sharma et al. 1998; Venturi et al. 2003; Weng et al. 2009), radiographic methods with radiopaque contrast media (Naoum et al. 2003), and scanning electron microscopy (Gilles and Reader 1990).

The aforementioned techniques have been used successfully over many years for the study of the root canal anatomy (Bergmans et al. 2001); however, their destructive nature produced irreversible changes in the specimens (Grover and Shetty 2012) and many artifacts (Perrini and Castagnola 1998) which, therefore, may not accurately reflect the canal morphology (Versiani et al. 2011a, 2012). Furthermore, these techniques did not allow for the three-dimensional analysis of the external and internal anatomy of the teeth at the same time (Grande et al. 2012). These inherent limitations have encouraged the search for new methods with improved possibilities (Bergmans et al. 2001).

Computer-assisted imaging was introduced in the field of endodontic research in 1986 by Mayo et al. . By combining different radiographic views of the tooth, a mathematically determined three-dimensional representation of the canals was obtained (Pao et al. 1984; Gullickson and Montgomery 1987; Dobó-Nagy et al. 2000). Some years later, an improved computerized method for visualization of the canal, based on the cross-sectional images of the root, was also introduced (Berutti 1993; Blašković-Šubat et al. 1995; Lyroudia et al. 1997a, b; Hegedus et al. 2000; Lyroudia et al. 2002). These methods allowed the development of 3D models as well as the quantitative measurements of some morphological parameters of the root canal; however, they were still destructive methods and the thickness of sections and material loss were found to influence the accuracy of the results (Hegedus et al. 2000).

In 1990, Tachibana and Matsumoto (1990) suggested the use of computed tomographic (CT) imaging in endodontics, but they concluded that CT had only a limited usefulness because of its low spatial resolution. Further improvements in the digital image systems allowed the study of the root canal anatomy using conventional medical CT (Gambill et al. 1996; Robinson et al. 2002; Cimilli et al. 2005; Reuben et al. 2008; Garg et al. 2013; Nayak and Singh 2013), magnetic resonance microscopy (Baumann et al. 1993; Baumann 1994; Baumann and Doll 1997; Appel and Baumann 2002; Tanasiewicz 2010; Idiyatullin et al. 2011; Sustercic and Sersa 2012), tuned-aperture computed tomography (Nance et al. 2000; Barton et al. 2003), optical coherence tomography (Shemesh et al. 2007), and cone-beam CT (Matherne et al. 2008; Huang et al. 2010; Michetti et al. 2010; Wang et al. 2010; Bauman et al. 2011; Lee et al. 2011; Neelakantan et al. 2011; Zhang et al. 2011a, b; Zheng et al. 2011; Han et al. 2012; Hassan et al. 2012; Kim et al. 2012; Seo et al. 2012; Shenoi and Ghule 2012; Szabo et al. 2012; Tian et al. 2012; Yu et al. 2012). These methods hindered their use for the detailed study of the internal anatomy mainly because of their insufficient spatial resolution and slice thickness for detailing the small size of the root canal structures (Nielsen et al. 1995; Dowker et al. 1997).

A few years later, high-resolution X-ray micro-computed tomography (micro-CT) was suggested to be used as an advanced tool for detailed endodontic research (Nielsen et al. 1995). This is a nondestructive reproducible accurate technique that could be applied quantitatively as well as qualitatively for two- and three-dimensional assessment of the root canal system (Peters et al. 2000; Versiani et al. 2011b; Siqueira et al. 2013; Versiani et al. 2013b; Keleş et al. 2014; Ordinola-Zapata et al. 2014; Marceliano-Alves et al. 2014; Versiani et al. 2015). Nowadays, despite the impossibility of employing micro-CT for in vivo human imaging, it has been considered the most important research tool for the study of root canal anatomy (Peters et al. 2000; Versiani et al. 2011a, 2012, 2013a). Morphological studies on the pulp canal space using this new technology have shown a wide variety of shapes and configurations of the root canal system (Peters et al. 2000; Peters et al. 2001; Versiani et al. 2011a, c; Villas-Boas et al. 2011; Versiani et al. 2012; Ordinola-Zapata et al. 2013; Versiani et al. 2013a, c; Ordinola-Zapata et al. 2014). Thus, the purpose of this chapter is to discuss the complexity of root canal system and understand its implications in biofilm disinfection.

2 Macroscopic Anatomy of the Root Canal System

2.1 Pulp Chamber

Didactically speaking, the root canal system may be divided into two parts: the pulp chamber, located within the anatomic dental crown, and the root canal, found inside the radicular portion of the tooth (Fig. 1a) (Woelfel and Scheid 2002; Hargreaves and Cohen 2011; Nanci and Ten Cate 2013).

The pulp chamber is a single cavity that occupies the center of the crown and resembles the shape of the crown surface (Bjørndal et al. 1999). In anterior teeth, the pulp chamber is contiguous with the root canal, but in posterior teeth it usually presents a square shape with six sides: the floor, the roof, and the axial walls. The walls are located in the four axial surfaces of the pulp chamber, and they are identified as mesial, distal, buccal, or lingual (palatal) walls. The pulp-chamber roof is located just below the occlusion or incisal ridge and usually presents projections or prominences associated to the cusps or mamelons, occupied by pulpal tissues, the so-called pulp horns (Woelfel and Scheid 2002). The floor of the pulp chamber is nearly flat in young teeth, later becoming convex. The natural anatomy of the pulp-chamber floor frequently indicates the location of the entrances of the root canals (orifices). The orifice is an opening leading from the pulp chamber into a root canal, especially in a tooth with multiple canals (AAE 2012) (Fig. 1b). However, the size of the pulp chamber may be reduced and changed in shape by apposition of secondary or tertiary dentin because of aging physiological process or as a pathological response to pulpal irritation such as caries, restorative treatment, occlusal loading, or abrasion (Peiris et al. 2008). As a consequence, calcifications may hide the canal orifices by altering their original configurations, blocking the access to the root canals (Hargreaves and Cohen 2011).

Accessory canals that connect the pulp chamber to the periodontal ligament in the furcation region of a multi-rooted tooth (Vertucci and Williams 1974) may also

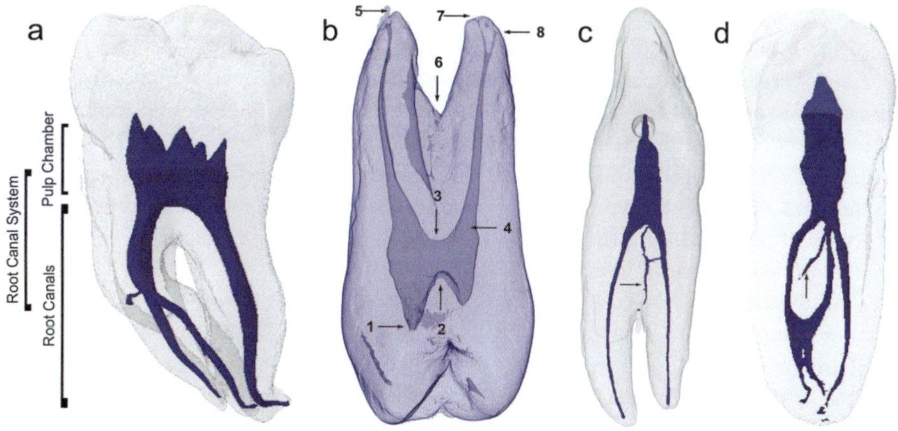

Fig. 1 (**a**) The root canal system is divided into the pulp chamber, located in the anatomic crown of the tooth, and the root canals, within the root. (**b**) Other important anatomical endodontic landmarks are the pulp horns (1), the pulpal roof (2), the pulpal floor (3), the canal orifice (4), the anatomic apex (5), the furcation (6), the apical foramen (7), the accessory foramina, and furcation canals (*arrows* in **c** and **d**)

occur. Furcation canals are derived from entrapment of periodontal vessels during the fusion of the parts of the diaphragm, which will become the floor of the pulp chamber (Cutright and Bhaskar 1969). They represent a through communication between the pulp canal space and the periodontal tissues and may be the cause of primary endodontic lesions in the interradicular region of multi-rooted teeth (Seltzer et al. 1967). Vertucci and Williams (1974) observed the presence of furcation canals in 13 % of mandibular first molars. In most of them, the canal extended from the center of the pulpal floor, but also from the mesial and distal aspects of the floor. Later, Vertucci and Anthony (1986) observed the presence of foramina on both the pulp-chamber floor and the furcation surface in 36 % of maxillary first molars, 12 % of maxillary second molars, 32 % of mandibular first molars, and 24 % of mandibular second molars. Recently, micro-computed tomographic studies have demonstrated the presence of furcation canals in two-rooted mandibular canines (Fig. 1c) (Versiani et al. 2011a) and three-rooted mandibular premolars (Fig. 1d) (Ordinola-Zapata et al. 2013).

Missed anatomy is one of the main causes of endodontic failure. In this way, special attention needs to be addressed to the anatomy of the pulp chamber in order to locate all the root canal orifices. After evaluating 500 pulp chambers of extracted teeth, Krasner and Rankow (2004) proposed some laws for aiding in the determination of the pulp-chamber position and the exact location and number of root canals in any individual tooth. They demonstrated that specific and consistent pulp-chamber floor and wall anatomy exist and proposed laws for assisting clinicians identify canal morphology. These laws are:

- *Law of centrality*: The floor of the pulp chamber is always located in the center of the tooth at the level of the cementoenamel junction (CEJ).
- *Law of concentricity*: The walls of the pulp chamber are always concentric to the external surface of the tooth at the level of the CEJ, i.e., the external root surface anatomy reflects the internal pulp-chamber anatomy.
- *Law of the CEJ*: The distance from the external surface of the clinical crown to the wall of the pulp chamber is the same throughout the circumference of the tooth at the level of the CEJ—the CEJ is the most consistent, repeatable landmark for locating the position of the pulp chamber.
- *Law of symmetry 1*: Except for maxillary molars, the orifices of the canals are equidistant from a line drawn in a mesial-distal direction, through the pulp-chamber floor.
- *Law of symmetry 2*: Except for the maxillary molars, the orifices of the canals lie on a line perpendicular to a line drawn in a mesial-distal direction across the center of the floor of the pulp chamber.
- *Law of color change*: The color of the pulp-chamber floor is always darker than the walls.
- *Law of orifice location 1*: The orifices of the root canals are always located at the junction of the walls and the floor.
- *Law of orifice location 2*: The orifices of the root canals are located at the angles in the floor-wall junction.
- *Law of orifice location 3*: The orifices of the root canals are located at the terminus of the root developmental fusion lines.

2.2 Root Canal

The root canal is the portion of the pulp space within the root of the tooth that tapers toward the apex, following the external outline of the root. The main root canal usually has a funnel shape and is ovoid in its cross section, having the greatest diameter at the orifice level. In general, canals are wider buccolingually than in the mesiodistal plane in longitudinal section. As previously mentioned, the anatomy of the root canal system is often complex, and the presence of a tapering canal and a single foramen is the exception rather than the rule (Vertucci 2005). Considering the large amount of dissimilarities in the classification of canal configuration, avoiding the comparison of the results among the reports, different classification systems have been proposed (Weine et al. 1969; Vertucci et al. 1974; Weine 1996; Gulabivala et al. 2001; Ng et al. 2001; Gulabivala et al. 2002; Sert and Bayirli 2004; Al-Qudah and Awawdeh 2006; Peiris et al. 2008; Gu et al. 2011; Kim et al. 2013; Leoni et al. 2013). These systems are based on the number of root canals that begin at the pulp-chamber floor, arise along the course of the canal, and open through an apical foramen (Vertucci 2005).

2.3 Root Canal System Configurations

Together with diagnosis and treatment planning, the knowledge of common root canal morphology and its frequent variations is a basic requirement for endodontic success (Vertucci 2005). Its significance is also highlighted by studies demonstrating that variations in the preoperative canal geometry, such as the presence of oval-shaped canals or isthmuses, had a greater effect on the changes that occurred during preparation than the instrumentation techniques themselves (Peters et al. 2001).

Weine et al. (1969) provided the first clinical classification of more than one canal system in a single root. He studied mesiobuccal roots of extracted maxillary first molars by the sectioning and radiographic methods and classified the canal configuration into four categories:

- Type I: a single canal from the pulp chamber to the apex (1 configuration)
- Type II: two separate canals leaving the chamber but merging short of the apex to form a single canal (2–1 configuration)
- Type III: two distinct canals from the pulp chamber to the apex (2 configuration)
- Type IV: a single canal leaving the chamber and dividing into two separate canals at the apex (1–2 configuration)

Vertucci et al. (1974) studied 200 cleared maxillary second premolars which had their pulp cavities stained with dye and found a much more complex systems, identifying eight different configurations:

- Type I: a single canal from the pulp chamber to the apex (1 configuration)
- Type II: two separate canals leaving the chamber but merging short of the apex to form a single canal (2–1 configuration)
- Type III: a single canal that divides into two and subsequently merges to exit as one (1–2–1 configuration)
- Type IV: two distinct canals from the pulp chamber to the apex (2 configuration)
- Type V: a single canal leaving the chamber and dividing into two separate canals at the apex (1–2 configuration)
- Type VI: two separate canals leaving the pulp chamber, merging in the body of the root, and dividing again into two distinct canals short from the apex (2–1–2 configuration)
- Type VII: a single canal that divides, merges, and exits into two distinct canals short from the apex (1–2–1–2 configuration)
- Type VIII: three distinct canals from the pulp chamber to the apex (3 configuration)

Despite the efforts to systematize the diversity of canal configurations, authors studying other groups of teeth in different racial groups observed additional types of root canal morphology (Gulabivala et al. 2001; Ng et al. 2001; Gulabivala et al. 2002; Sert and Bayirli 2004; Al-Qudah and Awawdeh 2006; Peiris

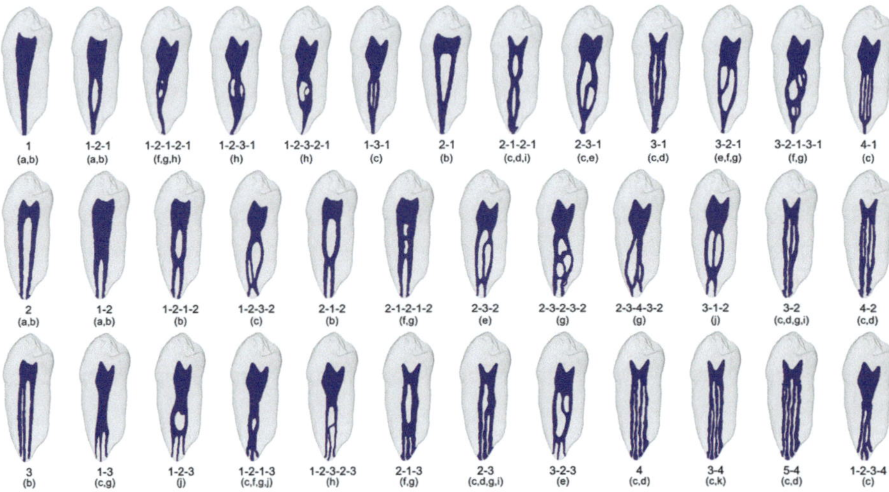

Fig. 2 Diagrammatic representation of root canal configurations, organized by the number of root canals at the apex, found by (**a**) Weine et al. (1969), (**b**) Vertucci et al. (1974), (**c**) Sert and Bayirli (2004), (**d**) Gulabivala et al. (2001), (**e**) Al-Qudah and Awawdeh (2006), (**f**) Gu et al. (2011), (**g**) Kim et al. (2013), (**h**) Leoni et al. (2014), (**i**) Ng et al. (2001), (**j**) Peiris et al. (2008), and (**k**) Gulabivala et al. (2002)

et al. 2008; Gu et al. 2011; Kim et al. 2013; Leoni et al. 2013). Figure 2 summarizes the varieties of root canal configurations found by different authors, organized by the number of root canals at the apex.

With the improvements of the imaging methods to evaluate the root canal anatomy, such as the micro-CT technology and the magnification approach in clinical practice, an increase in the reporting of complex anatomy is expected. To the clinician, it is important to be aware of the variability in the root canal configuration, because it calls the attention to their existence so that the anatomy may be properly recognized and treated (Vertucci 2005).

2.4 *Root Canal Anomalies*

Although statistics regarding the global prevalence of the anatomical configurations described in Fig. 2 are not available, it probably embraces 99 % of all variations. However, anomalous root and root canal morphology that do not correspond to any of those reported configurations can be found mostly associated with posterior teeth with varying frequency. Dental anomalies are formative defects due to genetic disturbances during the morphogenesis of the teeth. Anomalies may occur during the developmental stages of the tooth, which are manifested clinically in the later life once the tooth is fully formed (Woelfel and Scheid 2002; Hargreaves and Cohen 2011; Nanci and Ten Cate 2013). Failure to diagnose the teeth with

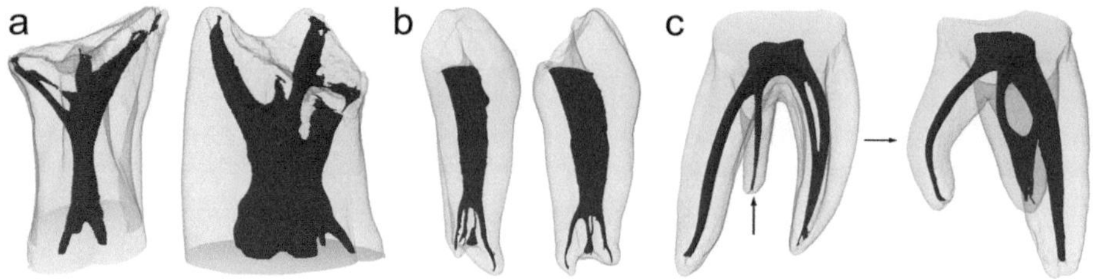

Fig. 3 Frontal and lateral views of molar (**a**) and premolar (**b**) taurodont teeth presenting a large pulp chamber with apical displacement of the pulpal floor and furcation of the roots; (**c**) frontal and lateral views of the roots of a mandibular first molar presenting a curved radix entomolaris (*arrows*)

anomalous anatomy may lead to misdiagnosis and a treatment plan that could cause permanent damage and tooth loss (Vertucci 2005). In this way, the clinician must be aware of the existence of some anatomical anomalies in order to implement an appropriate treatment plan. The most common anomalies with impact on endodontic practice include taurodontism, *dens invaginatus*, *radix entomolaris*, *radix paramolaris*, and C-shaped canals.

2.4.1 Taurodontism

Taurodontism is a dental morphologic variation in which the body of the tooth is enlarged and the roots are reduced in size. A taurodont tooth presents a large pulp chamber with apical displacement of the pulpal floor and furcation of the roots (Fig. 3a, b) (AAE 2012). The etiology of taurodontism is unclear. It is thought to be caused by the failure of Hertwig's epithelial sheath diaphragm to invaginate at the proper horizontal level, resulting in a tooth with normal dentin, short roots, elongated body, and enlarged pulp (Woelfel and Scheid 2002; Regezi et al. 2008; Nanci and Ten Cate 2013). The teeth involved are almost invariably molars or premolars. The condition can be uni- or bilateral and may affect single or multiple teeth (Sert and Bayrl 2004). Taurodontism has been graded according to its severity: least pronounced (hypotaurodontism), moderate (mesotaurodontism), and most severe (hypertaurodontism) (Shaw 1928). Clinically, the crowns of these teeth have normal characteristics; therefore, the diagnosis is radiological (Sert and Bayrl 2004). Owing to the complexity of the root canal anatomy and the proximity of the orifices to the root apex, the root canal filling procedures may be challenging. Because the pulp of a taurodont is usually voluminous, control of bleeding in case of pulpitis may take some time and effort compared to teeth with normal anatomy. Additional efforts such as application of ultrasonic combined with sodium hypochlorite as irrigant solution should be made to dissolve as much organic material as possible (Tsesis et al. 2003; Sert and Bayrl 2004; Metgud et al. 2009).

2.4.2 Dens Invaginatus/Dens Evaginatus

Dens invaginatus is a developmental defect resulting from invagination in the surface of the tooth crown before calcification has occurred. Clinically, it may appear as an accentuation of the lingual pit in the anterior teeth and, in its more severe form, gives a radiographic appearance of a tooth within a tooth, hence the term *dens in dente* (AAE 2012). Its etiology malformation is controversial and remains unclear. The affected teeth radiographically show an infolding of enamel and dentin which may extend deep into the pulp cavity and into the root and sometimes even reach the root apex (Hülsmann 1997). The most common associated clinical finding is an early pulpal involvement, explained by the existence of a canal extending from the invagination into the pulp (Ridell et al. 2001). The invagination also allows entry of irritants into an area which is separated from pulpal tissue by only a thin layer of enamel and dentin and presents a predisposition for the development of dental caries (Hülsmann 1997). Therefore, this condition must be recognized early and the tooth prophylactically restored (Metgud et al. 2009). The variability of its root canal system configuration is unlimited. Clinically, however, it can only be speculated upon from radiographs (Rotstein et al. 1987). In this way, the most commonly referred classification was proposed by Oehlers (1957a) who categorized this anatomic variation into three types:

- Type 1: The invagination is confined to the crown and does not extend beyond the amelocemental junction.
- Type 2: The invagination extends past the cementoenamel junction and does not involve the periradicular tissues, but may communicate with the dental pulp.
- Type 3: The invagination extends beyond the cementoenamel junction and may present a second apical foramen, with no immediate communication with the pulp.

In the literature, the reported prevalence of this anomaly varies from 0.25 (Poyton and Morgan 1966) to 10 % (Ruprecht et al. 1987), and the most affected teeth are maxillary lateral incisors, despite it may occur in any tooth (Oehlers 1957b; Hülsmann 1997). This high range frequency of *dens invaginatus* has been associated with the study design, sample size and composition, and diagnostic criteria (Hülsmann 1997; Ridell et al. 2001).

Dens evaginatus is an anomalous outgrowth of tooth structure resulting from the folding of the inner enamel epithelium into the stellate reticulum with the projection of structure exhibiting enamel, dentin, and pulp tissue (AAE 2012). It arises most frequently from the occlusal surface of involved posterior teeth, mainly premolars, and primarily from the lingual surface of associated anterior teeth (Oehlers et al. 1967; Levitan and Himel 2006). Its etiology remains unclear; however, it predominantly occurs in people of Asian descent with varying estimates reported from 0.5 (Kocsis et al. 2002) to 15 % (Merrill 1964), depending upon the population group studied. The presence of pulp within the cusp-like tubercle has great clinical significance. Because the tubercle may extend above the occlusal surface, malocclusion with the opposing tooth may cause abnormal wear or fracture of the tubercle, and this is how pulp exposure occurs (Levitan and Himel 2006). Subsequent pulpal inflammation or infection will most likely ensue. It is important for the clinician to be able to recognize and treat the entity soon after the affected teeth have erupted into the oral cavity in order to avoid the development of pathological conditions (Levitan and Himel 2006).

2.4.3 Radix Entomolaris/Radix Paramolaris

Radix entomolaris has been defined as a supernumerary root on a mandibular molar located distolingually, while *radix paramolaris* is an extra root located mesiobuccally (AAE 2012). The presence of these anatomical anomalies has been associated with certain populations with Mongoloid traits, such as Chinese, Eskimo, and American Indians (De Moor et al. 2004; Calberson et al. 2007). *Radix paramolaris* is a very rare structure, and its prevalence was found to be 0, 0.5, and 2 % for the mandibular first, second, and third molars, respectively (Carlsen and Alexandersen 1991), while *radix entomolaris* occurs with a higher frequency, ranging from 0.2 % (Tratman 1938) to 32 % (Turner 1971) of the studied sample. In radix molars, each root usually contains a single root canal. The orifice of the *radix entomolaris* is located disto- to mesiolingually from the main canal or canals of the distal root; the orifice of the *radix paramolaris* is located mesio- to distobuccally from the main mesial canals (Calberson et al. 2007). A dark line or groove from the main root canal on the pulp-chamber floor leads to these orifices (De Moor et al. 2004); however, they provide a limited practical aid for its identification in clinical practice (Wang et al. 2011). These anatomic variations present definite challenges to therapy because of their orifice inclination and root canal curvature (Fig. 3c). In this way, preoperative periapical radiographs at different horizontal angles or a CBCT exam are required to identify this additional root, which will also result in a modified opening cavity. An accurate diagnosis of these anatomic variations is important to avoid missed canals.

2.4.4 C-Shaped Canals

The C-shaped configuration was firstly reported in the endodontic literature by Cooke and Cox (1979) ; however, this canal configuration is well known since the early twentieth century (Keith and Knowles 1911). This anatomic variation is so named for the root and root canal cross-sectional shape of the capital letter “C” (Jafarzadeh and Wu 2007). Its main feature is the presence of one or more isthmuses connecting individual canals, which can change the cross-sectional and three-dimensional canal shape along the root (Melton et al. 1991; Jafarzadeh and Wu 2007; Fan et al. 2008). Typically, this configuration is found in teeth with fusion of the roots either on its buccal or lingual aspect and results from the failure

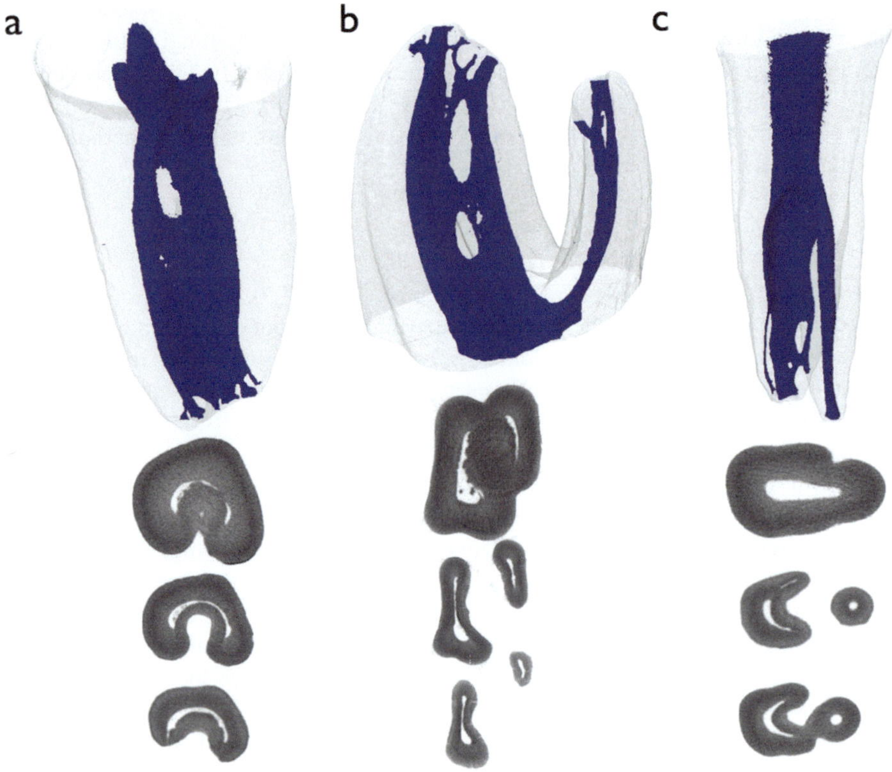

Fig. 4 Three-dimensional and cross-sectional micro-CT models of a C-shaped mandibular second molar (**a**), a C-shaped maxillary first molar with palatal and distobuccal radicular fusion (**b**), and a C-shaped mandibular first premolar (**c**)

of Hertwig's epithelial sheath to develop or fuse in the furcation area during the developing stage of the teeth (Woelfel and Scheid 2002; Nanci and Ten Cate 2013). Failure on the buccal side will result in a lingual groove, and the opposite cases would be possible (Nanci and Ten Cate 2013). In such teeth, the floor of the pulp chamber is frequently situated deeply and may assume an unusual anatomic appearance (Fan et al. 2008). Below the orifice level, the root structure of a C-shaped tooth can harbor a wide range of anatomic variations (Jafarzadeh and Wu 2007), which make it a challenge with respect to disinfection (Fig. 4a) (Solomonov et al. 2012). This variation may occur in different types of teeth (Boveda et al. 1999; Yilmaz et al. 2006; Cleghorn et al. 2008; Fan et al. 2008; Fan et al. 2012; Gu et al. 2013b); however, it is most commonly found in mandibular second molars (Fig. 4b) (Yang et al. 1988b; Manning 1990; Min et al. 2006; Rahimi et al. 2008; Zheng et al. 2011) with a reported prevalence ranging from 2.7 (Weine et al. 1988) to 44.5 % (Jin et al. 2006). There is significant ethnic variation in the frequency of C-shaped molars, being much more common in Asians than in Caucasians (Vertucci 2005). In Asian population, the reported prevalence was

10.6 % in Saudi Arabians (Al-Fouzan 2002), 19.14 % in Lebanese (Haddad et al. 1999), 31.5 % in Chinese (Yang et al. 1988b), and 44.5 % in Koreans (Jin et al. 2006). To date, only a few studies have addressed the efficacy of different systems in the preparation of C-shaped mandibular molar canals and showed a significant percentage of canal area unaffected by the instrumentation procedure (Cheung and Cheung 2008; Solomonov et al. 2012).

In 1991, Melton et al. proposed the first classification system for C-shaped canal configurations in mandibular second molars based on its cross-sectional shape:

- Category I: a continuous C-shaped canal running from the pulp chamber to the apex
- Category II: a semicolon-shaped orifice in which dentin separates a main C-shaped canal from one mesial distinct canal
- Category III: two or more discrete and separate canals which could join in the apical (subdivision I), middle (subdivision II), or coronal (subdivision III) thirds

It is important to point out that C-shaped mandibular molar teeth can present irregularities in their canal systems throughout the root, and the presence of these categories may vary from the pulp chamber to the apex (Jin et al. 2006). In this way, Fan et al. (2004) modified Melton's method and recommended to classify each portion of the same tooth using five categories:

- Category I: The shape was an uninterrupted "C" with no separation or division.
- Category II: The canal shape resembled a semicolon resulting from a discontinuation of the "C" outline.
- Category III: Two or three separate canals.
- Category IV: Only one round or oval canal in the cross section (normally found near the apex).
- Category V: No canal lumen (usually seen near the apex only).

Melton's classification (Melton et al. 1991) stated that categories II and III have separated canals, but no description was provided to differentiate them. In this modified classification, one of the canals in the category II would appear as an arc and would be more likely to extend into the "fused" area of the root where the dentin wall may be quite thin (Fan et al. 2004).

Based on radiographic features revealed by intraradicular contrast medium, three types of C-shaped canal systems of mandibular second molars were also described by Gao et al. (2006):

- Type I (merging type): Canals merge into one major canal before exiting at the apical foramen; partial dentin fusion area may appear in the coronal and (or) middle third of the root.
- Type II (symmetrical type): Separated mesial and distal canals are located at the mesial and distal parts of the root, respectively. From the buccolingual view, symmetry of the mesial and distal canals is present along the longitudinal axis of the root.
- Type III (asymmetrical type): Separate mesial and distal canals are evident. From a buccolingual view, the distal canal may have a large isthmus across the

furcation area, which commonly makes the mesial and distal canals asymmetrical.

Based on the shape of the pulp-chamber floor and the location of the dentin fusion, Min et al. (2006), using micro-CT technology, classified the pulp floor and orifice into four types:

- Type I: A peninsula-like floor with a continuous C-shaped orifice.
- Type II: A buccal, strip-like dentin connection exists between the peninsula-like floor and the buccal wall of the pulp chamber that separates the C-shaped groove into mesial and distal orifices. Sometimes the mesial orifice was separated into a mesiobuccal and a mesiolingual orifice by another strip-like dentin between the peninsula-like floor and the mesial wall of the pulp chamber.
- Type III: Only one mesial, strip-like dentin connection exists between the peninsula-like floor and the mesial wall, which separates the C-shaped groove into a small mesiolingual orifice and a large mesiobuccal-distal orifice. The mesiobuccal-distal orifice was formed by the merging of the mesiobuccal orifice and the distal orifice.
- Type IV: Non-C-shaped floors. One distal canal orifice and one oval or two round mesial canal orifices are present.

The presence of C-shaped canal anatomy has also been reported in third molars (Sidow et al. 2000), lateral incisors (Boveda et al. 1999; Gu 2011), mandibular first premolars (Cleghorn et al. 2008; Fan et al. 2008; Gu et al. 2013a, b; Ordinola-Zapata et al. 2013), mandibular first molars (Rice and Gilbert 1987), and maxillary first (Newton and McDonald 1984; De Moor 2002; Martins et al. 2013) and second (Yang et al. 1988a) molars. In maxillary molars, C-shaped canal configuration is a rare condition resulted from the fusion of the roots, and its prevalence appears to be below 1 % (De Moor 2002; Cleghorn et al. 2006; Martins et al. 2013). When the radicular fusion does not involve the mesiobuccal root, an additional canal in this root, the so-called MB2, can also be present (Fig. 4b). A recent review has identified three types of C-shaped configuration in maxillary molars, based on the fusion of the roots (Martins et al. 2013):

- Type A: fusion between palatal and distobuccal canals
- Type B: fusion between mesiobuccal and distobuccal canals
- Type C: fusion between two palatal canals

Mandibular first premolars present a variety of root canal configurations that include the presence of a C-shaped configuration system (Fig. 4c) (Vertucci 1984; Ordinola-Zapata et al. 2013). As in mandibular molars, C-shaped canal systems in mandibular first premolar vary among different ethnic groups with its prevalence being reported to range from 1 % (Yu et al. 2012) to 18 % (Lu et al. 2006) in different ethnic populations. This configuration has been highly associated with the presence of groove or concavity on the external root surface and Vertucci's type V configuration (Fan et al. 2012), i.e., a single canal that bifurcates at the middle third (Fan et al. 2008; Ordinola-Zapata et al. 2013). Radicular grooves in mandibular first premolars usually began 3 mm from the CEJ on the proximal lingual area of the middle root, not always extending to the root apex (Lu et al. 2006; Fan et al. 2008).

Preoperative diagnosis of C-shaped canals is complex, mainly because these unique anatomic features are not easily recognized on a traditional two-dimensional periapical radiograph (Solomonov et al. 2012). With the increased use of CBCT scanning, clinicians may be able to detect C-shaped canals before endodontic treatment. Nevertheless, even when recognized, the disinfection procedure will remain a challenge, mostly because of the isthmus areas. Irregular areas in a C-shaped canal that may house soft-tissue remnants or infected debris may escape thorough cleaning and may be a source of bleeding and severe pain (Fan et al. 2009). In this way, the use of dental microscope associated with sonic or ultrasonic instrumentation techniques may make treatment outcome more predictable (Vertucci 2005). Because of its challenging morphology, the C-shaped canal anatomy would increase the difficulty in root canal therapy and may account for the frequent occurrence of endodontic treatment failure (Lu et al. 2006).

2.5 *Isthmuses*

An isthmus is a narrow, ribbon-shaped communication between two root canals that may contain vital tissue, necrotic pulp, biofilms, or residual filling material (Fig. 5) (Weller et al. 1995; Vertucci 2005). It has been also reported as lateral interconnection or transverse anastomosis (Manning 1990). Any root that contains two or more root canals has the potential to contain an isthmus (Vertucci 2005). Researchers have also considered the presence of a partial isthmus, which was described as an incomplete communication with one or more patent openings

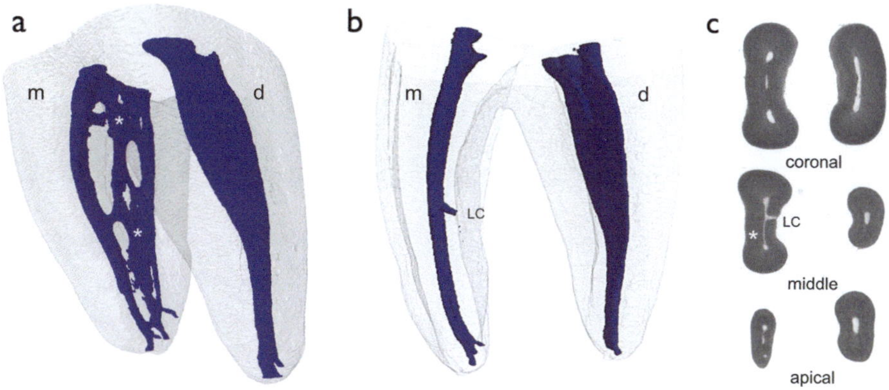

Fig. 5 (**a**) Lateral, (**b**) frontal, and (**c**) cross-sectional views of the mesial (m) and distal (d) roots of a micro-CT model of a mandibular first molar showing the presence of isthmuses (*) and a lateral canal (LC)

between two main canals (Weller et al. 1995). Studies in posterior teeth have shown the presence of necrotic pulp tissue and biofilms after chemomechanical preparation indicating that conventional disinfection methods have limited results in these hard-to-reach areas (Adcock et al. 2011; Endal et al. 2011). If surgery becomes necessary, the natural anatomy of the root canal is altered, and additional anatomic features such as an undebrided isthmus may become exposed and need to be addressed (Leoni et al. 2013). In this case, surgical microscopy and ultrasonic root-end preparation would help the clinician to better visualize the apex, incorporating both canals and the isthmus into the root-end preparation to ensure a proper debridement and sealing of the root canal system (Weller et al. 1995; Hsu and Kim 1997; Mauger et al. 1998; Adcock et al. 2011).

The knowledge about the morphology and prevalence of root canal isthmuses in posterior teeth is relevant for nonsurgical and surgical endodontic treatment (Hsu and Kim 1997). A summary of the isthmus prevalence in mandibular and maxillary molars found in laboratory studies using cross-sectioning and micro-CT methods is shown in Table 1.

Isthmuses may present different configurations, and its prevalence is dependent on the type of teeth, the root level, and the patient's age. Hsu and Kim (1997) classified the isthmus configuration into five types:

- Type I: Two canals with no notable communication.
- Type II: A hair-thin connection between the two main canals.
- Type III: Differs from type II because of the presence of three canals instead of two.
- Type IV: An isthmus with extended canals into the connection.
- Type V: There is a true connection or wide corridor of tissue between two main canals.

Fan et al. (2010), using micro-CT technology, described the configuration of the isthmuses in the mesial roots of mandibular molars into four categories:

- Type I (sheet connection): Narrow sheet and complete connection existing between two canals from the top to bottom of the isthmus. Sometimes, one or more small dentin fusions were discerned in the isthmus area.
- Type II (separate): Narrow but incomplete connection existing between two canals from the top to bottom of the isthmus.
- Type III (mixed): Incomplete isthmus existing above and/or below a complete isthmus.
- Type IV (cannular connection): Narrow cannular communication between two canals.

Isthmuses connecting multiple canals are the type of anatomical configuration that presents a clinical challenge directly related to the irrigating protocols because all preparation techniques often leave behind accumulated hard- and soft-tissue remnants as well as microorganisms in these hard-to-reach areas. The recognition and treatment of root canal isthmuses may be one factor that reduces the failure rate of endodontic treatment (Hsu and Kim 1997). Usually, a large amount of the

Table 1 Summary of laboratory studies on the prevalence and position of isthmuses in the mesial root canal system of mandibular molars and mesiobuccal (MB) root canal system of maxillary molars

Authors	Methodology	Conclusion
Weller et al. (1995)	Transverse serial 1-mm-thick sections from the apical 6 mm of 50 MB roots of maxillary first molars were examined under surgical operating microscope	The mean frequency of isthmuses was 30, 65, 90, 100, 82, and 81 % from 1- to 6-mm levels from the apex
Teixeira et al. (2003)	Transverse serial 1-mm-thick sections from the apical 6 mm of 50 MB roots of maxillary molars and 50 mesial roots of mandibular first molars were examined through a light microscope	Maxillary molars: the mean frequency of isthmuses was 11.4, 20, 31.5, and 23.6 % from 3 to 6 mm from the apex; mandibular molars: the mean % frequency of isthmuses was 6.6, 11.9, 20.5, 30.3, 33.3, and 32.4 % from 1 to 6 mm from the apex
Mannocci et al. (2005)	800 cross sections from the apical 5 mm of 20 mesial roots of mandibular first molars were investigated by means of micro-CT technology	The mean frequency of isthmuses was 17.2, 36.7, 50.2, 33, and 34.7 % from 1 to 5 mm from the apex
Jung et al. (2005)	Transverse serial 1-mm-thick sections from the apical 5 mm of 47 MB roots of maxillary first molars and 50 mesial roots of mandibular first molars were examined through microscope	Maxillary molars: the mean frequency of isthmuses was 52.7, 52.6, 63.2, and 44.7 % from 2 to 5 mm from the apex; mandibular molars: the mean frequency of isthmuses was 60, 80, 82.5, and 70 % from 2 to 6 mm from the apex
Degerness and Bowles (2008)	Transverse serial 0.44-mm-thick sections from the apex of 153 MB roots of maxillary first and second molars were examined through stereomicroscope	The mean percentage of sections with isthmus up to 5 mm from the apex was 8.5 (0.64 mm), 35.3 (1.62 mm), 43.8 (2.15 mm), 51.6 (3.12 mm), 60.2 (3.64 mm), 66.7 (4.58 mm), and 76.5 % (5.1 mm)
Fan et al. (2010)	Cross-sectional images from the apical 5 mm of mesial roots of mandibular first ($n = 70$) and second ($n = 56$) molars, respectively, were investigated by means of micro-CT technology	The frequency of isthmus in the apical 5 mm of the mesial roots was 85 %. Mandibular first molars had more isthmuses with separate and mixed types, whereas second molars had more isthmuses with sheet connections
Villas-Boas et al. (2011)	1.0-mm increment cross-sectional images from the apical 4 mm of mesial roots of 60 mandibular molars were investigated by means of micro-CT technology	The mean frequency of isthmuses was 45, 55, 71, and 68 % from 1 to 4 mm from the apex
Harris et al. (2013)	0.5-mm increment cross-sectional images from the apical 6 mm of mesial roots of 22 mandibular molars were investigated by means of micro-CT technology	Isthmuses were found along the length of all mesial roots. On average, isthmus started at 4-mm level and ended 8.1 mm from the apex

isthmus area is left unaffected by the endodontic procedures (Siqueira et al. 2013). Besides, the dentin debris created during canal preparation may prevent the antibacterial activity of irrigant solution such as sodium hypochlorite or chlorhexidine because they do not have a significant effect against inorganic compounds. These findings support the clinical observation that isthmuses in failure cases were not adequately filled, probably because of the presence of debris or organic tissue in these areas (von Arx 2005). Currently, with the advent of microscopic endodontic techniques, it is possible to visualize, identify, and treat most of the isthmus areas with thin ultrasonic tips in both surgical and nonsurgical endodontic procedures.

2.6 Accessory and Lateral Canals

Accessory canal is referred here as any branch of the main pulp canal or chamber that communicates with the external surface of the root, while a lateral canal is defined as an accessory canal located in the coronal or middle third of the root (Fig. 5b, c) (AAE 2012). They are formed after a localized fragmentation of the epithelial root sheath develops, leaving a small gap, or when blood vessels running from the dental sac through the dental papilla persist (Ricucci and Siqueira 2010; Nanci and Ten Cate 2013). Accessory canals comprise potential pathways through which bacteria and/or their products from the necrotic root canal might reach the periodontal ligament and cause disease, and likewise, bacteria from periodontal pockets might reach the pulp, being difficult to reach, clean, disinfect, and fill during the endodontic treatment (Ricucci and Siqueira 2010). De Deus (1975) studied the frequency, location, and direction of the accessory canals in 1140 teeth using the diaphanization technique and showed that 27.4 % of the sample ($n = 330$) had accessory canals, mostly in the apical area (17 %) and less frequently in the middle (8.8 %) and coronal (1.6 %) thirds. Vertucci (1984) evaluated 2400 teeth and also observed lower occurrence of canal ramifications in the middle (11.4 %) and coronal (6.3 %) thirds compared to the apical level (73.5 %). On the other hand, using histological sections, Ricucci and Siqueira (2010) observed a much higher overall frequency of canal ramifications (75 %). This may be explained due to differences in the sample origin and evaluation methods; however, in all of these studies (De Deus 1975; Vertucci 1984; Ricucci and Siqueira 2010), premolars and molars have showed the highest frequency of accessory canals. Recent micro-CT studies on the root canal anatomy of mandibular anterior teeth showed that lateral canals are rare and were observed only in 4 % of the canines (Leoni et al. 2013; Versiani et al. 2013a).

Lateral canals are not usually visible in preoperative radiographs, but its presence can be suspected when there is a localized thickening of the periodontal ligament or a lesion on the lateral surface of the root. Morphological studies have demonstrated that the largest diameter (10–200 μm) of the accessory foramina was up to three times smaller than the mean diameter reported for the main apical foramen in permanent maxillary and mandibular molars (Dammaschke et al. 2004).

While this difference might explain why apical periodontitis is observed far more than lateral periodontitis (Ricucci and Siqueira 2010), the presence of accessory foramina with large diameters may imply that an inflammatory process can spread from pulpal to periodontal tissues and vice versa (Dammaschke et al. 2004). Large and patent lateral foramina might allow a higher amount of bacteria to reach and contact a larger area of the lateral periodontal ligament to cause disease, while the quantity of bacterial irritants in small ramifications might be insufficient to induce significant disease to be radiographically discernible (Ricucci and Siqueira 2010).

According to Weine (1996), lateral lesions can be radiographically classified into three types:

- Type I: Lateral lesion with no apical lesion: as the infection progresses apically, it might reach a sufficiently large lateral canal to allow a substantial amount of bacteria and bacterial products to reach the lateral periodontium to cause inflammation.
- Type II: Separate lateral and apical lesions: if the pathological process advances without professional intervention, an apical periodontitis lesion can also be visible.
- Type III: Coalescence of lateral and apical lesions: in some cases, the type II condition can progress to the so-called "wraparound" lesion.

Actually, these types might represent different stages of the disease progression within the root canal space once the conditions of the tissue within lateral canals and apical ramifications reflect the conditions of the pulp in the main root canal (Ricucci and Siqueira 2010). Clinically, it is also relevant that lateral canals cannot be instrumented. In this way, its content can only be neutralized by means of effective irrigation with a suitable antimicrobial solution or with an additional use of an intracanal medication.

2.7 Apical Canal

A hallmark of the apical region is its variability and unpredictability, which have fed considerable controversy concerning the termination point for endodontic procedures and for the final size of enlarging instruments (Bergenholtz and Spångberg 2004; Vertucci 2005). The main root canal ends at the apical foramen (major foramen), which usually opens laterally on the root surface at a distance of between 0.2 and 3.8 mm from the anatomic apex (Gutierrez and Aguayo 1995). The apical foramen is the main apical opening in the root structure that communicates with the dental pulp and generally contains neural, vascular, and connective elements, while the anatomic apex is the tip or the end of the root as determined morphologically (AAE 2012). Apical foramen has been observed to coincide with the anatomic apex in a frequency that ranged from 6.7 to 46 % (Green 1956; Burch and Hulen 1972; Pineda and Kuttler 1972; Vertucci 1984; Mizutani et al. 1992). Its mean size varied from 0.21 to 0.39 mm (Morfis et al. 1994) and its shape can be asymmetrical

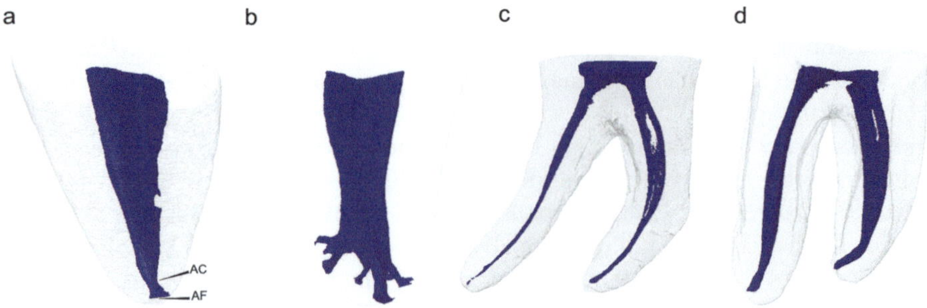

Fig. 6 Three-dimensional micro-CT models of root and root canals of different groups of teeth showing the presence of (**a**) apical constriction (AC), apical foramen (AF), and (**b**) apical delta, as well as, a gradual (**c**) and abrupt (**d**) apical curvature of the root canal

(Blašković-Šubat et al. 1992). The mesial root of mandibular molars, the maxillary premolar roots, and the mesiobuccal root of maxillary molars present the highest frequency of multiple apical foramina (Morfis et al. 1994). A previous study on the root apexes of all groups of permanent teeth has also demonstrated that the number of foramina on each root may vary from 1 to 16 (Gutierrez and Aguayo 1995).

The apical portion of the root canal having the narrowest diameter is called apical constriction (minor foramen) (AAE 2012). From the apical constriction, the canal widens as it approaches the apical foramen (Fig. 6a). The topography of the apical constriction is not constant (Dummer et al. 1984; Versiani et al. 2013a) and, when present, is usually located 0.5–1.5 mm from the center of the apical foramen (Vertucci 2005). The cementodentinal junction (CDJ) is the point at which the cemental surface terminates at or near the apex of a tooth and meets the dentin (AAE 2012). At this highly variable histological landmark that usually does not coincide with the apical constriction, pulp tissue ends and periodontal tissues begin (Ponce and Vilar Fernandez 2003). Despite canal cross sections have shown different shapes in different levels of the root in the same tooth, at the apical third, the canal used to be more round or slightly oval in shape in comparison with the middle and cervical thirds (Wu et al. 2000; Versiani et al. 2013a).

Another important configuration of the root canal at or near the apex is the apical delta, defined as a morphology in which the main canal divides into multiple accessory canals (Fig. 6b) (AAE 2012). In maxillary teeth, the percentage frequency of apical delta ranges from 1 % (central incisors) to 15.1 % (second premolars), while in mandibular teeth its frequency varies from 5 % (central incisors) to 14 % (distal root of first molars) (Vertucci 1984). Clinically speaking, the complex anatomical configuration of the apical delta, with several portals of exit which may harbor biofilms and infected dentinal tubules, makes disinfection procedure more challenging.

Knowledge of the root canal curvature is also an important factor for testing newly developed instruments and choosing the appropriate protocol for the cleaning and shaping procedures. Accidents reported after preparation of curved canals with different instruments including zip, separate instruments, ledge, and perforation, may considerably decrease the efficacy of the cleaning procedures at the apical third. Thus, preoperative determination of the degree of the canal curvature is of utmost importance. Nearly all root canals are curved in the apical third, particularly in a buccolingual direction, which is not evident on a standard radiograph (Vertucci 2005). Curvature may be a gradual curve of the entire canal (Fig. 6c) or a sharp curvature near the apex (Fig. 6d). Numerous methods have been proposed to determine, measure, and classify the root canal curvature (Backman et al. 1992; Cunningham and Senia 1992; Luiten et al. 1995; Nagy et al. 1995; Shearer et al. 1996; Weine 1996; Sert and Bayirli 2004), but the Schneider's method has been the most widely used. Schneider (1971) classified single-rooted permanent teeth according to the degree of curvature of the root, which was determined by firstly drawing a line parallel to the long axis of the canal and then a second line connecting the apical foramen to the point in the first line where the canal began to leave the long axis of the tooth. The angle formed by these two lines was the angle of curvature and its degree was classified as straight ($\leq 5°$), moderate (10° to 20°), or severe (25° to 70°). Later, Pruett et al. (1997) stated that two canals with similar angles measured by the Schneider method could have very different abruptness of curvatures, i.e., the shorter the length of curvature, the more abrupt the deviation at same the degree of curvature, hence representing a smaller radius of curvature. Thus, authors suggested that a more precise method for describing the canal curvature would be to consider both the degree and the radius of curvature.

Clinically, different angled views are necessary to determine the presence, direction, and severity of the root canal curvature. Schäfer et al. (2002) evaluated radiographically the degree and radius of curvature of 1163 root canals from all groups of teeth. The degree of curvature ranged from 0° to 75° and from 0° to 69° degrees in a clinical and proximal views, respectively. The highest median degree of curvature was observed in the clinical view of the mesiobuccal canal of maxillary molars and in the mesial canals of mandibular molars. The smallest median radii of curvature were 2.1 and 1.3 mm, observed in the clinical view of the palatal root of maxillary first molars and in the proximal view of mandibular first premolars, respectively. In several cases, the medians of the angles of proximal curvatures were greater than those of the clinical view curves. Additionally, a secondary curvature (S-shaped canal) was observed in 12.3 and 23.3 % of the maxillary and mandibular teeth, respectively.

3 Microscopic Anatomy of the Root Canal System

Dentin has been defined as a mineralized tissue that forms the bulk of the crown and root of the tooth, and its composition is approximately 67 % inorganic, 20 % organic, and 13 % water. It surrounds coronal and radicular pulp, forming the walls of the pulp chamber and root canals presenting porous irregularities known as dentinal tubules. Dentinal tubules are circular ducts in the dentin matrix that contains an odontoblastic process and fluid. A typical dentinal tubule has a diameter of 3–4 μm at its pulpal end and about 1 μm at the dentinoenamel or dentinocemental

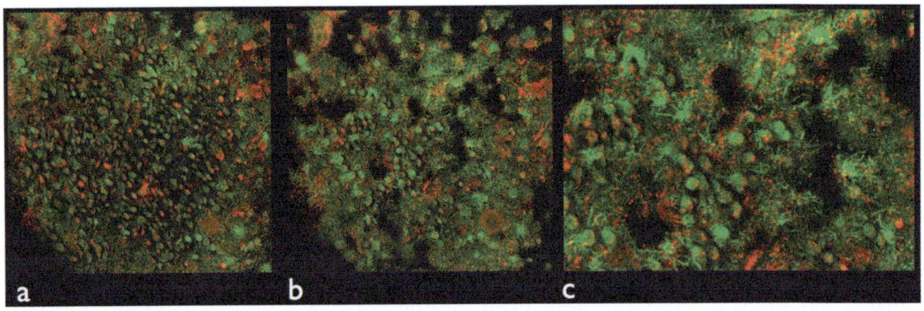

Fig. 7 Confocal laser scanning microscopy of biofilm-infected dentin. The images show the adherence of several morphotypes to the dentinal tubules and peritubular dentin. Dentin was labeled with the live/dead fluorescence kit: green represents live cells and red dead cells. (**a, b**) Deep and superficial layer of the attached biofilm (scanned area: 275 × 275 μm). Magnification of the infected dentin (**c**) shows a dense infection of bacteria organized in colonies of different morphotypes

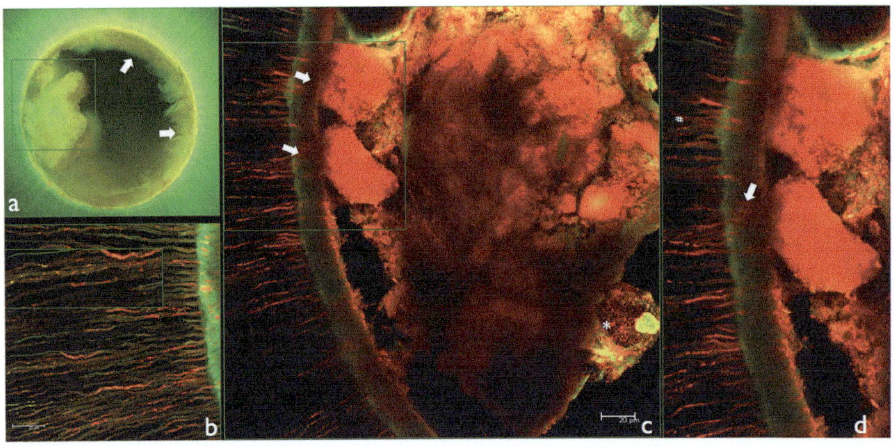

Fig. 8 Necrotic teeth of experimentally induced apical periodontitis in an animal model. The sample was stained using the acridine orange dye; the microbial RNA is usually stained in red/orange. (**a**) Panoramic view of the root canal content showing a large mass of necrotic tissue and organic structures attached to root canal walls (*arrows*). (**b**) Magnification of the dentin showing a dense infection within dentinal tubules. (**c**) Biofilm/dentin interface. A complex microbial amorphous structure heavily stained attached to the root canal wall (*square*). (**c**) Magnification of the dentin/biofilm interface (shown in **b**) revealing microbial cells invading the predentin and several dentinal tubules (*arrow*)

junction. The number of dentinal tubules ranges from 20,000 to 75,000 per square millimeter of dentin from the apex to the coronal level, respectively (AAE 2012; Nanci and Ten Cate 2013).

The dentin is usually exposed to the oral cavity by caries or fracture. These conditions expose dentin to the oral microorganisms, which may adhere to its surface, colonize, and invade the dentinal tubules (Figs. 7 and 8). Some factors can influence the infection of the dentin; basically, colonization is limited by dentin permeability and the bacteria ability to adhere to dentin substrate (Love and Jenkinson 2002). Morphological analysis of the dentinal tubules in the apical area showed that fine (300–700 mm in diameter) and micro (25–200 mm diameter) tubular branches, which run at 45° and 90° to the main tubules, respectively, were also frequently present. On the other hand, in some areas no tubules were observed (Mjör 2009). Additionally, in the presence of apical periodontitis, exposure of the dentinal tubules in the apical region may occur due to the development of external apical root resorption, which may favor the spreading of microorganisms, complicating the disinfection procedures at this level (Vertucci 2005).

4 Concluding Remarks

Outcomes of nonsurgical and surgical endodontic procedures are highly influenced by variations in complex canal anatomies. The high frequency of fins and communications between canals within the same root make it almost impossible for any mechanical or chemical technique to completely disinfect the root canal system. Some factors, such as aging, pathology, occlusion, and the secondary deposition of cement, can increase the aforementioned variations, thus, the purpose of the treatment must be toward reducing the level of contamination as much as possible and to entomb the remaining microorganisms. Careful interpretation of angled radiographs, proper access preparation, and a detailed exploration of the tooth, ideally under magnification, are essential prerequisites for a successful treatment outcome.

Spreading and flushing the irrigant throughout the canal space assume a pivotal role in treatment because it acts mechanically and chemically on remnants of necrotic pulp tissue and microbial biofilm communities. In order to circumvent limitations generated by the unpredictable anatomical configuration of the root canal, making cleaning and disinfection procedures more predictable, several instruments and techniques have been developed. Ideally, efficient irrigation solutions and protocols are required to provide fluid penetrability to such an extent as to accomplishing a microcirculation flow throughout the intricate root canal anatomy.

In laboratory-based studies, several experimental models have been used to understand the intracanal effect of irrigants by different irrigation protocols. It includes artificially created grooves, histological cross sections, computational fluid dynamics, and in vivo use of radiopaque solutions. These methodological approaches provide valuable information about the quality of cleaning and shaping procedures which cannot otherwise be obtained, but they are unable to show some critical factors, such as the volume of the solution or the root canal areas effectively touched by the irrigant.

An ideal experimental model should allow a reliable in situ volumetric quantitative evaluation of the root canal space, offering a deeper and comprehensive understanding on capabilities and limitations of different irrigation protocols.

Recently, a nondestructive experimental model that allows a two- and three-dimensional in situ quantification of several outcome parameters related to irrigation in the complex root canal space was proposed. These interesting aspects definitely open a new methodological appraisal to study the efficiency of irrigation procedures to combat microbial biofilm infections.

References

AAE (2012) Glossary of endodontics terms, 8th edn. American Association of Endodontists, Chicago

Adcock JM, Sidow SJ, Looney SW et al (2011) Histologic evaluation of canal and isthmus debridement efficacies of two different irrigant delivery techniques in a closed system. J Endod 37:544–548

Adloff P (1913) Das Durchsichtigmachen von Zähnen und unsere Wurzelfüllungsmethoden. Dtsch Monatsschr Zahnheifkd 31:445

Al-Fouzan KS (2002) C-shaped root canals in mandibular second molars in a Saudi Arabian population. Int Endod J 35:499–504

Al-Qudah AA, Awawdeh LA (2006) Root canal morphology of mandibular incisors in a Jordanian population. Int Endod J 39:873–877

Appel TR, Baumann MA (2002) Solid-state nuclear magnetic resonance microscopy demonstrating human dental anatomy. Oral Surg Oral Med Oral Pathol Oral Radiol Endod 94:256–261

Backman CA, Oswald RJ, Pitts DL (1992) A radiographic comparison of two root canal instrumentation techniques. J Endod 18:19–24

Baratto-Filho F, Fariniuk LF, Ferreira EL, Pécora JD, Cruz-Filho AM, Sousa-Neto MD (2002) Clinical and macroscopic study of maxillary molars with two palatal roots. Int Endod J 35:796–801

Barker BC, Lockett BC, Parsons KC (1969) The demonstration of root canal anatomy. Aust Dent J 14:37–41

Barker BC, Parsons KC, Mills PR, Williams GL (1973) Anatomy of root canals. I. Permanent incisors, canines and premolars. Aust Dent J 18:320–327

Barker BC, Parsons KC, Mills PR, Williams GL (1974a) Anatomy of root canals. III. Permanent mandibular molars. Aust Dent J 19:408–413

Barker BC, Parsons KC, Mills PR, Williams GL (1974b) Anatomy of root canals. II. Permanent maxillary molars. Aust Dent J 19:46–50

Barton DJ, Clark SJ, Eleazer PD, Scheetz JP, Farman AG (2003) Tuned-aperture computed tomography versus parallax analog and digital radiographic images in detecting second mesiobuccal canals in maxillary first molars. Oral Surg Oral Med Oral Pathol Oral Radiol Endod 96:223–228

Bauman R, Scarfe W, Clark S, Morelli J, Scheetz J, Farman A (2011) Ex vivo detection of mesiobuccal canals in maxillary molars using CBCT at four different isotropic voxel dimensions. Int Endod J 44:752–758

Baumann MA (1994) A new approach to demonstration of root canal anatomy. J Dent Educ 58:704–708

Baumann MA, Doll GM (1997) Spatial reproduction of the root canal system by magnetic resonance microscopy. J Endod 23:49–51

Baumann MA, Schwebel T, Kriete A (1993) Dental anatomy portrayed with microscopic volume investigations. Comput Med Imaging Graph 17:221–228

Bergenholtz G, Spångberg LS (2004) Controversies in endodontics. Crit Rev Oral Biol Med 15:99–114

Bergmans L, Van Cleynenbreugel J, Wevers M, Lambrechts P (2001) A methodology for quantitative evaluation of root canal instrumentation using microcomputed tomography. Int Endod J 34:390–398

Berutti E (1993) Computerized analysis of the instrumentation of the root canal system. J Endod 19:236–238

Bjørndal L, Carlsen O, Thuesen G, Darvann T, Kreiborg S (1999) External and internal macromorphology in 3D-reconstructed maxillary molars using computerized X-ray microtomography. Int Endod J 32:3–9

Black GV (1890) Descriptive anatomy of the human teeth. The Wilmington Dental Manufacturing Co, Philadelphia

Blašković-Šubat V, Maričić B, Sutalo J (1992) Asymmetry of the root canal foramen. Int Endod J 25:158–164

Blašković-Šubat V, Smojver B, Maričić B, Sutalo J (1995) A computerized method for the evaluation of root canal morphology. Int Endod J 28:290–296

Boveda C, Fajardo M, Millan B (1999) Root canal treatment of an invaginated maxillary lateral incisor with a C-shaped canal. Quintessence Int 30:707–711

Burch JG, Hulen S (1972) The relationship of the apical foramen to the anatomic apex of the tooth root. Oral Surg Oral Med Oral Pathol 34:262–268

Burger CL, Mork TO, Hutter JW, Nicoll B (1999) Direct digital radiography versus conventional radiography for estimation of canal length in curved canals. J Endod 25:260–263

Calberson FL, De Moor RJ, Deroose CA (2007) The radix entomolaris and paramolaris: clinical approach in endodontics. J Endod 33:58–63

Carabelli G (1842) Systematisches Handbuch der Zahnheilkunde II Anatomic des Mundes. Braumuller and Siedel, Vienna

Carlsen O, Alexandersen V (1991) Radix paramolaris in permanent mandibular molars: identification and morphology. Scan J Dent Res 99:189–195

Cheung LH, Cheung GS (2008) Evaluation of a rotary instrumentation method for C-shaped canals with micro-computed tomography. J Endod 34:1233–1238

Cimilli H, Cimilli T, Mumcu G, Kartal N, Wesselink P (2005) Spiral computed tomographic demonstration of C-shaped canals in mandibular second molars. Dentomaxillofac Radiol 34:164–167

Cleghorn BM, Christie WH, Dong CC (2006) Root and root canal morphology of the human permanent maxillary first molar: a literature review. J Endod 32:813–821

Cleghorn BM, Christie WH, Dong CC (2008) Anomalous mandibular premolars: a mandibular first premolar with three roots and a mandibular second premolar with a C-shaped canal system. Int Endod J 41:1005–1014

Cooke HG 3rd, Cox FL (1979) C-shaped canal configurations in mandibular molars. J Am Dent Assoc 99:836–839

Cunningham CJ, Senia ES (1992) A three-dimensional study of canal curvatures in the mesial roots of mandibular molars. J Endod 18:294–300

Cutright DE, Bhaskar SN (1969) Pulpal vasculature as demonstrated by a new method. Oral Surg Oral Med Oral Pathol 27:678–683

Dammaschke T, Witt M, Ott K, Schafer E (2004) Scanning electron microscopic investigation of incidence, location, and size of accessory foramina in primary and permanent molars. Quintessence Int 35:699–705

De Deus QD (1975) Frequency, location, and direction of the lateral, secondary, and accessory canals. J Endod 1:361–366

De Moor RJ (2002) C-shaped root canal configuration in maxillary first molars. Int Endod J 35:200–208

De Moor RJ, Deroose CA, Calberson FL (2004) The radix entomolaris in mandibular first molars: an endodontic challenge. Int Endod J 37:789–799

Degerness R, Bowles W (2008) Anatomic determination of the mesiobuccal root resection level in maxillary molars. J Endod 34:1182–1186

Dewey M (1916) Dental anatomy. C.V. Mosby Company, St. Louis
Dobó-Nagy C, Keszthelyi G, Szabó J, Sulyok P, Ledeczky G, Szabó J (2000) A computerized method for mathematical description of three-dimensional root canal axis. J Endod 26:639–643
Dowker SE, Davis GR, Elliott JC (1997) X-ray microtomography: nondestructive three-dimensional imaging for in vitro endodontic studies. Oral Surg Oral Med Oral Pathol Oral Radiol Endod 83:510–516
Dummer PM, McGinn JH, Rees DG (1984) The position and topography of the apical canal constriction and apical foramen. Int Endod J 17:192–198
Endal U, Shen Y, Knut A, Gao Y, Haapasalo M (2011) A high-resolution computed tomographic study of changes in root canal isthmus area by instrumentation and root filling. J Endod 37:223–227
Estrela C, Pereira HL, Pécora JD (1995) Radicular grooves in maxillary lateral incisor: case report. Braz Dent J 6:143–146
Fabra-Campos H (1989) Three canals in the mesial root of mandibular first permanent molars: a clinical study. Int Endod J 22:39–43
Fan B, Cheung GS, Fan M, Gutmann JL, Bian Z (2004) C-shaped canal system in mandibular second molars: part I–Anatomical features. J Endod 30:899–903
Fan B, Yang J, Gutmann JL, Fan M (2008) Root canal systems in mandibular first premolars with C-shaped root configurations. Part I: Microcomputed tomography mapping of the radicular groove and associated root canal cross-sections. J Endod 34:1337–1341
Fan B, Min Y, Lu G, Yang J, Cheung GS, Gutmann JL (2009) Negotiation of C-shaped canal systems in mandibular second molars. J Endod 35:1003–1008
Fan B, Pan Y, Gao Y, Fang F, Wu Q, Gutmann JL (2010) Three-dimensional morphologic analysis of isthmuses in the mesial roots of mandibular molars. J Endod 36:1866–1869
Fan B, Ye W, Xie E, Wu H, Gutmann JL (2012) Three-dimensional morphological analysis of C-shaped canals in mandibular first premolars in a Chinese population. Int Endod J 45:1035–1041
Fasoli G (1913) Sull'anatomia dei canali radicolari dei denti umani. Stomatologia 11:409–415
Ferraz JA, de Carvalho Junior JR, Saquy PC, Pécora JD, Sousa-Neto MD (2001) Dental anomaly: dens evaginatus (talon cusp). Braz Dent J 12:132–134
Fisher G (1908) Beitrage zur behandlung enkrankter zahne mit besonderer beucksichtigung der anatomie und pathologie der wurzelkanale. Georg Thiéme, Leipzig
Gambill JM, Alder M, del Rio CE (1996) Comparison of nickel-titanium and stainless steel hand-file instrumentation using computed tomography. J Endod 22:369–375
Gao Y, Fan B, Cheung GS, Gutmann JL, Fan M (2006) C-shaped canal system in mandibular second molars part IV: 3-D morphological analysis and transverse measurement. J Endod 32:1062–1065
Garg AK, Tewari RK, Agrawal N (2013) Prevalence of three-rooted mandibular first molars among Indians using SCT. Int J Dent 2013:183869
Gilles J, Reader A (1990) An SEM investigation of the mesiolingual canal in human maxillary first and second molars. Oral Surg Oral Med Oral Pathol 70:638–643
Grande NM, Plotino G, Gambarini G et al (2012) Present and future in the use of micro-CT scanner 3D analysis for the study of dental and root canal morphology. Ann Ist Super Sanita 48:26–34
Green D (1955) Morphology of pulp cavity of the permanent teeth. Oral Surg Oral Med Oral Pathol 8:743–759
Green D (1956) A stereomicroscopic study of the root apices of 400 maxillary and mandibular anterior teeth. Oral Surg Oral Med Oral Pathol 9:1224–1232
Green EN (1958) Microscopic investigation of root canal diameters. J Am Dent Assoc 57:636–644
Green D (1973) Double canals in single roots. Oral Surg Oral Med Oral Pathol 35:689–696
Grover C, Shetty N (2012) Methods to study root canal morphology: a review. Quintessence Int 6:171–182

Gu YC (2011) A micro-computed tomographic analysis of maxillary lateral incisors with radicular grooves. J Endod 37:789–792
Gu Y, Lee JK, Spangberg LS et al (2011) Minimum-intensity projection for in-depth morphology study of mesiobuccal root. Oral Surg Oral Med Oral Pathol Oral Radiol Endod 112:671–677
Gu YC, Zhang YP, Liao ZG, Fei XD (2013a) A micro-computed tomographic analysis of wall thickness of C-shaped canals in mandibular first premolars. J Endod 39:973–976
Gu Y, Zhang Y, Liao Z (2013b) Root and canal morphology of mandibular first premolars with radicular grooves. Arch Oral Biol 58:1609–1617
Guerisoli DM, de Souza RA, de Sousa Neto MD, Silva RG, Pécora JD (1998) External and internal anatomy of third molars. Braz Dent J 9:91–94
Gulabivala K, Aung TH, Alavi A, Ng YL (2001) Root and canal morphology of Burmese mandibular molars. Int Endod J 34:359–370
Gulabivala K, Opasanon A, Ng YL, Alavi A (2002) Root and canal morphology of Thai mandibular molars. Int Endod J 35:56–62
Gullickson DC, Montgomery S (1987) The study of root canal morphology using a digital image processing technique. J Endod 13:158–163
Gutierrez JH, Aguayo P (1995) Apical foraminal openings in human teeth. Number and location. Oral Surg Oral Med Oral Pathol Oral Radiol Endod 79:769–777
Gysi A, Röse C (1894) Sammlung von Mikrophotographien zur Veranschaulichung der mikroscopischen Struktur der Zähne des Menschen. Mikrophotographien der Zahnhistologie. Schweiz, Zürich
Haddad GY, Nehme WB, Ounsi HF (1999) Diagnosis, classification, and frequency of C-shaped canals in mandibular second molars in the Lebanese population. J Endod 25:268–271
Han X, Yang H, Li G, Yang L, Tian C, Wang Y (2012) A study of the distobuccal root canal orifice of the maxillary second molars in Chinese individuals evaluated by cone-beam computed tomography. J Appl Oral Sci 20:563–567
Hargreaves KM, Cohen S (2011) Cohen's pathways of the pulp, 10th edn. Mosby, St. Louis
Harris SP, Bowles WR, Fok A, McClanahan SB (2013) An anatomic investigation of the mandibular first molar using micro-computed tomography. J Endod 39:1374–1378
Hassan BA, Payam J, Juyanda B, van der Stelt P, Wesselink PR (2012) Influence of scan setting selections on root canal visibility with cone beam CT. Dentomaxillofac Radiol 41:645–648
Hegedus C, Flora-Nagy E, Martos R et al (2000) 3D reconstruction based on hard tissue microtome cross-section pictures in dentistry. Comput Methods Programs Biomed 63:77–84
Hess W (1917) Die pulpaamputation als selbstandige wurzelbehandlungs method. Georg Thiéme, Leipzig
Hess W, Zurcher E (1925) The anatomy of the root canals of the teeth of the permanent and deciduous dentitions. John Bale, Sons & Danielsson, Ltd., London
Hsu YY, Kim S (1997) The resected root surface. The issue of canal isthmuses. Dent Clin North Am 41:529–540
Huang CC, Chang YC, Chuang MC et al (2010) Evaluation of root and canal systems of mandibular first molars in Taiwanese individuals using cone-beam computed tomography. J Formos Med Assoc 109:303–308
Hülsmann M (1997) Dens invaginatus: aetiology, classification, prevalence, diagnosis, and treatment considerations. Int Endod J 30:79–90
Idiyatullin D, Corum C, Moeller S, Prasad HS, Garwood M, Nixdorf DR (2011) Dental magnetic resonance imaging: making the invisible visible. J Endod 37:745–752
Jafarzadeh H, Wu YN (2007) The C-shaped root canal configuration: a review. J Endod 33:517–523
Jin GC, Lee SJ, Roh BD (2006) Anatomical study of C-shaped canals in mandibular second molars by analysis of computed tomography. J Endod 32:10–13
Jung IY, Seo MA, Fouad AF et al (2005) Apical anatomy in mesial and mesiobuccal roots of permanent first molars. J Endod 31:364–368

Kaffe I, Kaufman A, Littner MM, Lazarson A (1985) Radiographic study of the root canal system of mandibular anterior teeth. Int Endod J 18:253–259
Keith A, Knowles FHS (1911) The description of teeth of palaeolithic man from Jersey. J Anat Physiol 46:12–27
Keleş A, Alcin H, Kamalak A, Versiani MA (2014) Micro-CT evaluation of root filling quality in ovalshaped canals. Int Endod J 47(12):1177–1184
Kerekes K, Tronstad L (1977a) Morphometric observations on the root canals of human molars. J Endod 3:114–118
Kerekes K, Tronstad L (1977b) Morphometric observations on root canals of human anterior teeth. J Endod 3:24–29
Kerekes K, Tronstad L (1977c) Morphometric observations on root canals of human premolars. J Endod 3:74–79
Kim Y, Lee SJ, Woo J (2012) Morphology of maxillary first and second molars analyzed by cone-beam computed tomography in a korean population: variations in the number of roots and canals and the incidence of fusion. J Endod 38:1063–1068
Kim Y, Chang SW, Lee JK et al (2013) A micro-computed tomography study of canal configuration of multiple-canalled mesiobuccal root of maxillary first molar. Clin Oral Investig 17:1541–1546
Kocsis G, Marcsik A, Kokai E, Kocsis K (2002) Supernumerary occlusal cusps on permanent human teeth. Acta Biol Szeged 46:71–82
Krasner P, Rankow HJ (2004) Anatomy of the pulp-chamber floor. J Endod 30:5–16
Lee JH, Kim KD, Lee JK et al (2011) Mesiobuccal root canal anatomy of Korean maxillary first and second molars by cone-beam computed tomography. Oral Surg Oral Med Oral Pathol Oral Radiol Endod 111:785–791
Leoni GB, Versiani MA, Pécora JD, Sousa-Neto MD (2014) Micro–computed tomographic analysis of the root canal morphology of mandibular incisors. J Endod 40(5):710–716
Levitan ME, Himel VT (2006) Dens evaginatus: literature review, pathophysiology, and comprehensive treatment regimen. J Endod 32:1–9
Love RM, Jenkinson HF (2002) Invasion of dentinal tubules by oral bacteria. Crit Rev Oral Biol Med 13:171–183
Lu TY, Yang SF, Pai SF (2006) Complicated root canal morphology of mandibular first premolar in a Chinese population using the cross section method. J Endod 32:932–936
Luiten DJ, Morgan LA, Baugartner JC, Marshall JG (1995) A comparison of four instrumentation techniques on apical canal transportation. J Endod 21:26–32
Lyroudia K, Mikrogeorgis G, Nikopoulos N, Samakovitis G, Molyvdas I, Pitas I (1997a) Computerized 3-D reconstruction of two "double teeth". Endod Dent Traumatol 13:218–222
Lyroudia K, Samakovitis G, Pitas I, Lambrianidis T, Molyvdas I, Mikrogeorgis G (1997b) 3D reconstruction of two C-shape mandibular molars. J Endod 23:101–104
Lyroudia K, Mikrogeorgis G, Bakaloudi P, Kechagias E, Nikolaidis N, Pitas I (2002) Virtual endodontics: three-dimensional tooth volume representations and their pulp cavity access. J Endod 28:599–602
Manning SA (1990) Root canal anatomy of mandibular second molars. Part II. C-shaped canals. Int Endod J 23:40–45
Mannocci F, Peru M, Sherriff M, Cook R, Pitt Ford TR (2005) The isthmuses of the mesial root of mandibular molars: a micro-computed tomographic study. Int Endod J 38:558–563
Marceliano-Alves MF, Sousa-Neto MD, Fidel SR, Steier L, Robinson JP, Pécora JD, Versiani MA (2014) Shaping ability of single-file reciprocating and heat-treated multifile rotary systems: a micro-CT study. Int Endod J. doi:10.1111/iej.12412 [Epub ahead of print]
Martins JN, Quaresma S, Quaresma MC, Frisbie-Teel J (2013) C-shaped maxillary permanent first molar: a case report and literature review. J Endod 39:1649–1653
Matherne RP, Angelopoulos C, Kulild JC, Tira D (2008) Use of cone-beam computed tomography to identify root canal systems in vitro. J Endod 34:87–89

Mauger MJ, Schindler WG, Walker WA 3rd (1998) An evaluation of canal morphology at different levels of root resection in mandibular incisors. J Endod 24:607–609
Mayo CV, Montgomery S, de Rio C (1986) A computerized method for evaluating root canal morphology. J Endod 12:2–7
Melton DC, Krell KV, Fuller MW (1991) Anatomical and histological features of C-shaped canals in mandibular second molars. J Endod 17:384–388
Merrill RG (1964) Occlusal anomalous tubercles on premolars of alaskan Eskimos and Indians. Oral Surg Oral Med Oral Pathol 17:484–496
Metgud S, Metgud R, Rani K (2009) Management of a patient with a taurodont, single-rooted molars associated with multiple dental anomalies: a spiral computerized tomography evaluation. Oral Surg Oral Med Oral Pathol Oral Radiol Endod 108:e81–e86
Meyer W, Scheele E (1955) Die Anatomie der Wurzelkanäle der oberen Frontzähne. Deutsch Zahnärztl 10:1041–1045
Michetti J, Maret D, Mallet JP, Diemer F (2010) Validation of cone beam computed tomography as a tool to explore root canal anatomy. J Endod 36:1187–1190
Min Y, Fan B, Cheung GS, Gutmann JL, Fan M (2006) C-shaped canal system in mandibular second molars Part III: the morphology of the pulp chamber floor. J Endod 32:1155–1159
Mizutani T, Ohno N, Nakamura H (1992) Anatomical study of the root apex in the maxillary anterior teeth. J Endod 18:344–347
Mjör IA (2009) Dentin permeability: the basis for understanding pulp reactions and adhesive technology. Braz Dent J 20:3–16
Moral H (1915) Uber das vorkommen eines vierten kanals in oberen molaren. Österr-Ungar Vjschr Zahnheilk 31:313–325
Morfis A, Sylaras SN, Georgopoulou M, Kernani M, Prountzos F (1994) Study of the apices of human permanent teeth with the use of a scanning electron microscope. Oral Surg Oral Med Oral Pathol 77:172–176
Mueller AH (1933) Anatomy of the root canals of the incisors, cuspids and bicuspids of the permanent teeth. J Am Dent Assoc 20:1361–1386
Mühlreiter E (1870) Anatomie des menschlichen Gebisses von med. Th. E. de Jonge Cohen. Arthur Felix, Leipzig
Nagy CD, Szabo J, Szabo J (1995) A mathematically based classification of root canal curvatures on natural human teeth. J Endod 21:557–560
Nance R, Tyndall D, Levin LG, Trope M (2000) Identification of root canals in molars by tuned-aperture computed tomography. Int Endod J 33:392–396
Nanci A, Ten Cate AR (2013) Ten Cate's oral histology: development, structure, and function, 8th edn. Elsevier, St. Louis
Naoum HJ, Love RM, Chandler NP, Herbison P (2003) Effect of X-ray beam angulation and intraradicular contrast medium on radiographic interpretation of lower first molar root canal anatomy. Int Endod J 36:12–19
Nattress BR, Martin DM (1991) Predictability of radiographic diagnosis of variations in root canal anatomy in mandibular incisor and premolar teeth. Int Endod J 24:58–62
Nayak BG, Singh I (2013) Using spiral computed tomography for endodontic management of a mandibular first molar with a middle mesial canal: a case report. Gen Dent 61:43–46
Neelakantan P, Subbarao C, Ahuja R, Subbarao CV, Gutmann JL (2011) Cone-beam computed tomography study of root and canal morphology of maxillary first and second molars in an Indian population. J Endod 36:1622–1627
Newton CW, McDonald S (1984) A C-shaped canal configuration in a maxillary first molar. J Endod 10:397–399
Ng YL, Aung TH, Alavi A, Gulabivala K (2001) Root and canal morphology of Burmese maxillary molars. Int Endod J 34:620–630
Nielsen RB, Alyassin AM, Peters DD, Carnes DL, Lancaster J (1995) Microcomputed tomography: an advanced system for detailed endodontic research. J Endod 21:561–568

Oehlers FA (1957a) Dens invaginatus (dilated composite odontome). I. Variations of the invagination process and associated anterior crown forms. Oral Surg Oral Med Oral Pathol 10:1204–1218, contd
Oehlers FA (1957b) Dens invaginatus (dilated composite odontome). II. Associated posterior crown forms and pathogenesis. Oral Surg Oral Med Oral Pathol 10:1302–1316
Oehlers FA, Lee KW, Lee EC (1967) Dens evaginatus (evaginated odontome). Its structure and responses to external stimuli. Dent Pract Dent Rec 17:239–244
Okumura T (1927) Anatomy of the root canals. J Am Dent Assoc 14:632–636
Ordinola-Zapata R, Bramante CM, Villas-Boas MH, Cavenago BC, Duarte MH, Versiani MA (2013) Morphologic micro–computed tomography analysis of mandibular premolars with three root canals. J Endod 39:1130–1135
Ordinola-Zapata R, Monteiro Bramante C, Gagliardi Minotti P, Cavalini Cavenago B, Gutmann JL, Moldauer BI, Versiani MA, Hungaro Duarte MA (2014) Micro-CT evaluation of C-shaped mandibular first premolars in a Brazilian subpopulation. Int Endod J. doi:10.1111/iej.12380. [Epub ahead of print]
Pao YC, Reinhardt RA, Krejci RF, Taylor DT (1984) Computer-graphics aided instruction of three-dimensional dental anatomy. J Dent Educ 48:315–317
Pécora JD, Saquy PC, de Souza JE, Sousa-Neto MD (1991) Endodontic treatment of a maxillary lateral incisor presenting dens invaginatus and transposition to the region of the canine–case report. Braz Dent J 2:5–8
Pécora JD, Saquy PC, Sousa-Neto MD, Woelfel JB (1992a) Root form and canal anatomy of maxillary first premolars. Braz Dent J 2:87–94
Pécora JD, Woelfel JB, Sousa Neto MD, Issa EP (1992b) Morphologic study of the maxillary molars. Part II: Internal anatomy. Braz Dent J 3:53–57
Pécora JD, Conrado CA, Zuccolotto WG, Sousa-Neto MD, Saquy PC (1993a) Root canal therapy of an anomalous maxillary central incisor: a case report. Endod Dent Traumatol 9:260–262
Pécora JD, Sousa Neto MD, Saquy PC (1993b) Internal anatomy, direction and number of roots and size of human mandibular canines. Braz Dent J 4:53–57
Pécora JD, Sousa-Neto MD, Saquy PC, Woelfel JB (1993c) In vitro study of root canal anatomy of maxillary second premolars. Braz Dent J 3:81–85
Peiris HR, Pitakotuwage TN, Takahashi M, Sasaki K, Kanazawa E (2008) Root canal morphology of mandibular permanent molars at different ages. Int Endod J 41:828–835
Perrini N, Castagnola L, Hess W, Keller's O (1998) Anatomical plates: studies on the anatomical structure of root canals in human dentition by a method of making the tooth substance transparent (1928). Altini Communicazioni Grafiche, Milan
Peters OA, Laib A, Ruegsegger P, Barbakow F (2000) Three-dimensional analysis of root canal geometry by high-resolution computed tomography. J Dent Res 79:1405–1409
Peters OA, Laib A, Gohring TN, Barbakow F (2001) Changes in root canal geometry after preparation assessed by high-resolution computed tomography. J Endod 27:1–6
Pineda F, Kuttler Y (1972) Mesiodistal and buccolingual roentgenographic investigation of 7275 root canals. Oral Surg Oral Med Oral Pathol 33:101–110
Ponce EH, Vilar Fernandez JA (2003) The cemento-dentino-canal junction, the apical foramen, and the apical constriction: evaluation by optical microscopy. J Endod 29:214–219
Poyton HG, Morgan GA (1966) Dens in dente. Dent Radiogr Photogr 39:27–33, passim
Preiswerk G (1903) Leherbuch und Atlas der Zahnheilkunde mit Einschluβ der Mund-Krankheiten. J. F. Lehmann, Munich
Pruett JP, Clement DJ, Carnes DL Jr (1997) Cyclic fatigue testing of nickel-titanium endodontic instruments. J Endod 23:77–85
Rahimi S, Shahi S, Lotfi M, Zand V, Abdolrahimi M, Es'haghi R (2008) Root canal configuration and the prevalence of C-shaped canals in mandibular second molars in an Iranian population. J Oral Sci 50:9–13
Regezi JA, Sciubba JJ, Jordan RCK (2008) Oral pathology: clinical pathologic correlations, 5th edn. Saunders/Elsevier, St. Louis

Reuben J, Velmurugan N, Kandaswamy D (2008) The evaluation of root canal morphology of the mandibular first molar in an Indian population using spiral computed tomography scan: an in vitro study. J Endod 34:212–215
Rice RT, Gilbert BO Jr (1987) An unusual canal configuration in a mandibular first molar. J Endod 13:515–515
Ricucci D, Siqueira JF Jr (2010) Fate of the tissue in lateral canals and apical ramifications in response to pathologic conditions and treatment procedures. J Endod 36:1–15
Ridell K, Mejare I, Matsson L (2001) Dens invaginatus: a retrospective study of prophylactic invagination treatment. Int J Paediatr Dent 11:92–97
Robertson D, Leeb IJ, McKee M, Brewer E (1980) A clearing technique for the study of root canal systems. J Endod 6:421–424
Robinson S, Czerny C, Gahleitner A, Bernhart T, Kainberger FM (2002) Dental CT evaluation of mandibular first premolar root configurations and canal variations. Oral Surg Oral Med Oral Pathol Oral Radiol Endod 93:328–332
Rocha LF, Sousa Neto MD, Fidel SR, da Costa WF, Pécora JD (1996) External and internal anatomy of mandibular molars. Braz Dent J 7:33–40
Rotstein I, Stabholz A, Heling I, Friedman S (1987) Clinical considerations in the treatment of dens invaginatus. Endod Dent Traumatol 3:249–254
Ruprecht A, Sastry KA, Batniji S, Lambourne A (1987) The clinical significance of dental invagination. J Pedod 11:176–181
Schäfer E, Diez C, Hoppe W, Tepel J (2002) Roentgenographic investigation of frequency and degree of canal curvatures in human permanent teeth. J Endod 28:211–216
Schneider SW (1971) A comparison of canal preparations in straight and curved root canals. Oral Surg Oral Med Oral Pathol 32:271–275
Seltzer S, Bender IB, Nazimov H, Sinai I (1967) Pulpitis-induced interradicular periodontal changes in experimental animals. J Periodontol 38:124–129
Seo DG, Gu Y, Yi YA et al (2012) A biometric study of C-shaped root canal systems in mandibular second molars using cone-beam computed tomography. Int Endod J 45:807–814
Sert S, Bayirli GS (2004) Evaluation of the root canal configurations of the mandibular and maxillary permanent teeth by gender in the Turkish population. J Endod 30:391–398
Sert S, Bayrl G (2004) Taurodontism in six molars: a case report. J Endod 30:601–602
Sharma R, Pécora JD, Lumley PJ, Walmsley AD (1998) The external and internal anatomy of human mandibular canine teeth with two roots. Endod Dent Traumatol 14:88–92
Shaw JC (1928) Taurodont teeth in South African races. J Anat 62:476–498
Shearer AC, Wasti F, Wilson NH (1996) The use of a radiopaque contrast medium in endodontic radiography. Int Endod J 29:95–98
Shemesh H, van Soest G, Wu MK, van der Sluis LW, Wesselink PR (2007) The ability of optical coherence tomography to characterize the root canal walls. J Endod 33:1369–1373
Shen Y, Gao Y, Lin J, Ma J, Wang Z, Haapasalo M (2012) Methods and models to study irrigation. Endod Topics 27:3–34
Shenoi RP, Ghule HM (2012) CBVT analysis of canal configuration of the mesio-buccal root of maxillary first permanent molar teeth: an in vitro study. Contemp Clin Dent 3:277–281
Sidow SJ, West LA, Liewehr FR, Loushine RJ (2000) Root canal morphology of human maxillary and mandibular third molars. J Endod 26:675–678
Siqueira JF Jr (2005) Reaction of periradicular tissues to root canal treatment: benefits and drawbacks. Endod Topics 10:123–147
Siqueira JF Jr, Alves FRF, Versiani MA et al (2013) Correlative bacteriologic and micro–computed tomographic analysis of mandibular molar mesial canals prepared by Self-Adjusting File, Reciproc, and Twisted File systems. J Endod 39:1044–1050
Solomonov M, Paque F, Fan B, Eilat Y, Berman LH (2012) The challenge of C-shaped canal systems: a comparative study of the self-adjusting file and ProTaper. J Endod 38:209–214
Sousa Neto MD, Paiva EP, Saquy PC, Pécora JD (1998) Treatment of dens invaginatus in supernumerary teeth. Aust Endod J 24:85–87

Sustercic D, Sersa I (2012) Human tooth pulp anatomy visualization by 3D magnetic resonance microscopy. Radiol Oncol 46:1–7

Sykaras SN (1971) Endodontic treatment of teeth with variations of the root canals. J Br Endod Soc 5:28–31

Szabo BT, Pataky L, Mikusi R, Fejerdy P, Dobo-Nagy C (2012) Comparative evaluation of cone-beam CT equipment with micro-CT in the visualization of root canal system. Ann Ist Super Sanita 48:49–52

Tachibana H, Matsumoto K (1990) Applicability of X-ray computerized tomography in endodontics. Endod Dent Traumatol 6:16–20

Tanasiewicz M (2010) Magnetic resonance imaging in human teeth internal space requirements for visualization of dental prosthetics. J Exp Clin Dent 2:19–24

Teixeira FB, Sano CL, Gomes BP, Zaia AA, Ferraz CC, Souza-Filho FJ (2003) A preliminary in vitro study of the incidence and position of the root canal isthmus in maxillary and mandibular first molars. Int Endod J 36:276–280

Tian YY, Guo B, Zhang R et al (2012) Root and canal morphology of maxillary first premolars in a Chinese subpopulation evaluated using cone-beam computed tomography. Int Endod J 45:996–1003

Tratman EK (1938) Three-rooted lower molars in man and their racial distribution. Br Dent J 64:264–274

Tsesis I, Shifman A, Kaufman AY (2003) Taurodontism: an endodontic challenge. Report of a case. J Endod 29:353–355

Turner CGN (1971) Three-rooted mandibular first permanent molars and the question of American Indian origins. Am J Phys Anthropol 34:229–241

Venturi M, Prati C, Capelli G, Falconi M, Breschi L (2003) A preliminary analysis of the morphology of lateral canals after root canal filling using a tooth-clearing technique. Int Endod J 36:54–63

Versiani MA, Pécora JD, Sousa-Neto MD (2011a) The anatomy of two-rooted mandibular canines determined using micro-computed tomography. Int Endod J 44:682–687

Versiani MA, Pécora JD, de Sousa-Neto MD (2011b) Flat-oval root canal preparation with self-adjusting file instrument: a micro-computed tomography study. J Endod 37(7):1002–1007

Versiani MA, Sousa-Neto MD, Pécora JD (2011c) Pulp pathosis in inlayed teeth of the ancient Mayas: a microcomputed tomography study. Int Endod J 44(11):1000–1004

Versiani MA, Pécora JD, Sousa-Neto MD (2012) Root and root canal morphology of four-rooted maxillary second molars: a micro-computed tomography study. J Endod 38:977–982

Versiani MA, Cristescu RC, Saquy PC, Pécora JD, de Sousa-Neto MD (2013a) Enamel pearls in permanent dentition: case report and micro-CT evaluation. Dentomaxillofac Radiol 42(6):20120332

Versiani MA, Leoni GB, Steier L, De-Deus G, Tassani S, Pécora JD, de Sousa-Neto MD (2013b) Microcomputed tomography study of oval-shaped canals prepared with the self-adjusting file, Reciproc, WaveOne, and ProTaper universal systems. J Endod 39(8):1060–1066

Versiani MA, Pécora JD, Sousa-Neto MD (2013c) Microcomputed tomography analysis of the root canal morphology of single-rooted mandibular canines. Int Endod J 46:800–807

Versiani MA, De-Deus G, Vera J, Souza E, Steier L, Pécora JD, Sousa-Neto MD (2015) 3D mapping of the irrigated areas of the root canal space using micro-computed tomography. Clin Oral Investig 19(4):859–866

Vertucci FJ, Williams RG (1974) Furcation canals in the human mandibular first molar. Oral Surg Oral Med Oral Pathol 38:308–314

Vertucci F, Seelig A, Gillis R (1974) Root canal morphology of the human maxillary second premolar. Oral Surg Oral Med Oral Pathol 38:456–464

Vertucci FJ (1984) Root canal anatomy of the human permanent teeth. Oral Surg Oral Med Oral Pathol 58:589–599

Vertucci FJ, Anthony RL (1986) A scanning electron microscopic investigation of accessory foramina in the furcation and pulp chamber floor of molar teeth. Oral Surg Oral Med Oral Pathol 62:319–326
Vertucci FJ (2005) Root canal morphology and its relationship to endodontic procedures. Endod Topics 10:3–29
Villas-Boas MH, Bernardineli N, Cavenago BC et al (2011) Micro-computed tomography study of the internal anatomy of mesial root canals of mandibular molars. J Endod 37:1682–1686
von Arx T (2005) Frequency and type of canal isthmuses in first molars detected by endoscopic inspection during periradicular surgery. Int Endod J 38:160–168
Wang Y, Zheng QH, Zhou XD et al (2010) Evaluation of the root and canal morphology of mandibular first permanent molars in a western Chinese population by cone-beam computed tomography. J Endod 36:1786–1789
Wang Q, Yu G, Zhou XD, Peters OA, Zheng QH, Huang DM (2011) Evaluation of x-ray projection angulation for successful radix entomolaris diagnosis in mandibular first molars in vitro. J Endod 37:1063–1068
Weine FS, Healey HJ, Gerstein H, Evanson L (1969) Canal configuration in the mesiobuccal root of the maxillary first molar and its endodontic significance. Oral Surg Oral Med Oral Pathol 28:419–425
Weine FS, Pasiewicz RA, Rice RT (1988) Canal configuration of the mandibular second molar using a clinically oriented in vitro method. J Endod 14:207–213
Weine FS (1996) Endodontic therapy, 5th edn. Mosby-Yearbook, St. Louis
Weller RN, Niemczyk SP, Kim S (1995) Incidence and position of the canal isthmus. Part 1. Mesiobuccal root of the maxillary first molar. J Endod 21:380–383
Weng XL, Yu SB, Zhao SL et al (2009) Root canal morphology of permanent maxillary teeth in the Han nationality in Chinese Guanzhong area: a new modified root canal staining technique. J Endod 35:651–656
Woelfel JB, Scheid RC (2002) Dental anatomy: its relevance to dentistry, 6th edn. Lippincott Williams & Wilkins, Philadelphia
Wu MK, R'Oris A, Barkis D, Wesselink PR (2000) Prevalence and extent of long oval canals in the apical third. Oral Surg Oral Med Oral Pathol Oral Radiol Endod 89:739–743
Yang ZP, Yang SF, Lin YC, Shay JC, Chi CY (1988a) C-shaped root canals in mandibular second molars in a Chinese population. Endod Dent Traumatol 4:160–163
Yang ZP, Yang SF, Lee G (1988b) The root and root canal anatomy of maxillary molars in a Chinese population. Endod Dent Traumatol 4:215–218
Yilmaz Z, Tuncel B, Serper A, Calt S (2006) C-shaped root canal in a maxillary first molar: a case report. Int Endod J 39:162–166
Yu X, Guo B, Li KZ et al (2012) Cone-beam computed tomography study of root and canal morphology of mandibular premolars in a western Chinese population. BMC Med Imaging 12:18
Zhang R, Yang H, Yu X, Wang H, Hu T, Dummer PM (2011a) Use of CBCT to identify the morphology of maxillary permanent molar teeth in a Chinese subpopulation. Int Endod J 44:162–169
Zhang R, Wang H, Tian YY, Yu X, Hu T, Dummer PM (2011b) Use of cone-beam computed tomography to evaluate root and canal morphology of mandibular molars in Chinese individuals. Int Endod J 44:990–999
Zheng Q, Zhang L, Zhou X et al (2011) C-shaped root canal system in mandibular second molars in a Chinese population evaluated by cone-beam computed tomography. Int Endod J 44:857–862

Part III
Outcome and Strategies of Treatment

Biofilm-Associated Infections in Root Canals: Treatment and Outcomes

Kishor Gulabivala and Yuan-Ling Ng

Abstract Root canal infections are essentially characterised by microbial biofilms that adhere to the root canal dentine and extend to the apical foramina and in some cases beyond. Primary objectives of root canal treatment are to eliminate these biofilms by using chemomechanical treatment protocols and to prevent reinfection. Desired outcomes of effective treatment are the discontinuation of destructive aspects of the host immune response and apical healing. This chapter reviews different treatment factors and their influence on the outcome of biofilm-associated infections in root canals.

1 Nature of the Biological Problem

The picture of the intraradicular microbiota is under continual review and improvement through new scientific insight, but it still remains relatively crude. The picture, at present, is synthesised from a variety of data sources that may be divided into two main streams of knowledge: microscopy (sometimes with in situ hybridisation) and diversity determination (through culture-dependent and culture-independent or molecular techniques). The distribution and diversity of the microbiota within the root canal system and dentine is unique to the individual tooth and notably variable. The state at any given time may represent the stage of infection with the bacterial biofilm extending up to and only sometimes beyond the apical canal terminus. Bacteria may proliferate beyond the apical foramina and into the periapical lesion, but this seems rare. The depth of penetration into dentine is also variable but confined to the area close to the root canal lumen.

Assuming all root canal infections originate from coronal breaches in tooth structure or exposure of natural communications, such as lateral canals, periapical lesion development would seem to be dependent upon initial microbial contamination of dentinal surfaces (root canal wall and dentinal tubules), successful initial

K. Gulabivala (✉) • Y.-L. Ng
Unit of Endodontology, UCL Eastman Dental Institute, University College London, London, UK
e-mail: k.gulabivala@ucl.ac.uk

L.E. Chávez de Paz et al. (eds.), *The Root Canal Biofilm*, Springer Series on Biofilms 9, DOI 10.1007/978-3-662-47415-0_8

colonisation (leading to an established microbial population), propagation and consolidation of its polymicrobial nature with generation of exopolymeric substances, succession of bacterial species within the growing biofilm community with establishment of anaerobic properties leading to critical colonisation and thence its definitive survival as an 'infection'. This progression may be tracked both temporally (Möller et al. 1981; Fabricius et al. 1982; Tani-Ishii et al. 1994) and spatially (Fabricius et al. 1982). Deep insight about the nature and course of apical advancement of the intraradicular biofilm is lacking, but inferences may be drawn from microbiota distribution and composition studies. The establishment of the microbiota within the apical canal complexities is associated with the opportunity for an imbalance to arise between microbiota and host, leading to ensuing apical pathosis. It is, however, not unknown that periapical host responses may commence earlier, whilst there is still intervening, viable and sometimes even healthy pulp tissue in the apical root anatomy (Moore 1967; Jordan et al. 1978; Byers et al. 1990; Caliskan 1995). The periapical tissue status in such cases may vary from frank periapical radiolucency to condensing periapical osteitis.

As the microbial biofilm gains further maturity and overwhelms the host mechanisms, the bioburden is said to be critical enough for a state of infection to exist. Bioburden is the presence of bacteria at the body tissue interface where microbes compete for a limited supply of oxygen and nutrients, creating a burden on the attempts at healing in the chronic inflammatory process. Bioburden is more than just the number of bacteria; it includes the diversity, virulence, and interaction of organisms (Daeschlein 2013). The progressive advancement of the bacterial biofilm in the root canal system from the coronal to the apical terminus is influenced by, as yet, undefined ecological factors in the root canal system. Speculative hypotheses about the mechanisms inherent in the process abound. They are based on ecological theory, seem plausible and are probably true, but positive proof remains elusive so far. Studies have, therefore, focused on the nature of microbial interactions within the consortia and with their environment. Interactions amongst microorganisms and their biotic (living) and abiotic (nonliving) surroundings are important in enabling their survival. The *relatively* restricted nature of the root canal microbiota suggests selective pressures (Fabricius et al. 1982). The positive and negative associations are thought to be due to nutritional interactions, local physiological conditions (Eh, pH), bacteriocins and bacterial coaggregation or physical attraction and binding. The classic 'infection continuum' described above as contamination, colonisation, critical colonisation and then infection may not reflect the full picture based on microbial/host relationship. An alternative description may be contamination, attachment, proliferation and critical concentration of signalling molecules rather than bacterial counts, followed by the subsequent natural adoption of a biofilm phenotype, which becomes progressively polymicrobial given opportunity (Jones and Kennedy 2012).

Microbial interactions within communities enable them to evolve depending on the local environmental variations, much like the differences between supragingival and subgingival plaque, which at only a few millimetres apart develop vastly differently depending on their salivary and serum nutritional sources, respectively.

A similar model is proposed for the root canal microbiota: coronal leakage may allow salivary ingress and facultative organisms to grow in the coronal part of the canal, whereas serum from the apical part of the canal may favour the growth of proteolytic bacteria at the root apex. The difference in the root canal system is that the nutritional sources are separated by a much greater distance in relative terms at 15–25 mm apart. This, therefore, also raises the question of what sustains biofilm progression in the mid-root. Evidence for different patterns of bacterial growth found in root canal systems comes from morphological and diversity studies (Richardson et al. 2009; Ozok et al. 2012).

The root canal environment has a unique natural history from the perspective of a nutritional source for bacteria in the human body, though it has not been characterised adequately in its necrotic, infected state from an ecological point of view. The pulp may provide a very rich supply of vital tissue and serum exudate during the stages of pulpal inflammation, but once it becomes necrotised, the nutritional supply is presumably rapidly exhausted as the environment is secluded by the dentine shell. The key fluid-based nutritional resources such as saliva, serum, blood and inflammatory exudate from pulpal tissues may dwindle. Where even minimal nutritional streams are available in the form of salivary leakage coronally and inflammatory exudate apically, bacterial communities may become established and reach maturity. As the initial habitat is altered by the primary colonisers, secondary invaders join and may replace them. According to classical ecological theory, succession ends when a relatively stable assembly of populations, called a climax community, is achieved. This concept has been difficult to apply to microbial communities in the general environment as random disturbances prevent the community from ever-reaching equilibrium although this may well be achieved in the secluded root canal environment in chronic cases. In the later stages (or perhaps throughout the process) of an enclosed root canal infection therefore, the bacteria presumably enter some sort of a starvation or dormant phase, although there is little published literature on this aspect in root canal sites. Although in animal models, periapical lesions may be induced in relatively short spans of time by direct inoculation of the pulp space (Gulabivala 2004), the actual amount of time taken may presumably vary enormously given the time taken for pulps to succumb from clinical evidence. The 'slow-burning' nature of a biofilm infection with starvation and dormancy states may play a significant part in the sustenance of chronic periapical disease (Costerton et al. 2003).

The overall ecological picture is one of a complex polymicrobial community that may function as one in response to its environment and presumably to treatment as well. The biofilm has a patchy and variable distribution through the root canal system. Depending upon the stage of infection, the colonisation may be dominant coronally, apically or in the middle (Fig. 1). The biofilm may have variable thickness from a few cells to hundreds (Fig. 2); the groups of organisms within each niche may vary and certainly each tooth will have its own unique diversity signature. Teeth with gross carious exposures may exhibit almost complete biofilm coverage of the root canal wall with prolific thickness (Fig. 3). Whilst it is tempting to speculate that enclosed root canal systems may exhibit restricted

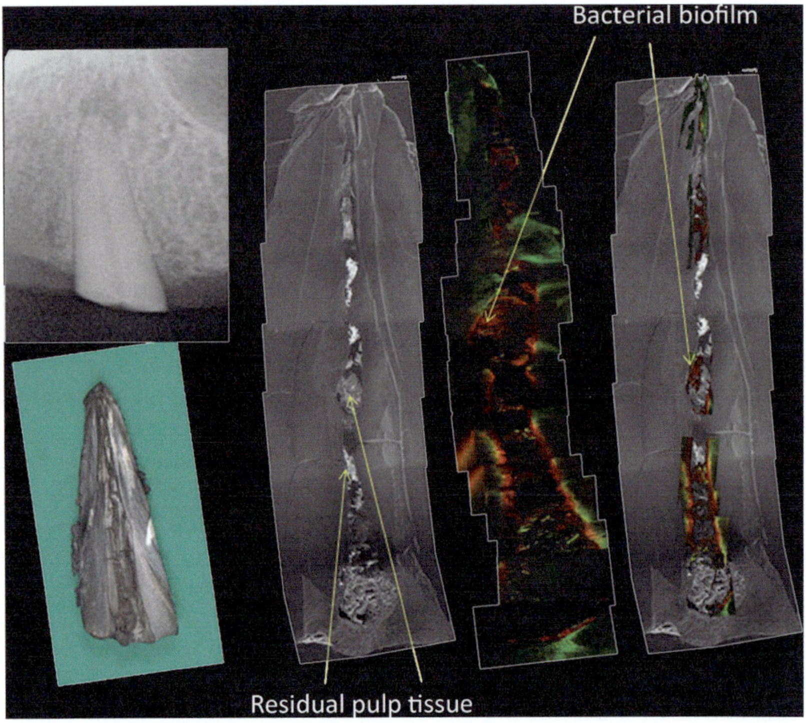

Fig. 1 Patchy distribution of the root canal biofilm, (**a**) radiographic image of root, (**b**) extracted split root, (**c**) composite SEM image of root, (**d**) composite CLSM image of root canal with fluorescent in situ hybridisation using bacterial probes, (**e**) SEM and CLSM superimposed images

microbiota with minimal diversity and teeth with open pulp chambers, wide microbial diversity, the truth is that such correlation is not proven.

The relative amount and composition of extracellular matrix (ECM) in biofilms within root canal systems (Fig. 4) are likely to be dictated by bacterial diversity and the microbiota's nutritional sources. However, insight about the ECM in root canal biofilms is currently lacking. It is important to determine these properties because they would presumably play a major role in biofilm eradication.

Where the resident infection propagates to the apical canal terminus with evidence of periapical suppuration, it is probable that PMNs will be found in the canal system, often lining the bacterial biofilm (Fig. 3) and indeed attempting to phagocytose the cells (Fig. 5). The extravascular life span of such cells is regarded as being 2–3 days, implying a very dynamic and nutritionally rich environment, in stark contrast to the notion of a stable ecosystem with minimal shift and disturbance (Richardson et al. 2009).

When the biofilm front extends beyond the confines of the root canal system apically, the bacteria involved are often mainly anaerobic, perhaps associated with foreign or dead (cellular or dentinal) material, to which defence cells have no or limited access. Given the richer nutritional source in this location, the biofilm may

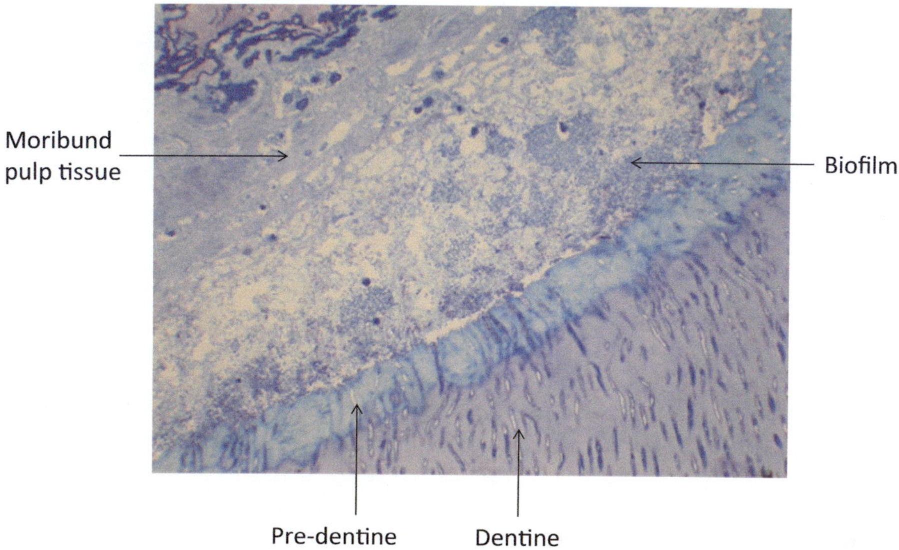

Fig. 2 Variable thickness of biofilm adherent to the root canal predentine with moribund pulp tissue towards the canal lumen

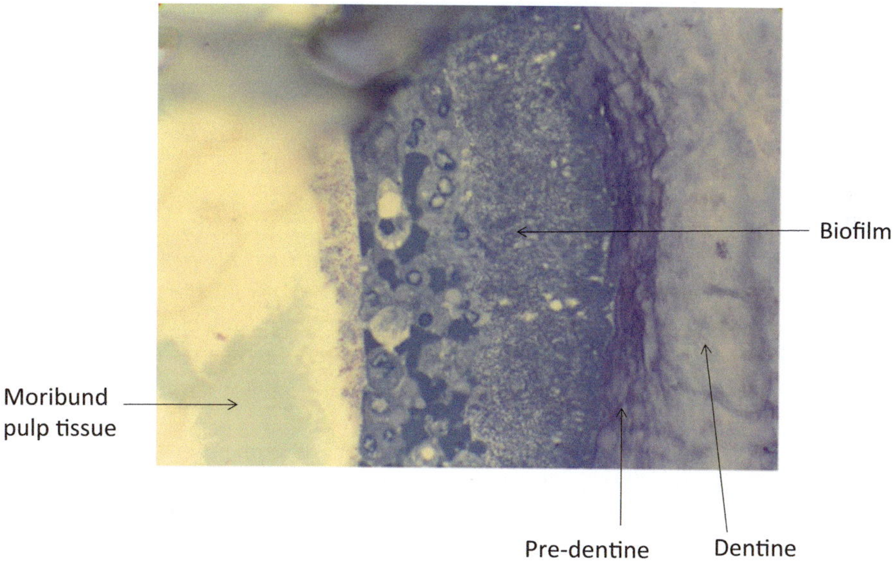

Fig. 3 Thick and rich biofilm layer coating canal wall in a cariously exposed tooth, with a PMN layer covering the biofilm towards the canal lumen aspect

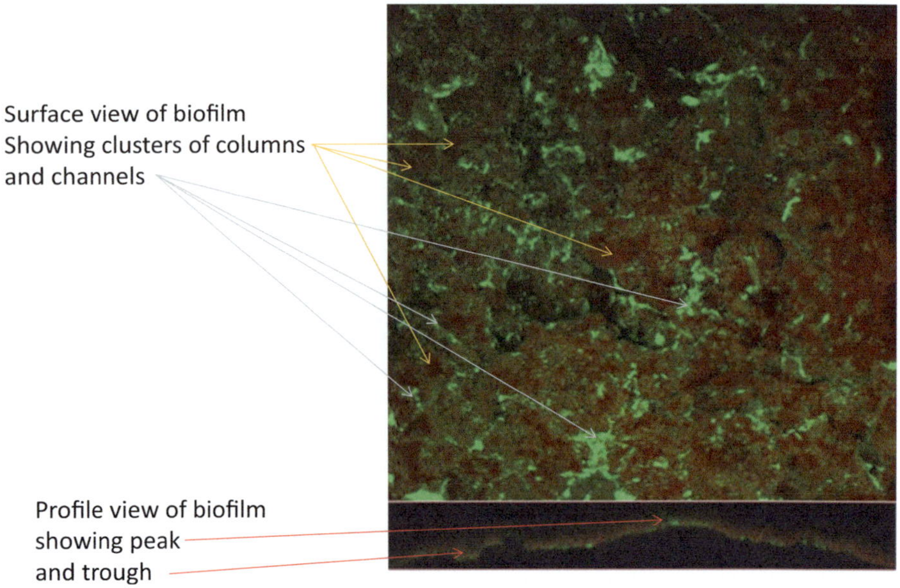

Fig. 4 Surface and cross-sectional views of the extracellular matrix (ECM) structure of a biofilm of a clinical strain using a dual staining method with the lectin distributed on the surface of the biofilm cells, field depth = 63 μm

exhibit a richer and thicker extracellular matrix covering and extending over the external surface of the root. Bacterial occupants of such plaques have been observed to be mainly coccal and rod forms, but fibrillar structures may also be present. Those implicated in periapical tissue invasion include *Actinomyces* species and *Propionibacterium propionicum*. Many others from the root canal microbiota have also been implicated, but their genuine presence as invaders of periapical tissues remains controversial because of the difficulties of obtaining uncontaminated samples. By definition, chronic periapical lesions *without suppuration* do not contain bacteria, whereas *suppurative lesions* must exhibit bacterial invasion of the periapical tissues. The problem inherent in such clinical classification lies in the difficulty of clinical discrimination between the two states. Whilst suppuration may be evident at a tissue level, it may not yet have manifested clinically.

The variations in the nature of intra- and extra-radicular infections, together with variations in host responses, probably result in the diverse clinical presentations of apical periodontitis. Given the association between infection and root canal treatment outcome, it is not surprising that clinical presentation (pain, swelling and sinus) might predict outcomes of treatment (Ng et al. 2011).

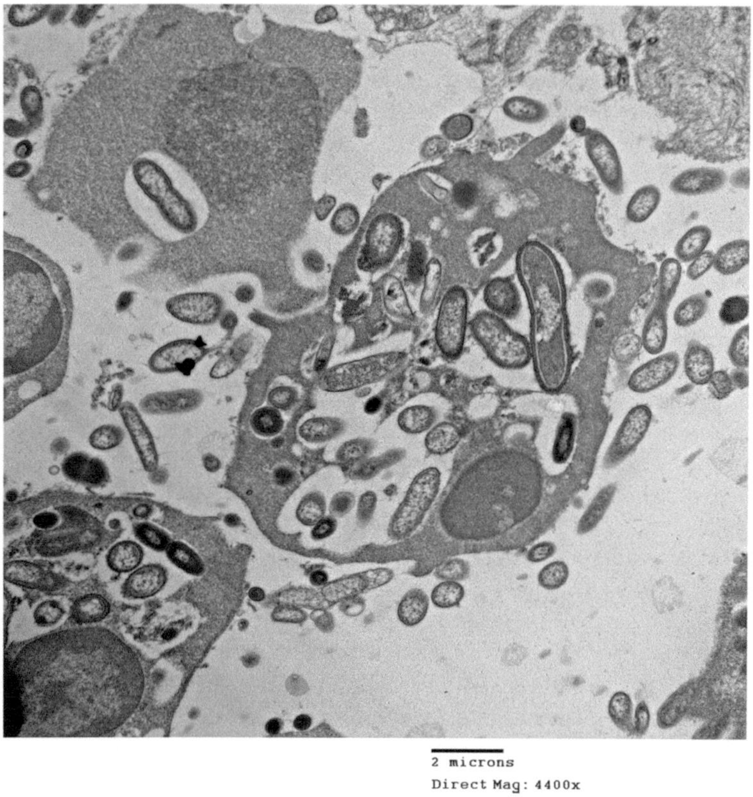

Fig. 5 PMN phagocytosis of root canal bacteria in situ

2 Nature of the Clinical Challenge

The previous section applied available biofilm concepts to develop a picture of biofilm physiology and distribution in root canal systems. Root canal infection then is essentially a bacterial biofilm coating the dentine surface (including the tubules) to variable depths and extending to the apical foramina and sometimes beyond. The biofilm is usually patchy and discontinuous; different populations of bacteria may communicate with each other through fluid films or columns by the medium of molecular messengers. Individual cells within this cooperative may respond to their immediate surroundings and neighbours by switching on or off relevant genes for survival or launching virulence gene expression. Nutrition depletion may slow down their metabolism and allow some to enter a dormant state. This may also make them more resistant to killing and uncultivable if sampled for culture. The bacterial population may therefore be diverse in terms of species and phenotypes (both biofilm and planktonic—although the true balance is unknown). This inter-reliance of the bacterial species is both therapeutically useful and potentially problematic.

The morphological complexity of the root canal system and the nature of microbial infection pitted against the limitations of contemporary instrumentation and irrigation techniques serve to characterise the severity of the clinical challenge. The morphological complexity of the root canal system is defined by variations in the shape and configurations of the root canal wall with narrow channel-like spaces and their blind-ending, surface- or space-connecting branches (Gulabivala et al. 2001). Its small size and volume also impose physical limitations on the resident fluid dynamics (Gulabivala et al. 2010). The previously described biofilm physiology imposes biological limitations on the ability of interventions to eradicate the infection (Abdullah et al. 2005; Bryce et al. 2009), depending on the agent used.

In a conceptually ideal model of biofilm eradication and wound dressing, a fluid agent would be delivered precisely to the apical termini of the root canal system without any canal enlargement, to disable the virulence of the biofilm and to disintegrate/detach and remove it. The perfectly debrided surfaces would then be rendered with a dressing delivered in fluid form to achieve perfect adaptation and prevent any repeat/subsequent microbial attachment or contamination. The empty space could be filled with a (removable) gel–sol precipitate to prevent bulk flow, ingress/egress of fluid or microbes. Such a treatment concept is a distant possibility that awaits development through multidisciplinary effort. In contrast, contemporary treatment techniques exhibit only a weak similarity to these idealised goals. Predictable and controlled delivery of fluids to the apical termini is currently not possible without some degree of canal enlargement by abrasive shaping, although the noninstrumentation technique did attempt to meet the basic principle of the approach (Lussi et al. 1995). Furthermore, the chemical activity of available fluid agents is limited in potency or by activity and therefore becomes spent on contact at a rate dictated by concentration, diffusion and mixing. The viscous domination of the root canal space compromises the delivery, flow and mixing of fluids throughout this space. Fresh unreacted fluid thus needs to be constantly circulated into the root canal system and spent reacted fluid must be moved out of the system without stagnation at the wall boundaries, a feat that is technologically challenging.

The sum total of the effect of chemomechanical intervention in the root canal system is therefore dictated by the combined interaction between root canal system shaping, nature of irrigant (type, concentration), irrigation regimen (method and rate of delivery, volume and concentration, method and frequency of agitation) and root canal system medication (type, activity and release profile of active ionic/molecular species, placement efficacy and chemical equilibria). Even within the confines of a putatively standardised chemomechanical protocol, intra-operator or interoperator variations would in all likelihood yield variations in outcome, even in controlled laboratory tests where anatomy and infection could be standardised. The outcomes of clinical/microbiological studies on root canal treatment would therefore be expected to reflect this complexity and variation. Such studies may reflect key unresolved challenges and point to possible pathways for optimisation.

The efficacy of root canal treatment has been evaluated using in vitro, ex vivo and in vivo study models. The outcome measures tested are diverse and include shaping efficacy and efficiency together with the maintenance of centring ratio, presence of residual root canal debris, extent of predentine removal, presence of smear layer, presence of residual biofilm, bacterial load reduction, altered bacterial diversity, periapical healing (radiographic or histological), tooth survival and patient reported outcomes. These measures may individually or collectively indicate the potential effectiveness and variability of the procedure, as well as any ceiling thresholds that may exist.

3 Rationale, Practical Concepts and Principles Underpinning Root Canal Treatment

3.1 Rationale

Based on the fact that periapical lesions develop as a result of the *interaction between* bacteria (and their products) and the host defences, it is clear that their prevention or resolution depends upon preventing or terminating this interaction. Given the relatively long history of root canal treatment and the relatively recent establishment of biological rationale (Gulabivala and Ng 2009), a series of different conceptual frameworks have been used over the decades and centuries to justify the approach used. Curiously, the principles of the approach have not altered for a very long time (Hall 1928), whilst technological advancements have made the procedures more efficient and amenable to the technical achievement of shaping and filling the root canal systems.

The treatment rationale for teeth without clinical evidence for apical periodontitis and for those with associated apical periodontitis is biologically different. In the former case, depending upon the precise clinical reason (elective, restorative, endodontic) for treatment, the main goal will be to prevent introduction of microbial infection into the apical anatomy, either from the environment or from the coronal part of the root canal system. The focus is therefore on prevention of microbial contamination/infection spread. On the whole, root canal treatment under such circumstances may have slightly different technical and biological challenges and is accompanied by evidence that a wide variety of protocols may suffice, giving high success rates with a relatively narrow range, as judged by continued clinical absence of apical periodontitis (Fig. 6).

Once the periapical lesion has become established, variations in the technical aspects of the chemomechanical protocol appear to have a greater impact on outcome (Fig. 7). Overall, the challenge is a different one because the purpose under these circumstances is not merely to prevent new contamination/infection but to eradicate the established bacterial biofilm and effect switching off of the host immune response coupled with accelerating wound healing. The challenge seems to

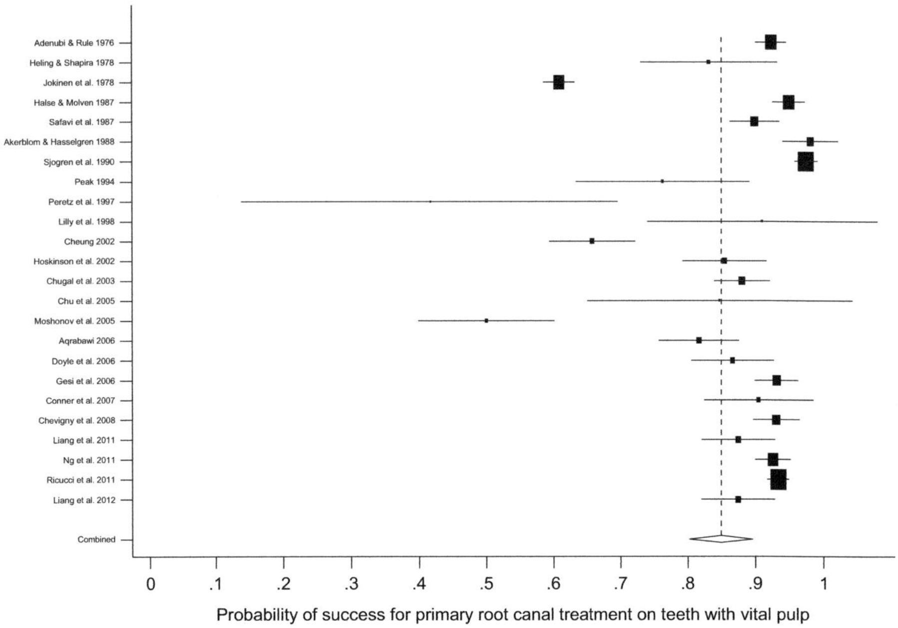

Fig. 6 Forest plot of clinical outcome data on presence/absence of apical radiolucency following treatment of teeth without prior apical periodontitis

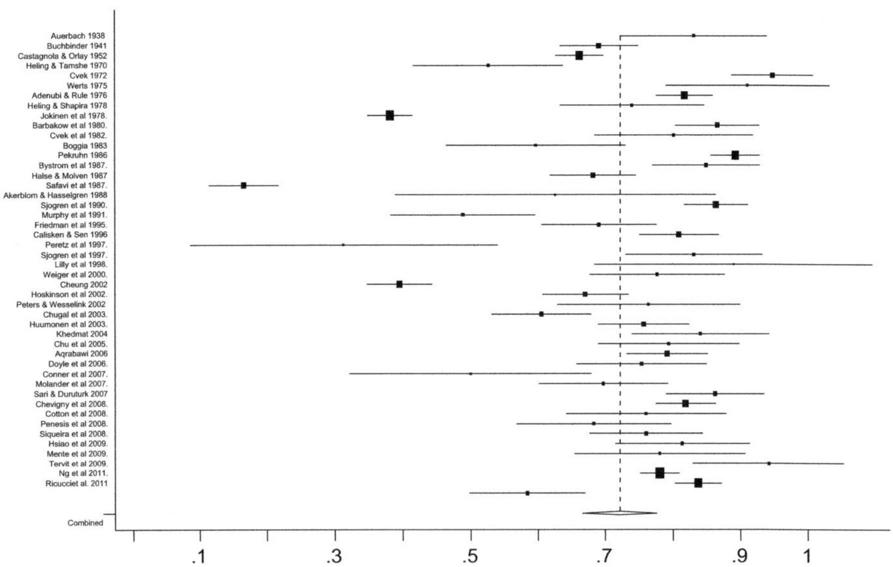

Fig. 7 Forest plot of clinical outcome data on presence/absence of apical radiolucency following treatment of teeth with prior apical periodontitis

be greater still if the periapical lesion is larger as it is associated with a microbiota of greater diversity. A number of protocols and conceptual models (Gulabivala and Ng 2009) have been used to achieve this general aim with an overall lower and broader range of success rates than in the case of teeth without apical periodontitis.

3.2 *Practical Concepts and Principles*

Root canal treatment is a multistep procedure, where each subsequent step is dependent on the effectiveness of the previous step. Root canal treatment is thought of in practical terms as comprising of the following steps:

Mechanical intraradicular preparation:

  Preceded by entry access and length determination

Chemical intraradicular preparation:

  Intra-appointment medication (root canal irrigation)
  Inter-appointment medication (root canal dressing)

Root canal system obturation

3.2.1 Principles of Mechanical Intraradicular Preparation

The first step involves creation of access to the root canal system by drilling a cavity from the occlusal surface to the root canal system: the *coronal access cavity*. Historically, mechanical preparation of the entire wall of the root canal system had been considered the most important means of bacterial biofilm removal. However, there has been a paradigm shift in the way this phase is now viewed, since it is now regarded as no more than an extension of the coronal access cavity into the root, that is, a *radicular access cavity* (Gulabivala and Stock 1995). This is principally because up to half of the surface of the root canal system surface remains uninstrumented, requiring the remainder to be cleansed of biofilm by means other than planing with metal instruments.

It has traditionally been held that canals should be prepared by controlled dentine removal so as to produce a regular taper with the minimum diameter at the apical constriction or canal terminus and the maximum diameter at the coronal end. The central axis of the canal should remain undeviated regardless of curvature, whilst the taper should be the minimal compatible with the delivery of irrigant and root filling material, so that root strength and integrity are preserved.

The apical diameter size and degree of taper selected as optimal have been the subject of much debate. The choice is usually based on personal preference and individual clinical experience rather than on sound scientific rationale. Widely tapered canals may allow better irrigant penetration, debridement and obturation, but these benefits are achieved at the expense of root strength and, possibly, long-term survival of the tooth. Advocates of narrowly prepared canals argue that a taper that allows irrigant penetration using narrow needles is sufficient for debridement and that obturation of such canals can be satisfactorily achieved with thermoplasticised gutta-percha techniques. Narrowly tapered preparations, if they allow adequate cleaning and obturation, are more desirable as they do not compromise root strength. Based on the irrigation requirements of the root canal system, teeth may be divided into those with simple or complex canal systems.

Simple Canal Systems (Types A, B and C)

In simple canal systems with narrow circular cross sections (Type A), preparation to a regularly tapered radicular access cavity may entirely (Fig. 8) encompass the original canal system. In such cases, biofilm removal may be achieved almost wholly by mechanical preparation with little reliance on the irrigant.

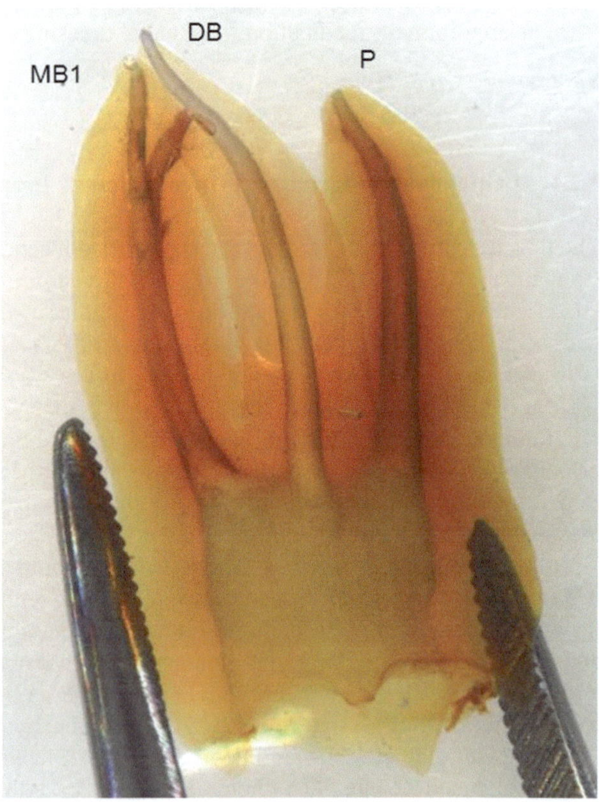

Fig. 8 Simple Type A canal system with canals so narrow and circular that complete wall debridement may be achieved by mechanical means alone without reliance on a chemical effect of the irrigant

Fig. 9 Simple Type B canal system with broad cross section that precludes complete canal wall debridement with mechanical means alone and requires reliance on a chemical effect of the irrigant

In contrast, in simple canal systems with wide non-circular but regular cross sections (Type B), such as canines, some premolars, distal and palatal roots of molars and teeth with 'C'-shaped canals, even preparation to a regularly tapered radicular access cavity may not allow complete biofilm debridement solely by mechanical preparation (Fig. 9). Access for irrigation is excellent in such teeth making canal enlargement by filing, redundant.

In further contrast, in simple canal systems with narrow circular cross sections and fins and ramifications extending off them (Type C), mechanical preparation would remove biofilms from the accessible walls but not from the uninstrumented accessory anatomy (Fig. 10). Examples of such anatomy may be drawn from all tooth types.

Complex Canal Systems

In teeth with more morphologically complex and irregular canal systems, the prepared radicular access cavity remains encompassed or surrounded by the original irregular canal system to a greater or lesser degree and may only be partly discernible or not at all. Examples may be drawn from mesial roots of mandibular or maxillary molars, mandibular incisors and some premolars, usually only when viewed proximally though (Fig. 11).

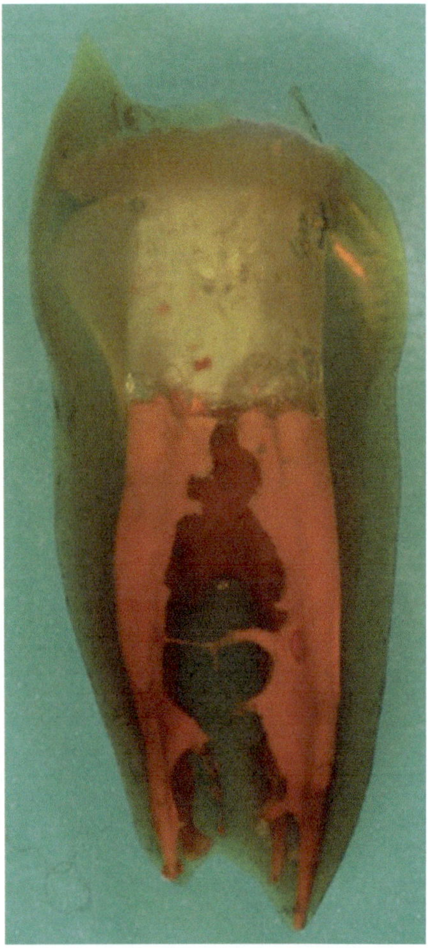

Fig. 10 Simple Type C canal system with narrow main canal and blind ending of communicating spaces extending the main canal at various angles that allows complete wall debridement of the main canal (as in Type A) but precludes wall debridement of the extensions without chemical means

3.2.2 Principles of Chemical Intraradicular Preparation (Root Canal Irrigation)

In order for the chemical fluid agent to reach all surfaces of the root canal system, a sufficient path must exist and the flow of the fluid irrigant must be facilitated. This requires that obstructions in the path have to be cleared of pulp tissue, dystrophic pulp calcification, pulp stones and bacterial biofilm. The mechanical shaping process may disrupt and displace such contents; any dentine chips generated from the shaping process would add a further element to the canal content. This heterogeneous mass of organic and inorganic tissue should ideally be flushed out of the canal system.

Pure flushing action is restricted by the viscously dominated environment of the root canal system, so an understanding of fluid dynamics is mandatory (Gulabivala et al. 2010). Strategies for delivering, mixing and replacing the irrigant must be effective. In addition, pure flushing action is only effective for loose debris; most of

a b

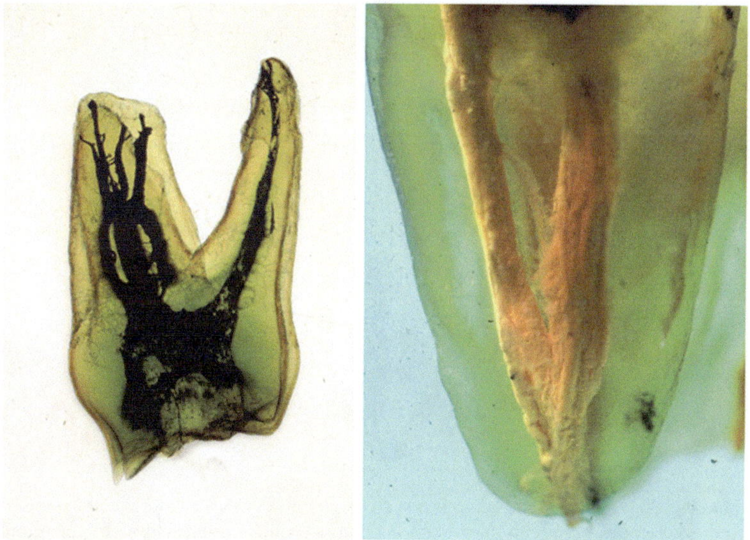

Fig. 11 Complex canal system whose overall wall geometry would remain predominantly unaltered by mechanical instrumentation and that would preclude debridement of the majority of the canal wall surface without the aid of chemical irrigation means: two separate examples, (**a**) unprepared and (**b**) root filled

the contents are likely to be attached or adherent and must therefore be loosened by agitation or chemical solvent action. Therefore, the act of flushing must exert a wall shear stress, be prolonged or repetitive and preferably also occur during mechanical preparation, although post-mechanical flushing is also considered advantageous.

The process of bacterial biofilm disruption by antibacterial fluids is complicated by the challenge of delivering the irrigant to the narrow, confined and distant apical anatomy and to the deepest layers of the multilayered microbial biofilm with its polysaccharide matrix protection. The first problem is overcome by enlarging and shaping the radicular access sufficiently to allow the irrigant to be delivered to the apical anatomy using an appropriately sized, narrow-gauge 'endodontic' needle, together with some form of agitation. The second problem is overcome by using a high enough concentration and/or sufficient volume of irrigant through continuous replenishment to provide sustained chemical potency. The combined action of mechanical and chemical cleaning is more efficient than either method alone in biofilm removal.

Once the mechanical shaping is completed, a final effort is made to reach all instrumented and, *particularly*, uninstrumented surfaces to allow chemical 'scrubbing' of the entire root canal wall to remove *biofilm from the uninstrumented surface* and any contaminated *smear layer from the instrumented surface* (Fig. 12). The strategies for delivering, agitating and mixing the irrigant into the complex anatomy from the *prepared* radicular access have increased multifold in recent years as this need became better understood. The approaches may be classified into (Gulabivala et al. 2010):

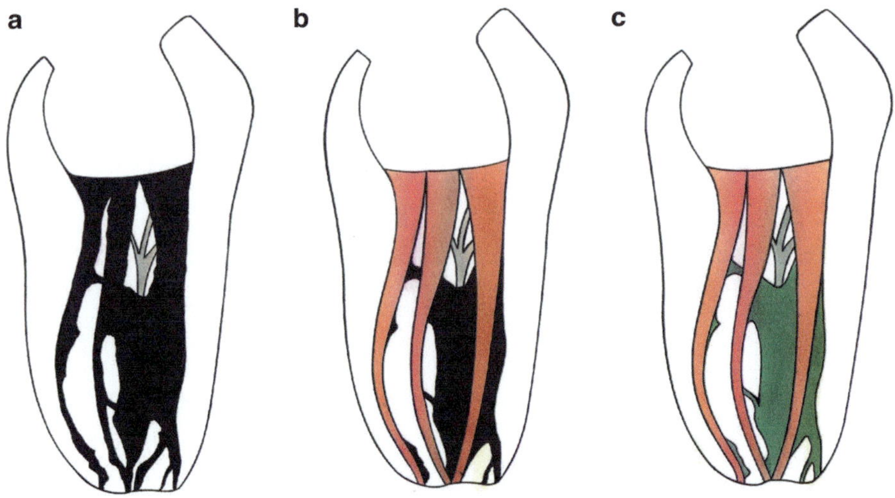

Fig. 12 (**a**) Unprepared complex root canal system, (**b**) instrumented root canal system revealing uninstrumented surfaces, (**c**) green uninstrumented area requires debridement by chemical irrigant delivered via the radicular preparation

- *Manual agitation*: using hand files propelled in a push–pull motion, a gutta-percha cone used in a push–pull motion
- *Sonic agitation*: using SonicAire® activated file, EndoActivator® and Vibringe® (vibrating needle attached to a syringe)
- *Ultrasonic agitation*: using Irrisafe file, ProUltra syringe (vibrating needle) and ultrasonically activated normal file
- *Pressure/vacuum agitation*: EndoVac, RinsEndo and NIT system
- Laser-induced agitation

The simplest method is to reciprocate a gutta-percha cone to agitate the irrigant. With this method, there is an increased risk of irrigant extrusion through the apical foramina. The optimal method for maximising wall shear stress and reducing apical extrusion is to use a gutta-percha cone that fits well apically but is slightly loose coronally.

Activation of irrigant with sonic or ultrasonic devices involves the agitation of either a syringe-delivered bolus of fluid in the canal or by simultaneous delivery (via a custom reservoir) and agitation. The sonic devices available for irrigant agitation include the Sonic Air Endo Handpiece (Micromega 1500, Besancon, France), which is driven by air pressure to produce vibration frequencies of 1500–3000 Hz and which also drives stainless steel files to aid simultaneous canal preparation. The EndoActivator® system (Advanced Endodontics, Santa Barbara, CA, USA) is electrically driven and works at the much lower frequencies of 33, 100 and 167 Hz. It was designed to use polymer tips of various sizes (ISO size 15, 25, 35) and tapers (0.02, 0.04) for agitation of the irrigant to avoid the potential risks of instrument separation associated with metal files, ledge formation and canal transportation. Preliminary data on debris and smear layer removal was promising (Ruddle 2007). Vibringe consists of a sonically activated device attached to a syringe, which creates agitation as the irrigant is delivered.

Much higher frequencies of vibration of the working tip (in the ultrasonic range, 20–40 kHz) are achieved with either magnetostrictive or piezoelectric devices. *Magnetostrictive transducers* produce an elliptical motion at the working tip, whilst *piezoelectric* transducers produce longitudinal or transverse linear motions. In either case, the oscillations are damped by contact of the file with the canal wall, also reducing the acoustic micro-streaming. Acoustic micro-streaming is enhanced by higher power output and greater flexibility of the file; therefore a small size is better. To maximise acoustic micro-streaming, it is best to use the ultrasonically energised file without contacting the canal walls; this is achieved by precurving the instrument and using it after canal preparation is complete, that is, the so-called but miss-labelled passive ultrasonic irrigation.

EndoVac® is a simple but ingenious device that delivers the irrigant into the coronal chamber but sucks the irrigant down into the canal through an attachment that passes to the apical end of the radicular preparation. There are two sizes of attachments; the smaller of the two is liable to blockage if the instructions are not followed. So far, the research has proved positive. Another device, the NIT®, also uses simultaneous delivery and suction, but this time through a single needle allowing separate flows in the inner and outer casings.

RinsEndo is an automated device operated by attachment to a handpiece coupling. It uses alternating positive and negative pressure changes to drive irrigant into the root canal system. It is moderately effective.

Apart from agitation, the exchange of irrigant relies on diffusion along a concentration gradient from the main radicular access, where the bolus is deposited towards the peripheral anatomical aspects. The efficacy of this also demands a higher delivered concentration, which could damage the dentine in the radicular preparation. If the antibacterial agents reach the bacteria in subinhibitory doses, there is a high chance of inducing resistance to them. This has been demonstrated not only for antibiotics but also for the antiseptic agents mentioned above.

In the majority of cases, these efforts are sufficient to eradicate the microbiota from the coronal aspects of the canal and also mostly from the middle aspects of the root canal system. However, the challenge lies in complete eradication from the apical part, where the bacterial biofilm usually remains virtually intact in the apical anatomy (Nair et al. 2005). It is therefore useful to use additional antibacterial measures that continue to work in the inter-appointment period (Vera et al. 2012). Overenthusiastic efforts to deliver the irrigant to the apical anatomy could result in irrigant extrusion. The consequences may range from minimal exacerbation of irritation to a severe sodium hypochlorite accident with tissue destruction.

3.2.3 Principles of Chemical Intraradicular Preparation (Root Canal Dressing)

The rationale for intra-canal medication between appointments is to help *complete* the aim of root canal system preparation, that is, to degrade *residual* microbial biofilm and organic tissue and to kill remaining bacteria that have escaped intra-appointment primary debridement. The medicament should, therefore, also prevent bacterial recolonisation of the root canal system from either those bacteria left behind after preparation or new invaders through lateral communications or the coronal access.

3.2.4 Principles of Root Canal System Obturation

Following debridement and medication of the root canal system, migration of microorganisms and their products from the oral cavity or root canal system to the periradicular tissues is prevented by obturating the space and sealing the access cavity permanently. The purpose is therefore to 'seal' the *entire* root canal system, including the coronal entry, whilst encouraging healing of the chronically inflamed periapical wound.

3.3 Conceptualisation of the Periapical Healing Process

The healing process after root canal treatment has not been deeply researched but can be conceptualised using the 'zones of Fish' in a chronic inflammatory lesion (Fish 1939). The removal of bacteria and their products should result in the reduction, if not (ideally) elimination, of the zones of infection and contamination. This allows the macrophages in the zone of irritation to invade the areas previously occupied by the zones of infection and contamination in order to remove dead cells and debris. This process also makes way for the osteoblasts and fibroblasts, together with new in-growing blood vessels and nerve fibres from the outermost and active zone of stimulation to proliferate into the zone of irritation. In this way, gradual healing may take place from the boundary of the lesion inwards until a normal periodontal ligament is established. Provided that the pluripotential cells in the periodontal tissues, in particular, have not been irreversibly damaged, ideal healing would eventually result in regeneration and the formation of cementum over the apical terminus, isolating the root canal system completely from the periapex, but this is not an inevitable end result. Incomplete removal of the infection would reduce but not eliminate the inflammatory area, resulting in suppression of the impetus for wound healing.

4 Evidence for Root Canal Treatment Efficacy

Studies conducted over many decades provide some insight about the relative efficacy of the various aspects or steps of the root canal treatment procedure(s).

4.1 *Effect of Chemomechanical Procedures on Intraradicular Microbiota*

The effect of different stages of root canal treatment on the intraradicular microbiota has been evaluated both qualitatively and quantitatively. Some studies merely report positive culture tests, whereas others have identified and quantified intraradicular microbiota before and after various stages of treatment, mostly using culture-dependent approaches and some using culture-independent approaches (Table 1).

The effect of the steps up to and including the ‘mechanical preparation’ of the canal(s) on the microbiota has been tested using only water or saline as the irrigant. Taken collectively, the studies show that negative cultures were achieved on average in 25 % of the cases (range 4.6–53 %). When sodium hypochlorite (concentration range 0.5–5.0 %) irrigation supplemented the steps up to ‘mechanical preparation’, the frequency of negative cultures immediately after debridement increased to an average of 75 % (range 25–98 %).

Most studies report culture reversals during the inter-appointment period when active antibacterial dressing is not used in the root canal system between appointments. The reversals are due to regrowth of residual bacteria or recontamination by bacterial leakage around the access cavity dressing.

Classical and well-controlled studies (Bystrom and Sundqvist 1981, 1983, 1985) evaluated the effect of various root canal treatment procedures on the microbiota, both qualitatively and quantitatively. They tested the effect of mechanical preparation, saline or sodium hypochlorite irrigation (0.5, 5.0, 5.0 % with EDTA), the addition of ultrasonic activation to the irrigation and calcium hydroxide dressing; each addition to the chemical canal preparation improved the antibacterial effect, reducing residual bacteria further. They found the antibacterial action to reduce the number of bacteria from an initial range of 10^2–10^8 cells to 10^2–10^3 fewer cells after initial debridement, further reducing down to no recoverable cells (from the prepared part of the root canal system) after inter-appointment dressing with calcium hydroxide. *It was notablethat the infection was more difficult to control when the diversity of the initial infection was greater, i.e., there were more species in greater abundance.* Calcium hydroxide dressing is also effective after mechanical preparation and irrigation with water. The efficacy of dressing with calcium hydroxide has recently become controversial because of emerging studies showing limited efficacy.

The collective antibacterial action during root canal treatment has not been shown to cause persistence of any particular species. Specific bacteria, therefore,

Table 1 Summary of studies evaluating the effect of root canal treatment procedures on the microbiota. For information on terms used within the table, refer the KEY at the end of the table

Study and year	Tx type M/I/D	Sample size	RD	Decont	Outcomes: Percentage of samples with bacterial presence/mean bacterial load: At baseline	After preparation ± irrigation	Next visit (after dressing±)	Persistent species (frequency)
Cultural								
Auerbach (1953)	M/I	60 teeth	✓	–	93 % (56/60)	Chlorinated soda (double strength: 22 % (12/56)	–	Not analysed
Stewart (1955)	M/I	50 teeth	✓	✓	–	3 %H_2O_2 and 0.5 % NaOCl: 6 % (3/50)	No dressing: 24 % (12/50)	Not analysed
Ingle and Zeldow (1958)	M	89 teeth	✓	✓	73 % (65/89)	H_2O: 70 % (62/89) Some initially –ve became +ve after Tx	–	Not analysed
Stewart et al. (1961)	M/I	77 teeth	✓	✓	100 % (77/77)	0.5 %NaOCl + Gly-oxide: 2 % (1/44) 0.5 % NaOCl + 3 %H_2O_2: 9 % (3/33)	No dressing: 0.5 %NaOCl + Gly-oxide: 34 % (15/44) 0.5 %NaOCl + 3 %H_2O_2: 39 % (17/33)	Not analysed
Melville and Slack (1961)	M/D	392 root canals	✓	✓	–	–	Eugenol beechwood creosote, tricresol, CMCP, or antibiotics: no data	695 pure isolates from 601 specimens: streptococci (36 %), lactobacilli (2 %), staphylococci (27 %), micrococci (11 %), *Neisseria* (6 %), G –ve rods (4 %), yeasts (7 %)
Nicholls (1962)	M/I	155 teeth	✓	✓	100 % (155/155)	Alkaline chloramine: 53 % (39/74) H_2O_2 and 2 %NaOCl: 50 % (30/60) H_2O and 2 %NaOCl: 71 % (15/21)	–	Not analysed

Grahnén and Krasse (1963)	M/I	97 teeth	✓	✓	77 % (75/97)	NaCl: 72 % (23/32) Biocept: 66 % (21/32) Nebacin: 36 % (12/33) Some initially –ve became +ve after Tx	No dressing: NaCl: 47 % (15/32) Biocept: 47 % (15/32) Nebacin: 18 % (6/33)	Not analysed
Engström (1964)	M/I/D	223 teeth	✓	✓	60 % (134/223), 20/134 +ve for enterococci (E) (21 % in RTxed teeth, 12 % in UnTxed teeth)	Biosept or iodophor, plus alcohol, chloroform and 0.5 % NaOCl: no data	5 % I_2 in 10 % IKI: 2nd visit: 43 % (58/134), 13/20 +ve E.f 3rd visit: 22 % (29/134), 5/20 +ve E.f 4th visit: 8 % (9/134), 2/20 +ve E.f 5th visit: 3 % (4/134), 2/20 +ve E.f 6th visit:2 % (3/134), 2/20 +ve E.f 7th visit:16 % (22/134), 3/20 +ve E.f	Enterococci persisted in canals at each visit with greater frequency than other organisms. Enterococci present more frequently in proximal areas of teeth with enterococci, 67 % vs. 31 %
Engström and Frostell (1964)	M/I/D	1170 samples	✓	✓	–	Biocept, alcohol, chloroform, 0.5 % NaOCl: no data	5 % I_2 in 10 % IKI: 25 % (332/1170) UnTxed with no pa:31 % (91/297) UnTxed with pa: 37 % (97/261) Txed with no pa: 21 (80/377) UnTxed with pa: 23 (54/235)	Before root filling, surviving species included: *Streptococcus, Staphylococcus, Enterococcus*, yeast, aerobic G +ve spore forming rods, *Micrococci*, diphtheroids, typical G +ve rods
Olgart (1969)	M/I/D	207 teeth	?	✓	72 % (149/207)	43 % (88/207) No difference between H_2O_2 and 0.5 %NaOCl vs. H_2O_2 and 1 % NaOCl	No dressing: 34 % (70/207)	*Streptococcus*, *Lactobacillus*, *Peptococcus*, *Peptostreptococcus*, *Veillonella*, *Fusobacterium*, *Corynebacterium*, *Eubacterium*, *Leptotrichia* All types persistent, simply dependent upon initial types

(continued)

Table 1 (continued)

Study and year	Tx type M/I/D	Sample size	RD	Decont	Outcomes: Percentage of samples with bacterial presence/mean bacterial load – At baseline	After preparation ± irrigation	Next visit (after dressing±)	Persistent species (frequency)
Goldman and Pearson (1969)	M/I	563 patients	?	?	–	NaOCl: 24 % (133/563)	–	242 organisms identified: *Enterococci* (most common), streptococci (64 %), staphylococci (17 %), *Neisseria*, *Lactobacillus*, *Micrococcus*, *Corynebacteria*, AnO_2 G +ve cocci
Myers et al. (1969)	?	214 Txed teeth	?	?	–	–	r culture: After 1 –ve: 26 % (28/108) After 2 –ve: 13 % (14/106) Reversal rate after 1 –ve culture = 26 %, after 2 –ve cultures = 13 %	*S. faecalis* (21 %), streptococci (52 %), staphylococci (8 %), Corynebacteria (4 %), unclassified G +ve rods (8 %)
Bence et al. (1973)	M/I	33 teeth	✓	✓	100 % (33/33)	Pre-irritation: 1st file: 93 %, enlargement with # 3 = 14 %, #4 = 11 %, #5 = 21 % (32 % of instruments showed +ve culture, regardless of size) 5.25 %NaOCl 48 h culture: 4 % dentine, 10 % pp 5 day culture: 8 % dentine, 26 % pp	No dressing: 8 % dentine, 12 % pp samples of teeth with negative culture after irrigation	Not analysed
Tsatsas et al. (1974)	M/I/D	205 teeth with a –ve culture after treatment**	?	?	–	–	Cresatin, CMCP, or polyantibiotic paste: 2nd culture: 24 % (49/205) 3rd culture: 80 % (39/49)	Streptococci (82 %), *Bacillus* (33 %), staphylococci (8 %), *Candida* (3 %), *Pseudomonas pyocyaneus* (3 %)

Mejàre (1975)	NA	612 cultures at RF with a –ve previous culture	✓	✓	–	–	r culture: 15 % (92/612), 27/92 yielded 29 isolates of enterococci	*S. faecalis*, subspecies *faecalis* (10), *zymogens* (3), *liquefaciens* (8), *atypical variate* (6). *S. faecium, var faecium* (1), *var durans* (1)
Akpata (1976)	M/D	20 extracted teeth	NA	✓	100 % (20/20)	NaCl: 65 % (13/20) Load = 200 CFUs/tooth	38 %CMCP: 20 % (2/10) Load = 50 CFUs/tooth When PP sample –ve, crushed tooth yielded –ve culture When PP +ve, crushed teeth yielded +ve or –ve cultures	Not analysed
Cvek et al. (1976)	M/I	108 teeth	✓	✓	NaCl group: 53 % (18/34) 0.5 % NaOCl group: 63 % (29/46) 5 % NaOCl group: 79 % (22/28)	NaCl: 83 % (15/18) 0.5 % NaOCl: 59 % (17/29) 5 % NaOCl: 68 % (15/22)	–	Staphylococci, *S. faecalis*, *Micrococcus*, *Peptostreptococcus*, *Lactobacillus*, *Corynebacteria*, *Propionibacterium*, *Actinomyces*, *Veillonella*, *Bacteroides* spp., *Eubacterium alactolyticum*, *Fusobacterium*, *Leptotrichia*, *Haemophilus*, *Bacillus subtilis*
Bystrom and Sundqvist (1981)	M/I	15 teeth	✓	✓	100 % 15/15, load = 10^5 (10^2–10^7) 89 strains in all. 1–10 strains/canal	Saline: 100 % (15/15)	No dressing: 47 % (7/15) (5th visit) Where initial bacteria load high, difficult to eliminate	No specific species were persistent Streptococci, *Peptostreptococcus*, *Arachnia*, *Eubacterium*, *Lactobacillus*, *Actinomyces* spp., *Fusobacterium*, *Bacteroides*, *Wolinella recta*, *Capnocytophaga*, *Enterobacter agglomerans*, *Veillonella parvula*, *Selenomonas sputigena*, *Eikenella corrodens*

(continued)

Table 1 (continued)

Study and year	Tx type M/I/D	Sample size	RD	Decont	Outcomes: Percentage of samples with bacterial presence/mean bacterial load – At baseline	After preparation ± irrigation	Next visit (after dressing±)	Persistent species (frequency)
Bystrom and Sundqvist (1983)	M/I	15 teeth	✓	✓	100 % (15/15); 169 strains isolated; 1–11 strains/canal; 2/canal if 10^5 cells and 8 if more cells	0.5 % NaOCl: 87 % (13/15)	No dressing: 20 % (3/15) (5th visit)	None of the species were more resistant to Tx Streptococci, *Peptostreptococcus, Arachnia, Eubacterium, Lactobacillus, Actinomyces* spp., *Fusobacterium, Bacteroides, Wolinella recta, Capnocytophaga, Enterobacter agglomerans, Veillonella parvula*
Bystrom and Sundqvist (1985)	M/I	60 teeth	✓	✓	100 % (60/60) Load = 10^5	0.5 % NaOCl: no data 5 % NaOCl: no data 5 % NaOCl + 15 % EDTA: no data	No dressing 0.5 % NaOCl: 12/20 (2nd visit); 8/20 load = 10^5 (3rd visit) 5 % NaOCl: 10/20 (2nd visit); 6/20, load = 10^6 (3rd visit) 5 % NaOCl + 15 % EDTA: 11/20 (2nd visit); 3/20, load = 10^5 (3rd visit)	44 strains recovered at 3rd visit; 80 % AnO_2. No indication that specific bacteria were resistant to Tx Streptococci, *Peptostreptococcus, Arachnia, Eubacterium, Lactobacillus, Actinomyces* spp., *Fusobacterium, Bacteroides, Wolinella recta, Capnocytophaga ochracea, Veillonella parvula*
Byström et al. (1985)	M/I/D	65 teeth	✓	✓	100 % (65/65) Load: CH gp: 10^2–10^7 CP/CMCP gp: 10^3–10^7	0.5 % NaOCl: no data 5.0 % NaOCl: no data	CH: 0/35 (1 month), 1/35 (2–4 days) CP/CMCP (2 weeks): 10/30, load = 10^5	Mostly AnO_2 G +ve bacteria *Bacteroides, Fusobacterium, Propionibacterium, Peptostreptococcus, Eubacterium, Streptococcus, Actinomyces, Lactobacillus, Bifidobacterium, Enterococcus* spp.

Sjögren and Sundqvist (1987)	M/I	31 teeth	✓	✓	100 % (31/31) Load = 10^4	0.5 % NaOCl plus ultrasonic debridement: no data	No dressing: 29 % (9/31) at 2nd visit 23 % (7/31) at 3rd visit	*Fusobacterium nucleatum*, *Arachnia propionica*, *Actinomyces*, *Streptococcus intermedius*, *Peptostreptococcus*, *Streptococcus milleri*
Koontongkaew et al. (1988)	M/I/D	15 teeth	✓	✓	100 % (15/15)	3 % H_2O_2/5.25 % NaOCl: no data	CMCP: 1 day dressing: 40 % (2/5) 3 day dressing: 20 % (1/5) 7 day dressing 10 % (1/10) No dressing: 60 % (3/5) after 1 day, 20 % (1/5) after 3 or 7 days	Not analysed
Reit and Dahlén (1988)	M/I/D	35 teeth	✓	✓	91 % (32/35) 1–7 strains/canal	0.5 % NaOCl: no data	CH: After 14 days: 23 % (8/35) After 21 days: 26 % (9/35)	After 21-day CH dressing: 12 strains were isolated from 9 canals Species included G +ve AnO_2 rods, *Enterococcus*, *Lactobacillus*
Cavalleri et al. (1989)	M/I	10 teeth	✓	?	–	5 % NaOCl: no data	–	Persistent species: *Streptococcus epidermidis* (40 %), *Aerococci* (30 %), *Bacteroides fragilis* (20 %) New species: *Klebsiella rhinoscleromatis* (20 %), *S. faecalis* (20 %), *Acinetobacter calcoaceticus* (20 %), *Actinomyces odontolyticus* (10 %)
Molander et al. (1990)	M/I/D	25 teeth	✓	✓	96 % (24/25) 81 strains isolated	0.04 % iodine: no data	Clindamycin: after 14 days: 16 % (4/25) After 21 days: 24 % (6/25)	Only two strains of enterococci in pre-Tx flora but post-Tx flora dominated by enterococci

(continued)

Table 1 (continued)

Study and year	Tx type M/I/D	Sample size	RD	Decont	Outcomes: Percentage of samples with bacterial presence/mean bacterial load: At baseline	After preparation ± irrigation	Next visit (after dressing±)	Persistent species (frequency)
Sjögren et al. (1991)	M/I/D	30 teeth	✓	✓	100 % (30/30) Load: 10 min samples, 10^3 (10^2–10^8) 7 day samples, 10^4 (10^2–10^6)	0.5 % NaOCl: 50 % (15/30) Load: 10^2–10^3	CH: 10 min: 50 % (6/12) at 1 week later 7 day: 0 % (0/18) (none after 1–5 weeks later without dressing)	*Fusobacterium*, *Bacteroides*, *E. faecalis*, *Actinomyces*, *Peptostreptococcus*, *Eubacterium*, *Lactobacillus*, *Wolinella recta*
Ørstavik et al. (1991)	M/I/D	23 teeth	✓	✓	96 % (22/23)	NaCl irrigation and enlarged to #20–25: 87 % (20/23); load = 10^4 further to #35–80. No data	CH: 34 % (8/23); load = 10^4 #35/40: 40 % (6/15) # > 40: 25 % (2/8)	Not analysed
Yared and Bou Dagher (1994)	M/I/D	60 teeth	✓	✓	100 % (60/60)	1 % NaOCl, enlarged to #25, 73 % (22/30) enlarged to #40, 23 % (7/30)	CH: 0 % (0/60)	Not analysed
Gomes et al. (1996)	M/I	42 root canals Primary Tx ($n = 15$) Secondary Tx ($n = 27$)	✓	✓	95 % (40/42); 1–9 strains/ tooth	2.5 % NaOCl: no data	Empty canal (7–10 days): 73 % (29/40); 1–8 strains/tooth	After instrumentation and dressing, Oblig AnO_2 (52 %), facultative AnO_2 (64 %); G +ve (74 %); G −ve (36 %) were found In 1° cases, no significant qualitative change in microbiota In 2° cases: prevalence compared with initial samples Decreased = *Peptostreptococcus* Increased = *Veillonella*, *Porphyromonas*, *Propionibacterium*, *Enterococcus* Same = staphylococci, *Bacteroides*, *Fusobacterium*

Sjögren et al. (1997	M/I	55 teeth (single canal)	✓	✓	100 % (55/55)	0.5 % NaOCl: 40 % (22/55)	–	After instrumentation 1–6 species per canal. Surviving species included: *Enterococcus*, *Streptococcus*, *Peptostreptococcus*, *Eubacterium*, *Actinomyces*, *Propionibacterium*, *Bacteroides*, *Prevotella*, *Fusobacterium*, *Campylobacter*
Siren et al. (1997)	?	80 +ve canal samples (40 with enterics, 40 with non-enterics)	✓	✓	–	–	–	Enteric gp; *E. faecalis* 24 (8 monoinfections), *Enterobacter cloacae* 5(2), *Enterobacter agglomerans* 1(1), *Enterobacter sakazakii* 1, *Klebsiella oxytoga* 1, *Acinetobacter* spp. 1, *Pseudomonas aeruginosa* 1(1) brackets = pure cultures. These occurred with *Streptococcus*, *Actinomyces*, *Micrococcus*, *Staphylococcus*, *Neisseria*, *P. micros*, *Fusobacterium*, *Prevotella*, *Campylobacter* and *Candida* Non-enteric gp; *Streptococcus*, *Actinomyces*, *Micrococcus*, *Lactobacillus*, *Neisseria*, *P. micros*, *Fusobacterium*, *Prevotella*, *Porphyromonas* and *Candida*; Assoc with clinical factors all significant

(continued)

Table 1 (continued)

Study and year	Tx type M/I/D	Sample size	RD	Decont	Outcomes: Percentage of samples with bacterial presence/mean bacterial load: At baseline	After preparation ± irrigation	Next visit (after dressing±)	Persistent species (frequency)
Waltimo et al. (1997)	?	967 canal samples (resistant cases)	✓	✓	–	–	No medication for 2 days before sampling: 72 % (692/967), 7 % (47/692) +ve for yeasts	48 isolates; all except one *Candida albicans, C. glabrata, C. guilliermondii, C. inconspicua*; exception was *Geotrichum candidum*; 6 (13 %) in pure culture and 41 (87 %) in mixed culture; of these facultative bacteria were G +ve in 39/41 samples; 1 had G −ve bacteria (*Capnocytophaga & Neisseria*). Most frequent G +ve bacteria were streptococci. AnO_2 bacteria (*P. micros & Fusobacterium*) were isolated from 12 canals (26 %)
Dalton et al. (1998)	M/I	46 teeth	✓	✓	100 % (46/46) Load = 10^5	NaCl + NiTi files, 68 % (15/22); NaCl + K files, 75 % (18/24) Load = 10^2	–	Not analysed
Reit et al. (1999)	M/I/D	50 teeth	✓	✓	84 % (42/50)	Enlarged to #35 (curved) or #50 (straight) with 0.5 % NaOCl: no data	5 % IKI (5–7 days): 44 % (22/50) Empty (7 days): 44 % (22/50)	After instrumentation and dressing, 58 % persistent bacteria were facultative. Surviving species included *Prevotella, Porphyromonas, Capnocytophaga, Eubacterium, Veillonella, Peptostreptococcus, Lactobacillus, Streptococcus, Enterococcus, Staphylococcus*
Peciuliene et al. (2000)	M/I/D	25 teeth		✓	80 % (20/25); 70 % (14/20) +ve *E. faecalis*	2.5 % NaOCl and 17 % EDTA: no data	Medication unknown: 28 % (7/25); 71 % (5/7) +ve *E. faecalis*	All residual *E. faecalis* were present in pure culture

Shuping et al. (2000)	M/I/D	42 teeth	✓	✓	98 % (41/42) Load = 10^5	1.25 % NaOCl: 38 % (16/42) Load = 10^1	CH: 8 % (3/40) Load = <10	Not analysed
Lana et al. (2001)	M/I/D	31 teeth	✓	✓	87 % (27/31)	2.5 % NaOCl: no data	CH: 13 % (4/31) Empty for 7 days: 23 % (7/31)	Yeasts, *Lactobacillus*, *Streptococcus*, facultative anaerobic G −ve rods. *Gemella* and *Pseudomonas* were found in only one canal
Peciuliene et al. (2001)	M/I/D	40 teeth	✓	✓	83 % (33/40)	2.5 % NaOCl and 17 % EDTA: 30 % (10/33)	50 % of sample: CH (10–14 days): 50 % of sample: 2 % I_2 in 10 % IKI (10 min):5 % (1/20)	After instrumentation and irrigation, *E. faecalis* was present in 6 teeth (5 as pure culture), other species included *Proteus mirabilis*; facultative or anaerobic G +ve bacteria
Peters and Wesselink (2002)	M/I/D	42 teeth	✓	✓	Instrumentation to #20: 100 % (42/42); load = 10^6, 4.6 species/canal	Enlarged to #35 with 2 % NaOCl: 23 % (10/42); load = 10^3, 2.8 species/canal	CH (4 weeks): 71 % (15/21); load = 10^3, 2.1 species/canal Further irrigation: 43 % (9/21); load = 10^2	Surviving species after instrumentation and dressing: *Fusobacterium, Prevotella, Bacteroides, Capnocytophaga, Eubacterium, Actinomyces, Bifidobacterium, Propionibacterium acnes, Veillonella, Peptostreptococcus, Gemella morbillorum, Streptococcus, Staphylococcus* Surviving species after final irrigation: *Fusobacterium, Prevotella, Capnocytophaga, Actinomyces, Propionibacterium acnes, Veillonella, Peptostreptococcus*

(continued)

Table 1 (continued)

Study and year	Tx type M/I/D	Sample size	RD	Decont	Outcomes			
					Percentage of samples with bacterial presence/mean bacterial load			Persistent species (frequency)
					At baseline	After preparation ± irrigation	Next visit (after dressing±)	
Card et al. (2002)	M/I	40 mandibular teeth/ canals	✓	✓	95 % (38/40)	1 % NaOCl Profile instrumentation (.04 taper): 0/13 of cuspids and bicuspids, 5/27 of mesio-buccal canals Further LightSpeed instrumentation to size 57.5–65: 3/27 mesio-buccal canals of molars Only 1/16 of those mesio-buccal canals with detectable communication with the mesio-lingual canals had +ve culture after the first preparation using ProFile instruments	No data	Not analysed
Chávez de Paz et al. (2003)	M/I/D	200 teeth	✓	✓	Initial samples were not taken	Type: unknown. No data	CH or IKI: 54 % (107/200) Tooth type, crown and pulpal status had no significant association Size of lesion, persistent pain, type of medication had significant association	After instrumentation and dressing, 235 strains isolated with 2 strains per case. Surviving species included: non-mutant or mutant group streptococci, *Enterococcus*, coagulase-negative staphylococci, *Peptostreptococcus*, *Lactobacillus*, *Bifidobacterium*, *Propionibacterium*, *Actinomyces*, *Eubacterium*, *Clostridium*, *Veillonella*, *Prevotella*, *Fusobacterium*, *Enterobacter* (lactose positive), *Porphyromonas*

Kvist et al. (2004)	M/I/D	96 teeth	✓	✓	98 % (94/96)	0.5 % NaOCl: 63 % (60/96)	CH (7 days): 36 % (16/44) IPI (10 min): 29 % (15/52) Number of treatment visits, bacterial load, number of strains in the initial samples had no significant effects	After instrumentation and dressing, 59 strains isolated. Surviving species included: fusiform rods, *Prevotella, Actinomyces*, Other G+ anaerobic rods, *Peptostreptococcus, Proteus, Lactobacillus,* G+ aerobic spore-forming rods, *Streptococcus, Enterococcus, Staphylococcus*
Chu et al. (2006)	M/I/D	88 canals	✓	✓	99 % (87/88)	0.5 % NaOCl: no data	CH, Septomixine forte, or Ledermix: 36 % (32/88) All canals showed significant reduction of bacterial load Exposure of pulp, tooth type, acute vs. chronic condition, size of lesion and type of medication had no significant effect	After instrumentation and dressing, 88 strains isolated. Surviving species included: *Campylobacter, Fusobacterium, Porphyromonas, Wolinella, Capnocytophaga, Kingella, Suttonella indologenes, Actinomyces, Bifidobacterium, Clostridium subterminale, Actinobacillus, Corynebacterium, Lactobacillus, Veillonella, Neisseria, Peptostreptococcus, Gemella, Leuconostoc, Staphylococcus, Streptococcus*
Paquette et al. (2007)	M/I/D	22 teeth (single canal)	✓	✓	100 % (22/22) Load = 10^5	2.5 % NaOCl: 68 % (15/22) Load = <10	2 % CHX: 45 % (10/22) Load = 10^2	Not analysed
Siqueira et al. (2007a)	M/I/D	11 teeth (single rooted)	✓	✓	100 % (11/11) Load = 10^4–10^7	2.5 % NaOCl: 55 % (6/11) Load = 1–10^4	CH/CPMC: 9 % (1/11) Load = 1–10^2	After instrumentation and irrigation; 11 isolates; most prevalent streptococci. After instrumentation and dressing, only 1 strain (*Propionibacterium acnes*) isolated

(continued)

Table 1 (continued)

Study and year	Tx type M/I/D	Sample size	RD	Decont	Outcomes: Percentage of samples with bacterial presence/mean bacterial load — At baseline	After preparation ± irrigation	Next visit (after dressing±)	Persistent species (frequency)
Siqueira et al. (2007b)	M/I/D	11 teeth (single rooted)	✓	✓	100 % (11/11) Load = 10^3–10^8	2.5 % NaOCl: 45 % (5/11) Load = 1–10^7	CH: 18 % (2/11) Load = 1–10^5	After instrumentation and irrigation: 7 isolates (amongst them streptococci and staphylococci). After instrumentation and dressing, only 2 strains (*Fusobacterium nucleatum*, *Lactococcus garvieae*) isolated
Vianna et al. (2007)	M/I/D	24 teeth (single root)	✓	✓	100 % (24/24) Load = 10^5	Saline + 2 %CHX gel: 33 % (8/24) Load = 10^2 (in positive cases)	2 %CHX, CH or mixture: 54 % (13/24) Load = 10^2 (in positive cases) Type of medication had no significant effect.	After instrumentation and dressing, 30 strains isolated. Surviving species included: *Aerococcus, Gemella, Peptostreptococcus, Neisseria, Veillonella, Capnocytophaga, Actinomyces, Bifidobacterium, Clostridium, Lactobacillus, Propionibacterium*
Wang et al. (2007)	M/I/D	43 canals	✓	✓	91 % (39/43) Load = 10^5	Saline + 2 %CHX gel: 8 % (4/39), Load = <10	2 %CHX + CH: 8 % (3/36) Load = <10 Size of apical preparation (40 vs. 60) had no significant effect	Not analysed
Markvart et al. (2012)	M/I	24 teeth			88 % (21/24)	2.5 % NaOCl: 63 % (15/24)	Further 17 % EDTA irrigation and 10 min 5 % IKI medication: 50 % (12/24) Box preparation (#60): 67 % (8/12) Cone preparation (#25–30): 33 % (4/12); no significant difference	After instrumentation and dressing, surviving species included: *Fusobacterium, Prevotella*, Gram +ve anaerobic rods, *Lactobacillus, Streptococcus, Enterococcus, Staphylococcus*

Xavier et al. (2013)	M/I/D	48 teeth (single canal)	✓	✓	100 % (40/40)	1 % NaOCl: 75 % (9/12) 2 % CHX: 75 % (9/12) No significant difference between NaOCl and CHX	CH: 75 % (18/24)	Not analysed
Molecular								
Rolph et al. (2001)	M/I/D	41 teeth 15 unTx 26 RTxed	✓	✓	UnTx: 75 % (6/8) culture/PCR RTx: 45 % (5/11) culture, 91 % (10/11) PCR	4 % NaOCl and 15 % EDTA: no data	CH: UnTx: 43 % (3/7) culture, 71 % (5/7) PCR RTx: 27 % (4/15) culture, 47 % (7/15) PCR	Not analysed
Sakamoto et al. (2007)	M/I/D	15 teeth	✓	✓	100 % (15/15) qPCR Load = 10^7	2.5 % NaOCl and 17 % EDTC: 67 % (10/15) Load = 10^4	CH/CMCP: 67 % (10/15) Load = 10^4	After instrumentation and dressing, surviving species included: *Streptococcus, Fusobacterium, Neisseria, Prevotella, Propionibacterium, Rothia, Veillonella,* uncultured *Lautropia*
Rocas and Siqueira (2011)	M/I/D	24 teeth	✓	✓	100 % (24/24) qPCR	2.5 % NaOCl: 54 % (13/24)	CH or CH/CMCP: 38 % (9/24) No significant difference between CH and CH + CMCP	After instrumentation and dressing, 18 taxa isolated in the CaOH group and 3 in the CaOH/CMCP group (the finding may be confounded by the difference in the initial samples). Surviving species included: *Selenomonas, Propionibacterium, Porphyromonas, Actinomyces, Bacteroidetes, Filifactor, Dialister, Parvimonas, Streptococcus, Enterococcus, Pyramidobacter, Treponema, Prevotella, Fusobacterium, Tannerella, Peptostreptococcus, Campylobacter, Solobacterium, Eikenella, Olsenella*

(continued)

Table 1 (continued)

Study and year	Tx type M/I/D	Sample size	RD	Decont	Outcomes: Percentage of samples with bacterial presence/mean bacterial load — At baseline	After preparation ± irrigation	Next visit (after dressing±)	Persistent species (frequency)
Paiva et al. (2013)	M/I/D	14 teeth for PCR, 12 teeth qPCR	✓	✓	100 % (12/12) qPCR Load = 10^6	2.5 % NaOCl, 17 % EDTA/2 % CHX: 50 % (6/12) Load = 10^4	CH/CHX: 42 % (5/12) Load = 10^3 (included –ve samples)	After instrumentation and dressing, 13 taxa isolated. Surviving species included: *Eubacterium, Fusobacterium, Campylobacter, Prevotella, Corynebacterium, Clostridium, Hydrogenophilus, Massilia*, uncultured actinobacterium and 2 uncultured bacteria
Provenzano et al. (2013)	M/I	12 teeth	✓	✓	100 % (12/12) PCR	2.5 % NaOCl: 67 % (4/6) 2 % CHX: 83 % (5/6)	–	Not reported

M mechanical preparation, *I* irrigation, *D* dressing, *RD* rubber dam, *NV* non-vital, *V* vital, *Tx* treatment, *UnTx* untreated, *CH* calcium hydroxide, *NaOCl* sodium hypochlorite, *PP* paper point sample, *RTx* retreated, *G −ve* Gram negative, *Rfed* root filled, *r culture* culture test before obturation, *CFU* colony forming unit, *−ve* no growth after culture, *+ve* growth after culture, *Oblig* obligate, *CP* camphorated phenol, *CMCP* camphorated monochlorophenol were not implicated in persistent infections during primary (first attempt) root canal treatment. In contrast, data from secondary (second attempt) root canal treatment (Table 2) showed that certain species were more prevalent after biomechanical procedures than others suggesting that they may be more resistant to treatment protocols, contrary to the previous view. The persistent species often included *Enterococcus faecalis*, *Streptococcus* species, *Staphylococcus* species, *Lactobacillus* species, *Propionibacterium* species, *Actinomyces* species, yeasts and other Gram-positive bacteria.

Even though most longitudinal studies of the root canal microbiota do not definitively show resistance of particular species, other studies suggest that Gram-positive bacteria are found with an unexpectedly high frequency in posttreatment cultures. In further monkey-model experiments from Möller's group (Möller et al. 2004), facultative bacteria were found to be more resistant to chemomechanical treatment than anaerobic species from a 4-strain infection (*Streptococcus milleri*, *Peptostreptococcus anaerobius*, *Prevotella oralis*, *Fusobacterium nucleatum*). In addition, the survival rate of a 5-strain infection including *Enterococcus faecalis* was higher.

4.2 Effect of Obturation Procedures on Intraradicular Microbiota

The role of the root filling appears to be a subsidiary one in helping to control residual infection, since absence of a root filling makes no apparent difference to the healing (Klevant and Eggink 1983; Sabeti et al. 2006). Obturation of infected but already prepared canals resulted in some degree of effect in controlling the remaining infection, presumably by incarcerating residual microbiota (Katebzadeh et al. 1999, 2000). Apart from aiding reduction of the bioburden, the root filling may also serve to help control periapical wound healing and regeneration of the normal anatomical architecture of the periodontal ligament.

4.3 Effect of Persistent Bacteria on Root Canal Treatment Outcome

Numerous studies (Table 3), mostly historical, have analysed the effect of preobturation culture test on root canal treatment outcome. The original rationale for the test was established at the beginning of the last century following the demise in root canal treatment procedures due to the *focal infection* debate (Gulabivala and Ng 2009). In essence, given the obsession with potential systemic complications putatively due to residual root canal infection, the culture test was introduced to reassure the clinician and patient alike that the treatment had in some demonstrable way eliminated the infection. This proved a sufficiently effective quality control

Table 2 Summary of studies evaluating the microbiota in root-treated teeth with periapical disease. For information on terms used within the table, refer the KEY at the end of the table

Study and year	Sample: Size/Pa/Pain/RF quality and exposure	Previous treatment details: I/D time to RTx	Harvest method: RF removal: Mechanical	Harvest method: RF removal: Solvent	Harvest method: Canal sampling	Microbial findings: +ve samples	Microbial findings: Bacterial load and species recovered from more than 2 samples and other findings
Cultural							
Engström (1964)	113 teeth/54 with Pa/Pain?/RF quality?	I/D: none given Time to RTx:?	✓	$CHCl_3$ in some	Filed to apex, NaCl and PP	43/113 (38 %)	*E. faecalis* found in 8 % of all samples (39 % of teeth with Pa, 37 % of teeth without Pa) and 20.9 % of positive samples (24 % of teeth with Pa, 18 % of teeth without Pa). Main enterococcus was *S. faecalis*
Moller (1966)	264 teeth/261 with pa/Pain?/RF quality unknown and had no exposure to oral cavity	I/D: none given Time to RTx:?	?	✓	H_2O, H_2O + meat infusion, or VMGI and PP	120/264 (45 %)	$FacAnO_2$: *Streptococcus* including *Enterococcus* (19 % *E. faecalis*), *Lactobacillus, Micrococcus, Staphylococcus, Neisseria, Pseudomonas, Proteus, Escherichia, Bacillus* AnO_2: *Peptostreptococcus, Veillonella, Bacteroides, Fusobacterium, Lactobacillus, Actinomyces, Leptotrichia, Candida*
Gomes et al. (1996)	21 teeth/12 with Pa/13 with pain/RF quality?	I/D: none given Time to RTx:?	?	? washed with H_2O	PP moisten with NaCl	–	$FacAnO_2$: *Propionibacterium* (most frequently recovered), *Streptococcus, Lactobacillus, Actinomycetes* AnO_2: *Prevotella, Peptostreptococcus, Eubacterium*
Molander et al. (1998)	120 teeth/100 with Pa/no pain/RF within 4 mm of apex	I/D: none given	GG and H files	$CHCl_3$ in 21 cases	Filed to #25 at WL	68/100 (Pa)	Pa group: 117 strains; 1–2 strains in most (85 %); v. heavy growth in 20 teeth; Predominance of G +ve facultative AnO_2 (69 %); Species included: *Enterococcus* (32), *Streptococcus* (14), *Lactobacillus* (11), *Escherichia* (8), G +ve anaerobic rods (8), *Staphylococcus* (7), *Fusobacterium* (5), *Prevotella* (5), *Peptostreptococcus* (4), *Propionibacterium* (4), *Lactobacillus* (3), *Candida* (3), *Actinomyces* (2), *Enterobacter* (2), *Klebsiella* (2); chloroform had major negative impact on recovery

		Time to RTx: 4^+ years			VMGA I and PP	20-Sep (No Pa)	No Pa group : 13 strains; 8 strains v. sparse growth (listed); length of RF, presence of post, size of lesion and tooth type did not influence recovery
Sundqvist et al. (1998)	54 teeth/Pa/no pain/RF good except one	I/D: none given	Burs and files	✗	Filed, NaCl and PP (sample 1)	24/54	In 19, single species present; when *E. faecalis* present (9), it was sole organism; species included *E. faecalis* (9), *Streptococcus* spp. (6), *P. micros* (2), *Actinomyces* (3), *Propionibacterium* (2), *Bacteroides gracilis* (3), *Candida albicans* (2)
		Time to RTx: 4^+ years			Sealed PP in canal (sample 2)	(44 %)	
						20/54 (sample 1 & 2)	
Peciuliene et al. (2000)	25 teeth/Pa/No pain/ RFquality?	I/D:?/CH not used	✓	✗	Same as Molander et al. (1998)	20/25	*E. faecalis* in 14 samples; 5 of these were in pure culture. Other species not reported
		Time to RTx:?					
Peciuliene et al. (2001)	40 teeth/Pa/No pain/ RF unsatisfactory	I/D: none given	✓	✗	Accessed to 1 mm from apex	33/40	Bacterial load: 10^1–10^7 CFUs
		Time to RTx: 5^+ years					Microbiological exam focussed on yeasts, G −ve enterics and *E. faecalis*
							Prevalence in canals: *E. faecalis* (21), *C. albicans* (6), G −ve enteric (3), other bacteria (17)
							Yeast: 50 % with *E. faecalis*; 50 % with other bacteria; *E. faecalis*, 52 % as pure culture; 48 % with other bacteria but as dominant isolate in all except one sample. No difference in lesion size by bacterial groups
Hancock et al. (2001)	54 teeth/Pa/Pain?/RF within 4 mm of apex; 21 had good, 28 had inadequate, 5 had poor restorations	I/D:?/4 had CH;	GG and H files	✗	Filed to #35 at 0.2–2 mm from apex; LDTM and PP +4 mm of last 3–5 F	34/54	PP: mean 1.69 spp./tooth (14 cases 1 sp., 14 cases 2 spp., 2 cases 3 spp., 1 had 4 spp.); 57 strains recovered *E. faecalis* (10 with 6 as pure culture), *Peptostreptococcus* (9), *Actinomyces* (8), *Streptococcus* (7), *Prevotella* (4), *Staphylococcus* (3), *Corynebacterium* (3), *Fusobacterium* (3), *Lactobacillus* (2), *Eubacterium* (2), *Porphyromonas* (2)

(continued)

Table 2 (continued)

Study and year	Sample	Previous treatment details	Harvest method			Microbial findings	
	Size/Pa/Pain/RF quality and exposure	I/D time to RTx	RF removal: Mechanical	RF removal: Solvent	Canal sampling	+ve samples	Bacterial load and species recovered from more than 2 samples and other findings
		Time to RTx: 3^{+} years 4 had fracture diagnosed after sampling				(33/54 by PP)	F: mean 1.5 spp./tooth (17 cases 1 sp.; 8 cases 2 spp.; 3 cases 3 spp.); 42 strains recovered with predominance of G +ve AnO_2 (88 %): *E. faecalis* (9 with 6 as pure culture), *Actinomyces* (9), *Streptococcus* (6), *Corynebacterium* (3), *Staphylococcus* (2), *Lactobacillus* (2), *Peptostreptococcus* (2), *Eubacterium* (2), *Prevotella* (2), *Porphyromonas* (2)
						(28/54 by F)	RF ended > 2 mm from apex; Pa lesions exceeding 5 mm were significantly associated with increase in bacterial recovery
Cheung and Ho (2001)	24 teeth/Pa/No pain/RF quality and exposure to oral cavity not standardised	I/D: none given	GG and H files	✗	Filed to #25 at 0.5–1 mm from apex, RTF and PP	12/18 (6 excluded)	Bacterial load: $10^{0}-10^{5}$ (median = 10^{3} CFU/mL)
		Time to RTx: 4^{+} years					22 species recovered (1–3 spp./tooth): *Streptococcus* (6), *Pseudomonas* (4), coagulase –ve staphylococci (4), *Gemella morbillorum* (2), *Neisseria* (2), *Peptostreptococcus* (3), *Candida albicans* (2)
							Poor RFs were associated with highest counts (10^{3}–10^{5}) and spp./canal (3–6)
Egan et al. (2002)	25 teeth/Pa/Pain?/RF quality and exposure to oral cavity not standardised	I/D: none given	H files	$CHCl_3$ in 3 cases	Filed to WL, NaCl and PP	4/25 +ve for yeasts	Yeasts were 14× more likely to be present in canals if already present in saliva
		Time to RTx:?					

Pinheiro et al. (2003)	60 teeth/Pa/6 with pain/38 with poor RF and 60 with poor restoration	I/D: none given	GG and files	✗	NaCl and PP	51/60	1–9 spp./canal (28 cases 1 sp., 8 cases 2 spp., 15 cases 3+ spp.)
		Time to RTx:?					118 isolates belonging to 37 species: 57 % were facultative anaerobes, 43 % were obligate anaerobes; Gram +ve species (83 %) dominant
							Bacterial species included *Enterococcus* (27 with 18 as pure culture), *Streptococcus* (17), *Peptostreptococcus* (13), *Actinomyces* (10), *Prevotella* (10), *Propionibacterium* (5), *Gemella* (4), *Veillonella* (4), *Fusobacterium* (3), *Lactobacillus* (2), *Staphylococcus* (2), *Candida* (2)
							Significant associations were found between history of pain and polymicrobial infection and anaerobes (*Prevotella, Fusobacterium*); spontaneous pain and *Peptostreptococcus*; tenderness to percussion and *P. intermedia*/*P. nigrescens*; sinus and *Streptococcus* and *Actinomyces*; coronally unsealed and *Streptococcus* and *Candida*; suboptimal RF and polymicrobial infections
Molecular/Cultural							
Kalfas et al. (2001)	2 teeth (anterior)/Pa in one case/pain/no RF but dressed with CH for 7 years in one case	I/D: none given	NA	NA	Filed, NaCl and PP	2 reports of persistent infection	The isolates which were similar to each other were identified as *Actinomyces radicidentis* by 16S gene sequencing; cells had a coccoid morphology
		Time to RTx:?					
Hashimura et al. (2001)	12 teeth/6 with Pa/3 with pain/RF quality?	I/D: none given	?	?	Debris from F	7/12 (specific species)	Species specific PCR: *Slackia exigua* (3), *Mogibacterium timidum* (3), *Eubacterium saphenum* (2)
		Time to RTx:?					

(continued)

Table 2 (continued)

Study and year	Sample	Previous treatment details	Harvest method			Microbial findings	
	Size/Pa/Pain/RF quality and exposure	I/D time to RTx	RF removal: Mechanical	RF removal: Solvent	Canal sampling	+ve samples	Bacterial load and species recovered from more than 2 samples and other findings
Rolph et al. (2001)	26 teeth/Pa/no pain/RF quality?	I/D: none given	ISO 35 orifice shaper	✗	Filed to WL, NaCl and PP	26-Nov	Comparison of culture and PCR +ve;
	Tx judged to have failed	Time to RTx:?				5/26 (culture)	Isolates from 9 refractory cases: *Actinomyces israelii, Eubacterium, Lactobacillus acidophilus, Peptostreptococcus* spp., *Porphyromonas, Streptococcus, Veillonella, Propionibacterium*; no *E. faecalis*
						10/26 (PCR)	Bacterial clones from 5 cases: *Streptococcus mitis, gordonii, intermedius, salivarius*; *Prevotella oris, nigrescens*; *Selenomonas infelix*; *Capnocytophaga gingivalis*; *Cytophaga* sp.; *Dialister* spp.; *Eubacterium yurii, brachy*; *Gemella haemolysins*; *Veillonella dispar*; *Peptostreptococcus* spp.; *Fusobacterium nucleatum*; *Mogibacterium* spp.; *Propionibacterium acnes; Solobacterium moorei; Micrococcus; Rothia dentocariosa*
Sakamoto et al. (2008)	9 teeth/Pa/no pain/RF within 2 mm of apex and had no direct exposure to oral cavity	I/D: none given	GG and files	✗	Filed to #15 at 1 mm from apex	9/9 (Taxa)	74 taxa identified (mean 10 taxa/canal); most dominant taxon: *Bacteroidetes**, *Synergistes, Flavobacteriaceae, Saprospiraceae**, *Pseudoramibacter alactolyticus, Corynebacterium durum, Pseudomonas aeruginosa**, *Prevotella nigrescens*. Only 11 taxa were found in more than one case: the above taxa with * and *Prevotella oris, Prevotella baroniae, Streptococcus mutans, Synergistes, Dialister, Enterococcus faecalis, Flavobacteriaceae, Peptostreptococcus*
		Time to RTx: 5^+ years			NaCl and PP + #15-file + retrieved RF		

Rocas and Siqueira (2008)	42 teeth/Pa/no pain/RF within 4 mm of apex and had no direct exposure to oral cavity.	I/D: none given	Same as Sakamoto et al. (2008)			34/34 (Taxa)	24/28 of target taxa identified with 1–12 per sample; only 5 cases harboured 5 target taxa; most frequently detected taxa included: *Propionibacterium acnes* (22), *Fusobacterium nucleatum* subspecies *nucleatum* (10), *Streptococcus* (7), *Propionibacterium acidifaciens* (6), *Pseudoramibacter alactolyticus* (6), *Enterococcus faecalis* (5), *Tannerella forsythia* (5)
		Time to RTx:2^{+} years					
Anderson et al. (2012)	21 teeth/Pa/no pain/RF within 4 mm of apex and had no direct exposure to oral cavity	I/D: none given	GG, NiTi and K files	✗	File #35 at 0.5–2 mm from apex	7/21 (culture)	Culture: 14 species (1–7 per sample) (*Enterococcus, Faecalis**, *Parvimonas micra, Proteus hauseri/vulgaris, Streptococcus oralis, S. salivarius, Lactobacillus fermentum, Actinomyces oris, Neisseria elongata, Dialister invisus, Streptococcus mutans, S. parasanguinis, Propionibacterium acnes**, *Rothia dentocariosa, Corynebacterium minutissimum, Rummeliibacillus stabekisii*) with the *species recovered in 2 samples
		Time to RTx:2^{+} years			NaCl and PP	7/21 (DNA) different subsets	Cloning: 14 taxa (*Enterococcus gallinarum/casseliflavus, Candida parapsilosis, Lactobacillus gasseri, Proteus hauseri/vulgaris, Streptococcus mutans, Peptostreptococcus stomatis, Selenomonas* sp., *Olsenella profusa, Delftia* sp., *Exiguobacterium aurantiacum, Pantoea agglomerans*, uncultured *Neisseria clone, Phocaeicola abscessus*) with none of the taxa identified in 2+ samples
							Majority of taxa *Firmicutes* (14 taxa), followed in rank order by *Actinobacteria, Proteobacteria* and *Bacteroidetes*

(continued)

Table 2 (continued)

Study and year	Sample	Previous treatment details	Harvest method			Microbial findings	
	Size/Pa/Pain/RF quality and exposure	I/D time to RTx	RF removal: Mechanical	RF removal: Solvent	Canal sampling	+ve samples	Bacterial load and species recovered from more than 2 samples and other findings
Wang et al. (2012)	58 teeth/Pa/6 with pain/RF quality and exposure to oral cavity not standardised	I/D: none given	NiTi files	✗	Patency achieved	54/55 (culture)	Culture: *E. faecalis* plus 1–4 species per canal, which increased with poor RF
		Time to RTx: ?			NaCl and PP		Species-specific PCR: *E. faecalis* in 38 % of canals which had no significant association with tooth location, clinical symptoms, restoration type and status, number of canals, quality of RF
Anderson et al. (2013)	50 teeth/Pa/25 with pain/RF within 4 mm of apex and had no direct exposure to oral cavity	I/D: none given	GG and NiTi files	✗	Filed to 0.5–2 mm from apex	40/50	450+ sequences/sample, 14 phyla and 277 genera identified; 741 species-level OTUs assigned using pyrosequencing
		Time to RTx:2[+] years			NaCl and PP	17 from symptomatic	Most abundant phyla in rank order: *Firmicutes, Proteobacteria, Actinobacteria, Bacteroidetes, Fusobacteria*
						23 from asymptomatic	Most abundant genera: *Streptococcus, Prevotella, Lactobacillus, Kocuria, Neisseria, Acinetobacter, Atopobium, Rothia, Pseudomonas, Propionibacterium, Schlegelella, Enterococcus, Phocaeicola, Leptotrichia, Fusobacterium, Enterobacter, Veillonella*, TM7 genera incertae sedis, *Haemophilus, Pseudoramibacter, Sphingomonas, Paracoccus*, unclassified *Pasteurellaceae, Pyramidobacter, Cellulosimicrobium*

Ex-vivo					
Fukushima et al. (1990)	21 extracted teeth/?/no pain/RF quality unknown but had no direct exposure to oral cavity	I/D: none given	Stored in RTF under AnO_2		Interpret with caution, because (a) teeth frozen (b) decontamination not verified; 62 % +ve for bacteria (52 % mixed cultures; 10 % pure cultures); AnO_2 bacteria in all except one; facultatives in 38 %; G +ve rods and cocci dominant; G +ve cocci dominant in 3 cases; G +ve bacteria: *Eubacterium, Lactobacillus, Propionibacterium, Actinomyces, Peptostreptococcus, Streptococcus*; G −ve bacteria: *Bacteroides, Fusobacterium, Selenomonas, Veillonella*
		Tx to extraction:?	Apex cut at 5 mm, frozen, longitudinal sectioned, one part fragmented into RTF		
Adib et al. (2004)	8 extracted teeth (/Pa/?/RF quality varied but had no direct exposure to oral cavity	I/D: none given	Tooth sectioned at crown/restoration level, root split longitudinally	8-Aug	6–41 species and 7–77 isolates per tooth
		Tx to extraction: 4^+ years or frank evidence of failure	Sample taken from dentine, restoration or GP using PP soaked in RTF		Predominance of $FacAnO_2$ (75 %), followed by AnO_2 (17 %), aerobes (5.6 %), *Candida* (2.4 %)
					G +ve (82 %) was more prevalent than G −ve (15 %)
					$FacAnO_2$: *Staphylococcus* (6), *Streptococcus* (6), *Enterococcus* (5), *Actinomyces* (8)
					AnO_2: *Peptostreptococcus* (7), *Pseudomonas* (7)

Rubber dam isolation was used in all studies; sampling field was decontaminated using various protocols in all studies except Hashimura et al. (2001) in which such information was not provide

M mechanical preparation, *I* irrigation, *D* dressing, *Pa* periapical lesion, *Tx* treatment, *UnTx* untreated, *ReTx* retreatment, *RFed* root filled, *G −ve* Gram negative, *PP* paper point sample, *F* file sample, *RTF* reduced transport fluid, *LDTM* liquid dental transport medium, *1°* primary cases, *2°* secondary cases, *GP* gutta-percha, AnO_2 anaerobic, $FacAnO_2$ facultative anaerobe, $CHCl_3$ chloroform, ? no information, ✓ applies, ✗ does not apply

Table 3 Summary of studies evaluating effect of preobturation culture test on treatment outcome (success). For information on terms used within the table, refer the KEY at the end of the table

Study and year	Sample size	RD	Decont	Treatment details	Success with −ve culture	Success with +ve culture	Qualifying comments and other findings
Rhein et al. (1926)	340 −ve; 152 +ve	?	?	Electrolytic medication	94 %	85 %	Data from Zeldow and Ingle (1963)
Buchbinder (1941)	1930–1936: 94 −ve 1936–1940: 76 −ve All with Pa	✓	?	Mechanical preparation; Dakin's solution, creosote–iodine and eugenol; 1930–1936 obturated by clinical criteria; 1936–1940 obturated by culture test	1930–1936: 63 % 1936–1940: 78 %		Strict criteria; overall success rate 69 %; reports result with negative (−ve) and without (?) culture test; initially single −ve required and then 2 −ve cultures required
Morse and Yates (1941)	265 −ve 237 pa	✓	I_2 tincture	Mechanical preparation; chlorinated soda Dakin's solution; 2–3 visits	93 %	–	All RFed after −ve culture; loose criteria followed; positive cultures had single or mixed flora: *Streptococcus* (α,β,γ), *Staphylococcus*, *Lactobacillus*, *Bacillus subtilis*, yeasts
Abramson (1961)	97 −ve; 38 +ve	✓	?	?	97 %	84 %	Data from Zeldow and Ingle 1963
Ingle (1961)	162 −ve	?	?	?	95 %	–	Data from Zeldow and Ingle (1963)
Oliet (1962)	31 −ve, 67 +ve	✓	Metaphen	No details; RF single cone or lateral condensation; some silver points	84 %	55 %	Loose criteria
Glick -	819 −ve	?	?	?	96 %	–	Data from Zeldow and Ingle (1963)
Zeldow and Ingle (1963)	14 −ve; 42 +ve single canals	?	Aseptic regimen	Mechanical preparation and irrigation with sterile water; RF at 2nd visit; regardless of culture test if no clinical S/S; no AnO_2	93 %	83 %	Loose criteria; 2+-yr follow-up; +ve cultures reduced success rates by 10 %
Seltzer et al. (1963)	1835 −ve, 500 +ve	?	?	Mechanical preparation and irrigation? Single cone or lateral condensation	Overall 84 % No pa 93 % Pa 76 %	Overall 82 % No pa 92 % Pa 75 %	Loose criteria and 6-month follow-up. Microbes in teeth when cultures +ve *Bacillus subtilis*, yeasts, diphtheroids, *E. coli*, *Flavobacterium*, *Micrococcus*, *Neisseria*, *Proteus*, *Pseudomonas aeruginosa*, *Staphylococcus*, *Streptococcus*

Frostell (1963)	138 −ve, 114 +ve, 252 including RFed teeth	✓	30 % H_2O_2 10 % I_2 tincture	Reaming and 1 % cetylpyridinium chloride; if narrow canals, EDTA; widened to H file 2–3; irrigation with chloroform; EtOH, Dakin's solution; dressing with 5 % I_2 in 10 % KI; sampled at every visit	85 %	69 %	Strict criteria; 4–5-year follow-up; percentage failures by residual bacterial species (1-year outcome): staphylococci (6/22, 28 %); enterococci (6/11, 55 %); streptococci (18/20, 90 %) Residual bacteria: *Staphylococcus, Streptococcus* (α,β,γ), *Enterococcus, Veillonella, Lactobacillus*, diphtheroids, coliform organisms, yeasts
Bender et al. (1964)	493 −ve, 213 +ve	?	?	Mechanical preparation and irrigation? Single cone/lateral condensation; no details of microbiological procedures	@ 6 m 84 % No pa 93 % Pa 76 % @ 2 years 82 % No pa 88 % Pa 77 %	@ 6 m 82 % No pa 92 % Pa 75 % @ 2 years 82 % No pa 91 % Pa 77 %	Loose criteria, 6-month and 2-year follow-up; subset of data shows that at 6 months and lat cond, signif diff between +ve (73 %) and −ve (83 %) cultures; but after 2 years no difference; suggests that this was due to slower healing
Engstrom (1964)	169 −ve, 137 +ve Txed by students 1956–1959	?	?	Details not given	Overall 83 % No pa 95 % Pa 73 %	Overall 69 % No pa 81 % Pa 59 %	Strict criteria, 4–5-year follow-up residual bacteria: *Staphylococcus, Streptococcus* (α,β,γ), *Enterococcus, Veillonella, Lactobacillus*, diphtheroids, coliform bacteria, yeasts, *Pseudomonas* Failure associated with *Streptococcus, Veillonella*, uncertain healing associated with AnO_2*Streptococcus, Enterococcus*, yeasts
Engstrom and Lundberg (1965)	80 −ve, 49 +ve No pa	✓	Aseptic technique	Pulp extirpation with Hedstrom files; irrigation with Biosept or Iodopax (0.02 % I_2); dressing with 2 % I_2 in 10 % KI; RF at 2nd visit; culture before RF	@ 4 years 85 %	@ 4 years 61 %	Strict criteria, 3.5–4-year follow-up; bacteria not identified
Oliet and Sorin (1969)	195 −ve, 165 +ve	✓	Tincture of metaphen	Mechanical preparation, irrigation, medication and absence of S/S; culture taken at preceding visit +ve; final sample taken and root canals filled with GP and Ag points or both	Overall 94 % No pa 95 % Pa 91 %	Overall 81 % No pa 83 % Pa 80 %	Criteria: healed or retrogressed; 6-month to 1-year follow-up; results analysed by tooth type, age, RF material and duration of follow-up; analysis showed differences in healing were not

(continued)

Table 3 (continued)

Study and year	Sample size	RD	Decont	Treatment details	Success with −ve culture	Success with +ve culture	Qualifying comments and other findings
							influenced by diagnosis, tooth type, age, RF material, duration after Tx; culture test was the most significant factor
Heling and Shapira (1978)	60 −ve, 20 +ve 118 teeth	✓	2.5 % I_2 tincture	Chemomechanical preparation with 15 % EDTA, irrigation with 5 % NaOCl & 3 % H_2O_2, CPC dressing and Obturation with GP and AH26; 2 −ve cultures before RF; Sample just before RF	Overall 80 % @ 1–3 years 74 % @ 4–5 years 86 % No Pa 82 % Pa 78 %	Overall 70 % @ 1–3 years 73 % @ 4–5 years 60 % No Pa 80 % Pa 67 %	Strict criteria, 1–5-year follow-up Other results: By culture −ve/+ve for: overfilled 60 %/50 %; underfilled 90 %/80 %; flush filled 89 %/60 %; with restoration 83 %/75 %; without restoration 71 %/67 %
Eggink (1982)	1007 −ve, 144 +ve Previously UnTxed and ReTxed teeth; classified by size of Pa	✓	Xylene 6 % I_2 tincture	Extirpate pulp or RF; initial culture at WL; medication with CPC; 2nd visit: reaming and irrigation with saline; no Rf without −ve culture result; follow-up at 1, 2, 3 years	40 %	36 %	Strict criteria, 1–3-year follow-up; evaluated relation between initial culture (no. of bacteria) and size of pa lesion; the pattern in ReTxed teeth less regular; found a relation between initial infection and healing but did not find a relation between r culture and healing; suggested the relation may be between no. of bacteria and outcome
Sjogren et al. (1997)	33 −ve, 22 +ve All with Pa; 53 teeth recalled	✓	Moller's protocol	Narrow canals enlarged to size 20 to WL; endosonics in coronal portion with 0.5 % NaOCl, NaS_2O_3 and culture test; then sample just before RF	94 %	68 %	Strict criteria, 5-year follow-up Number of persistent bacteria; 1–6; 93 % AnO_2; number of bacteria low 10^2–10^3; number and diversity of remaining bacterial cells; no indication that certain species were resistant to Tx; most common persistent bacteria: *Fusobacterium, Peptostreptococcus, Eubacterium, Prevotella, Actinomyces*

Peters and Wesselink (2002)	30 −ve, 8 +ve All with Pa	✓	80 % ethanol	Chemomechanical preparation to 35–60 apical size, 2 % NaOCl irrigation; samples before canal enlargement, after debridement, after removal of dressing material, and before RF.	@ 4–5 years 74 % Single visit 79 % Multi-visit 69 %	@ 4–5 years 88 % Single visit 86 % Multi-visit 100 %	Strict criteria, 4–5-year follow-up; residual bacterial load: $<10^2$ to 10^4
Waltimo et al. (2005)	44 −ve, 6 +ve All with Pa	✓	0.12 % CHX	Chemomechanical preparation with 2 % NaOCl irrigation, ± $Ca(OH)_2$; samples before canal enlargement, after debridement, and after removal of dressing material before RF	@1 year: −1.53 change in PAI	@1 year: −0.79 change in PAI	At completion of Tx, 4/20 (20 %) of single visit group, 0/18 (0 %) of two visits with $Ca(OH)_2$ group, and 2/12 (17 %) of two visits with no dressing group still had bacteria. Gram-positive facultative rods were the most dominant; bacteria identified: *Streptococcus* (α), *Peptostreptococcus, Veillonella, Prevotella, E. faecalis, Bacteroides*
Molander et al. (2007)	61 −ve, 27 +ve All with asymptomatic pa	✓	Moller's protocol	Chemomechanical preparation with 0.5 %NaOCl irrigation, dressed with 5 %IKI for 10 min or CaOH for 1 week. Samples after medication before RF	@ 2 years 80 %	@ 2 years 44 %	Strict criteria, 2-year follow-up; culture result did not have significant association with periapical healing
Tervit et al. (2009)	7 −ve, 10 +ve All with pa	✓	Moller's protocol	Chemomechanical preparation with 2.5 % NaOCl irrigation, dressed with 2 %CHX for 1–2 weeks. Samples before and after canal enlargement, after medication before RF	@ 2 years 86 %	@ 2 years 100 %	Strict criteria, 2-year follow-up; residual bacterial load before RF: 10^3–10^4

RD rubber dam, *Pa* periapical lesion, *Tx* treatment, *UnTx* untreated, *RTxed* root treated, *ReTx* retreated, *CaOH* calcium hydroxide, H_2O_2 hydrogen peroxide, *NaOCl* sodium hypochlorite, ? unknown, *EtOH* alcohol, *PP* paper point sample, I^2*tinc* iodine tincture, NaS_2O_3 sodium thiosulfate, *S/S* signs and symptoms, *Appt* appointment, *lat cond* lateral condensation, *RFed* root filled, *GP* gutta-percha, *Ag* silver point, AnO_2 anaerobic, *KI* potassium iodide, *r culture* before obturation, *CHX* chlorhexidine gluconate, *loose criteria* reduction in size of periapical lesion accepted as success, *strict criteria* complete healing only successful measure for the antimicrobial aspect of the treatment for that time period, and it gained acceptance.

The test was performed following completion of the chemomechanical phase of root canal treatment. It involved the deposition of a sampling fluid into the canal and filing of the dentinal surface to release dentine shavings. The canal fluid, together with any dentine shavings and bacterial biofilm, was soaked up using paper points. The paper points were immediately transferred to a sterile bottle containing bacterial culture broth, which was incubated at 37 °C. The root canal system would be sealed with an antibacterial agent until the next visit, when a negative culture result would signal readiness for obturation and a positive result, further debridement with another culture test. The data from such tests are diverse because of variations in protocols tested, as well as the sampling procedures, yet some interesting findings emerge.

Taken collectively, a preobturation negative culture result may increase treatment success (as judged by radiographic reduction/elimination of the periapical lesion) by an average of 12 % (range 0–26 %). Despite this, a number of factors led to the gradual abandoning of the culture test in clinical practice. One criticism was that numerous factors could have potentially accounted for the treatment outcome but were not all considered in these studies. One large study (Seltzer et al. 1963) *in particular* contributed to the demise of the culture test, but even their study showed a 10 % difference in success in favour of the negative culture test *when periapical disease was present*. The outcome is even worse when a positive culture test result combines with the presence of a periapical lesion (Fig. 13).

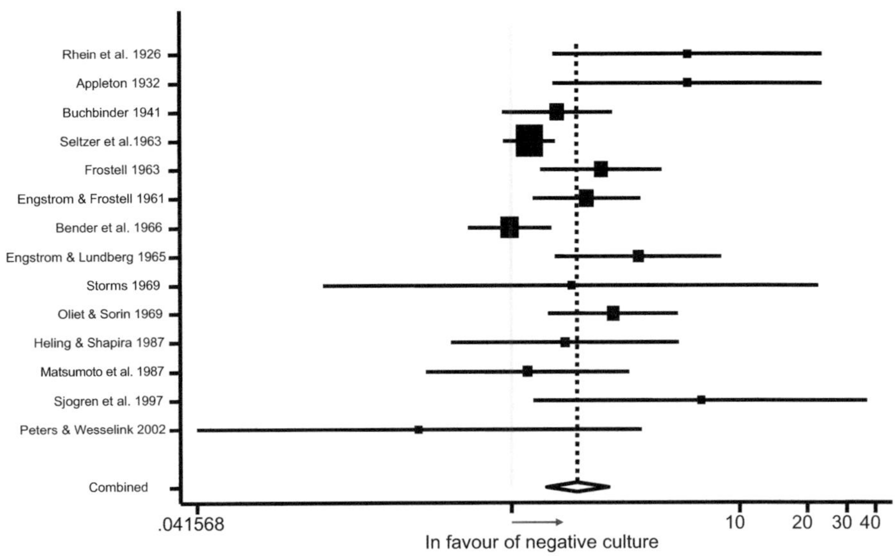

Fig. 13 Effect of negative vs. positive culture on root canal treatment outcome (included studies with retreatment cases) (OR = 1.9: 95 % CI 1.4, 2.6)

The bacteria in preobturation cultures include *Enterococcus*, *Streptococcus*, *Staphylococcus*, *Lactobacillus*, *Veillonella*, *Pseudomonas* and *Fusobacterium* species and yeasts. Some studies found no relationship between individual species and treatment failure, but others have. Whilst the overall failure rate for cases with positive cultures was 31 %, that for teeth with *Enterococcus* species was 55 % and for teeth with *Streptococcus* species was 90 % (Frostell 1963). In another study, good quality root canal treatment on 54 teeth with asymptomatic periapical disease gave an overall success rate of 74 %, but teeth with *Enterococcus faecalis* had a success rate of 66 % (Sundqvist et al. 1998). These associations cannot be regarded as cause–effect and a relationship should also be sought between numbers of bacteria and treatment outcome. The success rate for teeth with no bacteria was 80 %, whilst that for teeth with bacteria in the canal before obturation was 33 %.

A more recent monkey-model study (Fabricius et al. 2006) used 4- or 5-strain infection model to test the effect of debridement and obturation procedures on outcome. When bacteria remained after chemomechanical debridement, 79 % of the root canals were associated with non-healed periapical lesions, compared with 28 % when no bacteria were found to remain. Combinations of several residual bacterial species were more frequently related to non-healed lesions than were single strains. When no bacteria remained at the end of chemomechanical debridement, healing occurred independently of the quality of the root filling. In contrast, when bacteria remained, there was a greater correlation with non-healing in poor quality root fillings than in technically well-performed fillings. In root canals where bacteria were found after removal of the root filling, 97 % had not healed, compared with 18 % for those root canals with no bacteria detected upon removal of root filling. The study emphasises the importance of reducing bacteria below detection limits (at least from the part of the canal system accessible for sampling) before permanent root filling in order to achieve optimal healing conditions for the periapical tissues. It also reinforces the view that obturation does indeed play a role when there is residual infection. The ability to properly seal a surface still coated in biofilm must be questioned, yet even such an act of incarceration appears to inflict some suppression of the microbiota.

Regardless of the technique for obtaining a culture, the use of a negative culture to inform progress of treatment has a positive impact on treatment outcome. The association of specific species with treatment failure is not well established, but the identity of the small group of species isolated from positive cultures is relatively constant and may hold answers to treatment resistance and failure. It is, however, important to account for the other factors that influence root canal treatment outcome. Bacterial load reduction in the sample-able part of the root canal system alone may sometimes show no obvious correlation with periapical healing (Bystrom et al. 1987). Other factors also may need to be taken into account.

4.4 Factors Affecting Outcome of Root Canal Treatment

Clinical judgement of the outcome of treatment is based on the absence of signs of infection and inflammation, such as pain, tenderness to percussion of the tooth, tenderness to palpation of the related soft tissues, absence of swelling and sinus and radiographic demonstration of healing of the periapical lesion (if sufficient time has lapsed), with regeneration of a completely normal periodontal ligament space.

Absence of signs and symptoms of periapical disease but a persistence of a periapical radiographic radiolucency may indicate either healing by fibrous repair or persistent chronic inflammation. Only time and acute exacerbation may identify the latter, whereas the former should remain asymptomatic.

A systematic review and meta-analysis (Ng et al. 2007, 2008) of the factors affecting root canal treatment outcome revealed the following: the mean success rate was 83 % when a vital pulpectomy was performed as there was no established infection; this reduced to 72 % when the root canal treatment procedure was aimed at eradicating an established infection associated with a periapical lesion.

The individual factors having a major impact on root canal treatment outcome were:

- Presence and size of preoperative periapical lesion
- Apical extent of root canal treatment (instrumentation and root filling) in relation to radiographic apex
- Outcome of culture test prior to obturation
- Quality of root canal treatment judged by radiographic appearance of root filling
- Quality of the final coronal restoration

The factors having minimal effect on root canal treatment outcome were:

- Age of patient
- Gender of patient
- General health of patient
- Type of treatment technique (preparation, irrigation and obturation material and technique) other than length control

The improvements in techniques of mechanical and chemical canal preparation have not resulted in increases in success rates over the last century (Fig. 14). It is notable that all the factors having a strong influence on treatment outcome may be associated in some way with root canal infection. Further improvements in root canal treatment outcomes may therefore be obtained by understanding the nature of the root canal infection (especially that in the complex apical anatomy) and the manner in which the microbiota is altered by treatment.

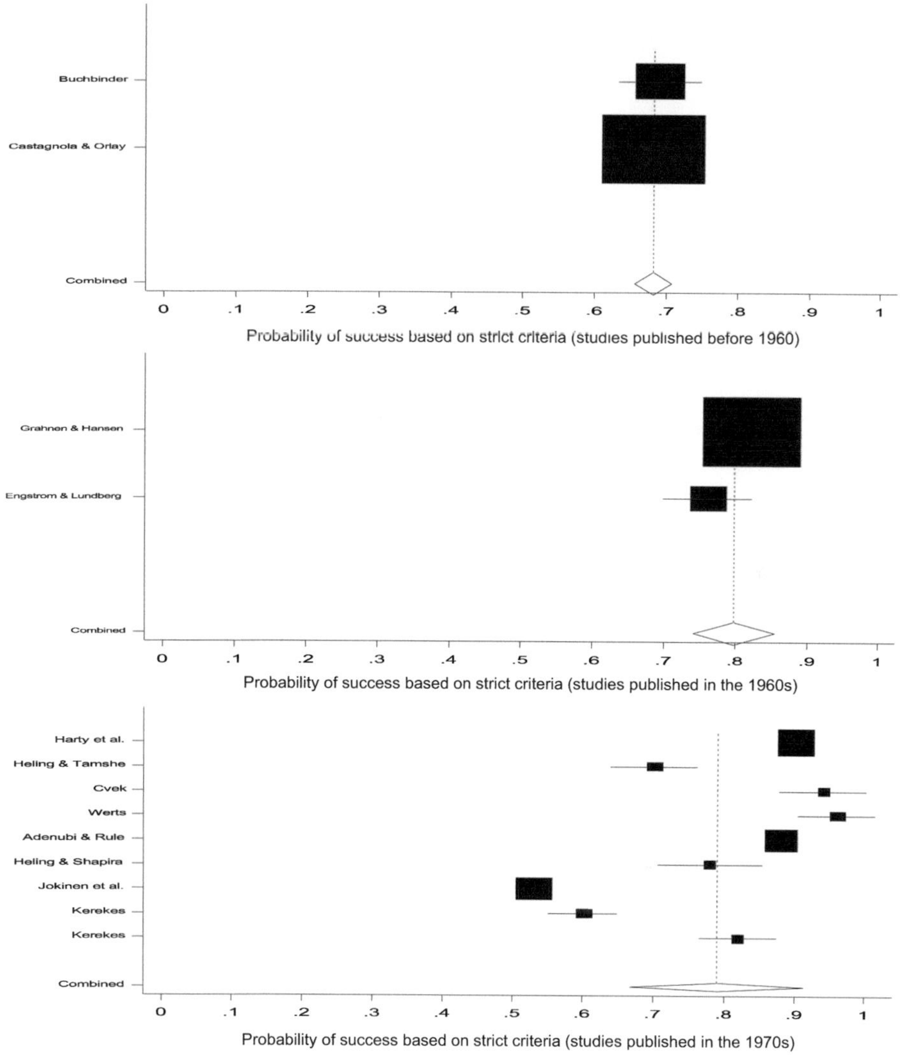

Fig. 14 (continued)

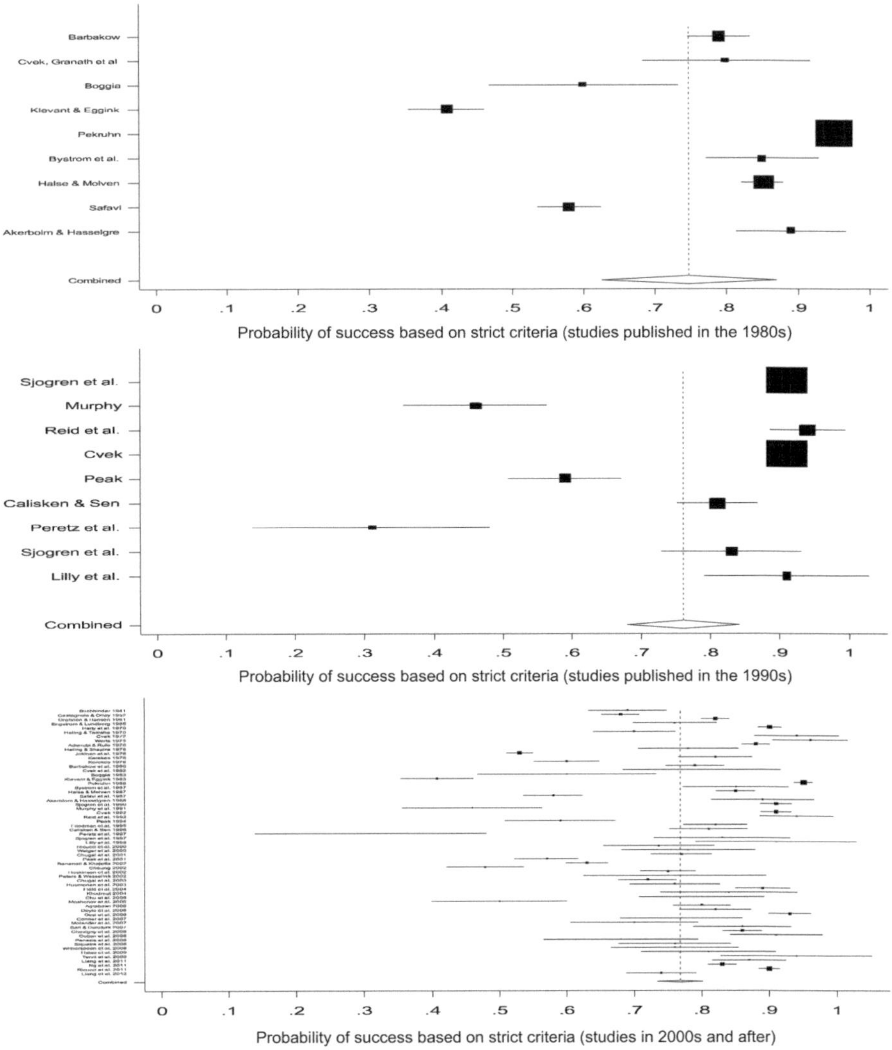

Fig. 14 Forest plots of root canal treatment outcomes, stratified by decades, showing a lack of improvement in outcomes

5 Root Canal Treatment Failure and Persisting Infection

When guideline standard root canal treatment is performed, the failure rate is 17–28 %. When the treatment is *technically* below guideline standard, as described for a high proportion of root canal treatments performed in general practice across the world (Fig. 15), the success rates are lower. The causes of root canal treatment failure (persistent periapical inflammation) may be summarised as follows according to Nair (2004):

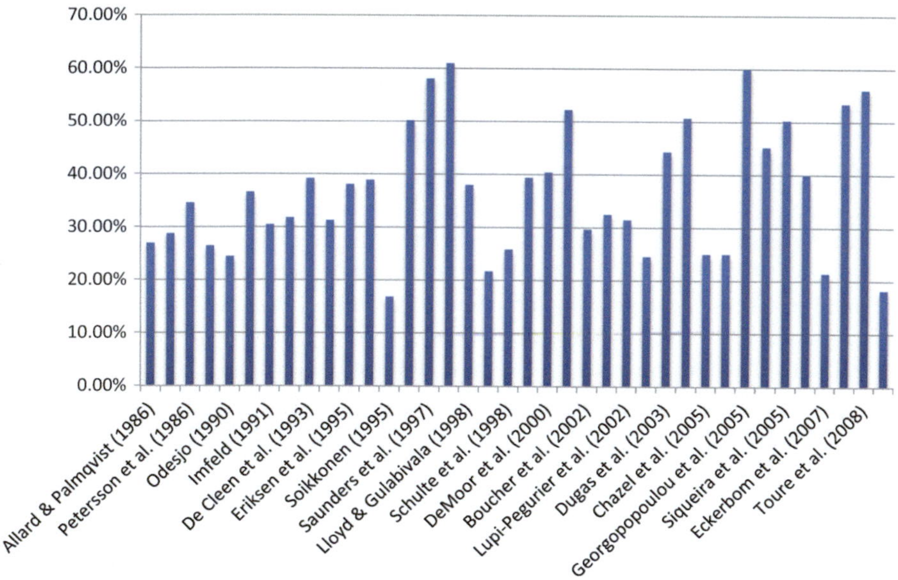

Fig. 15 Prevalence of periapical radiolucencies associated with root-treated teeth

Microbial

- Intraradicular (persistent or new infection)
- Extra-radicular (pre-existing or precipitated by treatment)

Non-microbial (persistent periapical inflammation/pathosis or new inflammation/pathosis)

- True cyst
- Foreign body reaction
- Fibrous healing
- Initial misdiagnosis

5.1 *Intraradicular Microbiota Associated with Failed Root Canal Treatment*

Microbial species recovered from root-treated teeth with persistent periapical disease show a different spectrum compared to that in untreated teeth. The microbiota is dominated by Gram-positive bacteria, many of which are coccoid facultative anaerobes. The retrieval of bacterial samples from obturated root canal systems is compromised by the need to remove root-filling material first, which may kill the bacteria present. The most frequently identified species are *Enterococcus faecalis*, *Propionibacterium* species, *Streptococcus* species, *Lactobacillus* species, *Peptostreptococcus* species and yeasts.

The species recovered from root-treated teeth reside in accessory canals, in dentinal tubules or in the main canal alongside the root filling. They are a subset of those found in untreated teeth, though the diversity and load are reduced. Unlike untreated teeth, treated teeth appear to contain few mixed cultures; often only three, two or one *cultivable* species are found with a mean of only 1.7 species per tooth. Teeth with poor root fillings have the highest bacterial counts (10^3–10^5) with a maximum of 3–6 *cultivable* species per canal and a diversity that resembles that of untreated teeth, perhaps reflecting poor treatment technique from a biological perspective. *Enterococcus faecalis* is the commonest species and when it is present in small numbers in the primary infection, it is easily eliminated, but if it infects in large numbers, it is difficult to eradicate.

Although specific bacterial species have not been implicated as being resistant to treatment in longitudinal studies, the presence of specific groups of bacteria in root-treated teeth suggests that other than being survivors from the pre-existing infection, they may be contaminants introduced during treatment. This may be due to inadequate tooth isolation, poor asepsis, leakage of the access dressing or access cavity being left open for drainage. It may also be the case that the presence of *E. faecalis* in the preliminary infection makes the infection more difficult to eradicate as a whole (Fabricius et al. 2006). If a particularly recalcitrant strain survives in the presence of poor obturation, it may colonise and contaminate the major part of the root canal system and root filling surface (Fig. 16).

5.2 Extra-radicular Microbiota Associated with Failed Root Canal Treatment

Obtaining a sample from the periapical tissues without contamination is notoriously difficult. Approaches include either sampling through the root canal or directly through the soft tissues, either method being susceptible to contamination. Many studies that have obtained periapical tissue have taken scrapings, which may include microorganisms from the apical part of the root canal system.

The bacterial front may extend beyond the apical foramen and into the periapical lesion. It is possible that the establishment of infection in this site may lead to its change to adapt to the new site. Such a periapical infection may then seed new infection into the root canal system and become a source of treatment resistance. This has been a topical area of research interest and controversy. The problem of its study is compounded by the assertion that extraction of teeth causes pumping motion that may move the bacterial front and artificially alter the morphological relationships existing in vivo (Dahlen and Möller 1992). Therefore not only can periapical tissues be contaminated from the root canal, but the root canal may be contaminated from the periapical tissues. Such movement of bacteria due to bulk flow of fluid caused by pressure changes could indeed occur but is less likely when the coronal part of the tooth is intact (Kapalas et al. 2011).

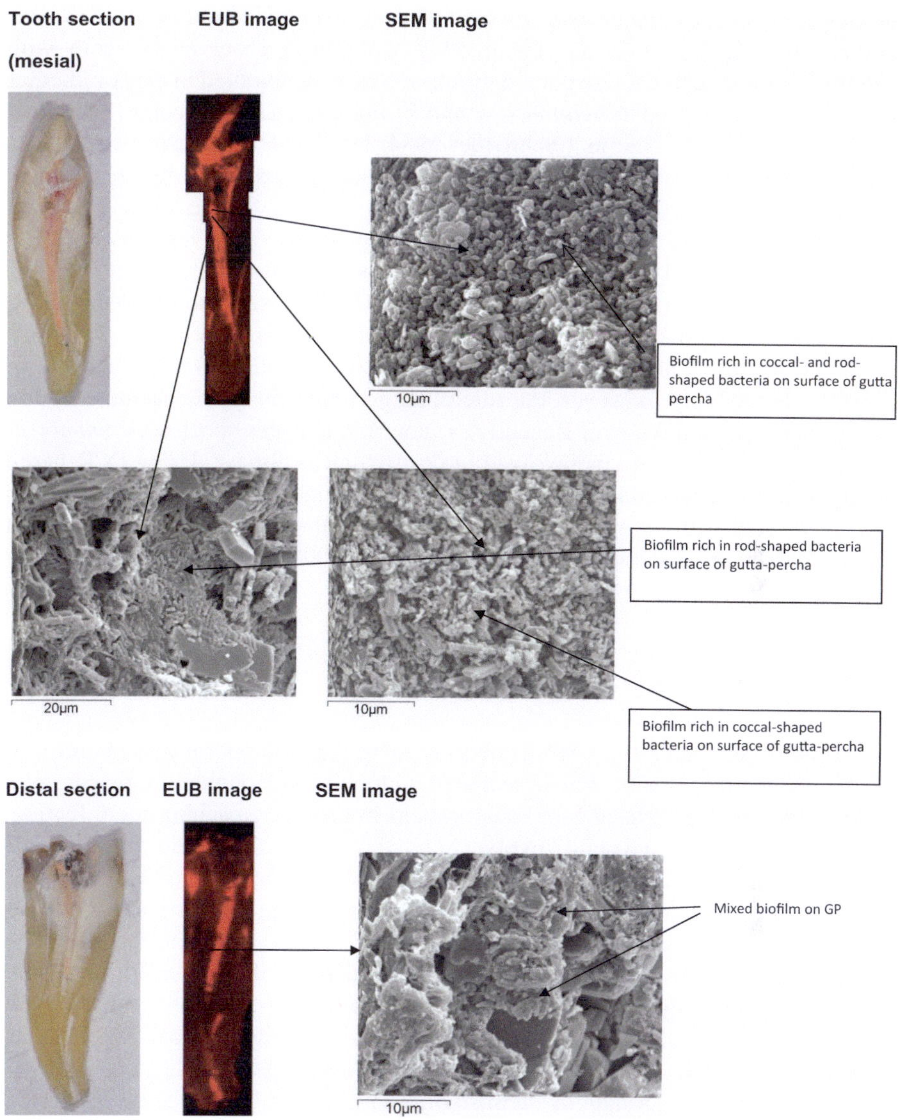

Fig. 16 Correlative images of root-treated tooth with persistent disease (images of split tooth), in situ hybridisation of bacterial presence by EUB probe (red fluorescence, and SEM images confirming bacterial morphotypes)

Periapical abscesses associated with untreated teeth appear to be dominated by *Streptococcus*, *Peptostreptococcus* and *Bacteroides* species, a finding consistent with the presence of these groups in root canals of teeth associated with symptoms. The profile of bacteria found in periapical tissues in cases of 'extra-radicular infection' is somewhat different. In addition to the above groups, they include

Actinomyces, *Propionibacterium*, *Fusobacterium*, *Prevotella* and *Staphylococcus* species; these groups overlap considerably with those observed in teeth with symptoms. Some regard the two main groups of bacteria involved in extra-radicular infections to be confined to *Actinomyces* and *Propionibacterium* species (Sundqvist and Reuterving 1980; Nair and Schroeder 1984; Nair et al. 1990a, b).

A. israelii is a repeated culprit in therapy-resistant cases and is by far the most common species involved in actinomycosis. *A. israelii* is the most prevalent *Actinomyces* species isolated from human abscesses; however, *Actinomyces gerencseriae* (formerly *A. israelii* serotype II) is also prevalent, found in 56 % and 25 % of human abscesses, respectively. Using checkerboard DNA–DNA hybridisation analysis of root canal samples from teeth diagnosed with periapical abscesses, *A. israelii* and *A. gerencseriae* have been reported in 14.8 % and 7.4 % of samples, respectively; however, the role of *A. gerencseriae* in persistent infection after root filling is unknown. Recently, a new *Actinomyces* species, *Actinomyces radicidentis*, was found to be involved in posttreatment disease. Using PCR-based detection, it has been shown to be present in untreated root canal infections and root-filled teeth with chronic apical periodontitis, although its prevalence in both types of infection was low (Figdor and Gulabivala 2008).

6 How Root Canal Treatment Works

In summary, root canal treatment is nothing more than the microbicidal control of an infected wound, albeit a wound secluded within the dark depths and reaches of a volume and space restricted root canal system in a tooth. Reaching the infection, specifically the infected wound, involves some technical skill. Indeed, there is some correlation between the technical standard of treatment and biological outcome; however, the relationship is not absolute. This must be because the technical manipulation does not have a direct and sole bearing on changing the nature of the resident microbial infection. The survival or demise of the resident microbial population is probably dependent upon the fundamental laws of biofilm physiology. It is a community of cells with inter-reliance and interdependence amongst its constituent members. Interference with the environment, even in a nonspecific way, may therefore facilitate the demise of the more fastidious species, triggering a chain reaction akin to a 'domino effect', which by deprivation of nutrients and stimulants (such as quorum sensing) from their neighbouring partner species allows a more global killing effect. Root canal treatment therefore probably works by a combination of direct and indirect killing effects. The importance of indirect killing is underestimated in modern root canal treatment and that is why knowledge of biofilm physiology may be therapeutically highly advantageous.

Conversely, biofilm existence also confers a range of defensive survival strategies. If a sufficiently effective attempt is *not made* to eradicate the biofilm, then its innate tendency would be to detect the changes and adapt to facilitate survival. Recognition that root canal infection is a collection of ECM-enclosed bacteria that behaves in the manner of an 'intelligent', multicellular organism (*biofilm*) is important in devising strategies for treatment because bacteria in biofilms are more resistant to killing. This is attributed to different factors including the observations that:

- The exopolysaccharide in which the bacteria are embedded may restrict diffusion of the antibacterial agents to the cells.
- Different layers of cells may similarly act as barriers to diffusion.
- Slower-growing bacterial cells, persister cells, metabolically inactive cells or dormant cells are more resistant to killing.
- Cells may exhibit specific resistance mechanisms.
- Biofilm phenotypes may be inherently more resistant.

The interdependence amongst different bacterial species with their environment must be the key to the success of root canal treatment as we know it. The treatment procedures (mechanical and chemical) essentially interfere with the environment, killing some bacteria and, by a domino effect, indirectly killing other species by altering the nutritional, physiological and toxic balance. The surviving bacteria are usually those physiologically 'hardy' enough to resist or adapt to changes in the environment induced by the treatment and capable of living independently of other species in the unique nutrition-depleted conditions. The organisms are described as 'hardy' because of their capacity to adapt. This means that a half-hearted first attempt at root canal treatment may result in a more recalcitrant infection to eradicate at the next attempt. It is therefore biologically most sensible to launch the most comprehensive effort at eradicating the infection at the first attempt.

6.1 Degradation of the Microbial Biofilm

The microbial biofilm contains an aggregate of microbes and extracellular polysaccharide matrix (ECPM), which are both highly adherent to the canal wall. The multiple bacterial cell layers covered by the additional thickness of ECPM, which exceeds the bacterial cell volume, render biofilm removal a difficult task. Removal is certainly facilitated by organic tissue solvents such as sodium hypochlorite and chelating agents such as EDTA. The EDTA helps by chelating and sequestering heavy metal ions which normally act as bridges to bond bacteria together in the biofilm. EDTA is routinely used in microbiology laboratories to wash bacterial cells free of the extracellular polysaccharide and to separate them. EDTA is likely to play an important role in degrading the biofilm on the uninstrumented walls of the root canal system; this is therefore a more significant function than smear layer removal.

Table 4 Relative efficacy of root canal irrigant agents in their different properties

Action/agent	Sodium hypochlorite	Iodine	Chlorhexidine	EDTA
Bacterial killing	+++++	+++++	+++	+
Dissolution	+++++	+	–	–
Biofilm penetration	+++++	++++	++	++
Biofilm break-up	+++++	–	–	++++

+ relative degree of potency, – no potency

6.2 *Antibacterial Action Against the Root Canal Microbiota*

Perhaps the most significant action of the irrigant must be its ability to kill all elements of the root canal microbiota. A number of different antibacterial agents have been used, which exhibit a range of actions against the various bacteria (Table 4).

It is a view that since biofilm exhibits multiple mechanisms of resistance or survival, multiple avenues of attack need to be used to facilitate its optimal eradication (Wolcott et al. 2009), known as multiple concurrent therapies. However, the approach of using multiple chemical agents against root canal isolates has not proven effective. The temptation to use multiple agents should be avoided because whilst at best they may potentiate each other (although the literature does not currently support this suggestion) or at worst they may neutralise each other. An example of the latter is the precipitate formed by mixing chlorhexidine with sodium hypochlorite. The latter combination has been shown to reduce the success rates of root canal treatment. The conceptual approach makes sense, but the right combination of therapies needs to be found.

6.3 *Derivation of Guidelines*

Optimal conditions for root canal treatment, as far as possible based on available best evidence, have been grouped into guidelines for root canal treatment by various organisations (European Society of Endodontology 2006; Canadian Academy of Endodontics 2006). Under such conditions of operation, it may be stated that using appropriate instruments in conjunction with adequate technical skill, the canals may be enlarged and predictably shaped, maintaining their original curvature and patency of the apical termini. Using appropriate concurrent irrigation regimens, the coronal and middle parts of root canal system are amenable to relatively effective and predictable debridement, whilst the peripherally extending fins and complex anatomy may remain contaminated to varying extents. Likewise, the apical and more complex anatomy is the most difficult to debride predictably. Under less than optimal conditions of debridement (non-guideline standards), the coronal and middle parts of the root canal system may remain at least partly contaminated, whilst the apical part will remain predictably contaminated. Despite such residual apical contamination by biofilms (Nair et al. 2005), clinical success rates, as measured by radiographic detection of residual periapical lesions, remain relatively high (Ng et al. 2007, 2011). It can only be surmised that total bacterial eradication in the apical anatomy is fortunately unnecessary in the majority of cases.

7 Nature of Future Challenges

The studies by Nair et al. (2005) and Vera et al. (2012) confirm the trends in residual apical infection during contemporary root canal treatment procedures. This implies that residual infection in the apical anatomy is the norm following completion of root canal treatment and that an ongoing interaction beyond the end of treatment, between the residual infection, root filling material and host defences, plays a definitive role in determining the final biological outcome. This explains several clinical observations about periapical healing outcomes. It explains why, despite variations and changes in the technical aspects of the chemomechanical protocol, the success rates of root canal treatment have not improved over the last century (Fig. 14). It also explains why the success rates are so sensitive to the apical length of root canal debridement (Fig. 17). It further explains why the periapical lesion can take so long to heal after termination of the treatment procedure (Fig. 18).

Matsumiya and Kitamura (1960) made very interesting observations from an animal study, in which artificially infected root canal systems were chemomechanically debrided and dressed with calcium hydroxide. Over the course of time, they observed that periapical healing occurred in parallel with a concurrent reduction in the bacterial presence. Their interpretation was not that periapical healing was facilitated by a gradual bacterial demise but that the bacterial demise was facilitated by periapical healing. This insightful statement probably indicates part of the truth. The nature of infection and its activity (expressed genes), the nature of host response and its activity (expressed genes) and any ongoing effect of the root canal filling (and prior chemomechanical treatment) collectively and interactively determine the final outcome over time. The outcome per individual tooth is therefore unpredictable save that we know about 70–80 % will ultimately result in regeneration to various degrees. Future challenges to improving success rates therefore depend upon effective research along the lines of residual biofilm physiology, host response and genetics, treatment effects on microbiota and host and the physics of fluid dynamics.

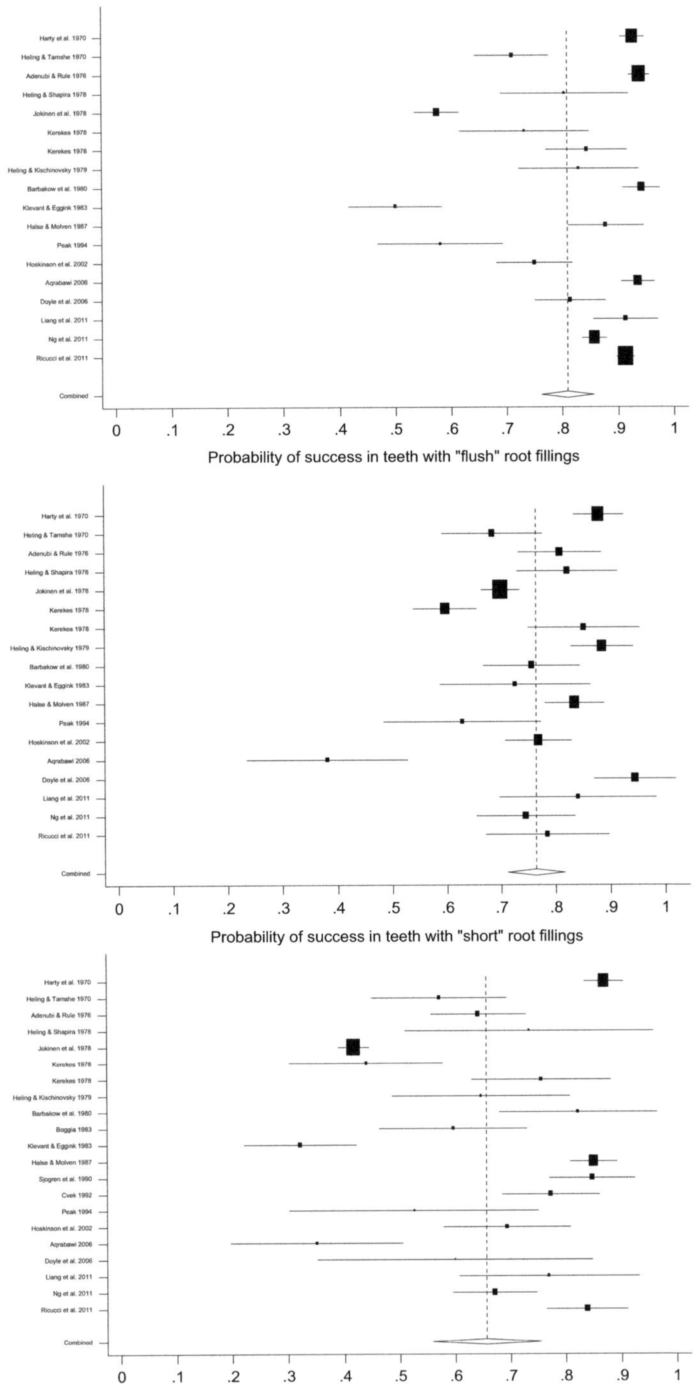

Fig. 17 Forest plots of root canal treatment outcome data (based on strict criteria) partitioned by apical extent of root filling

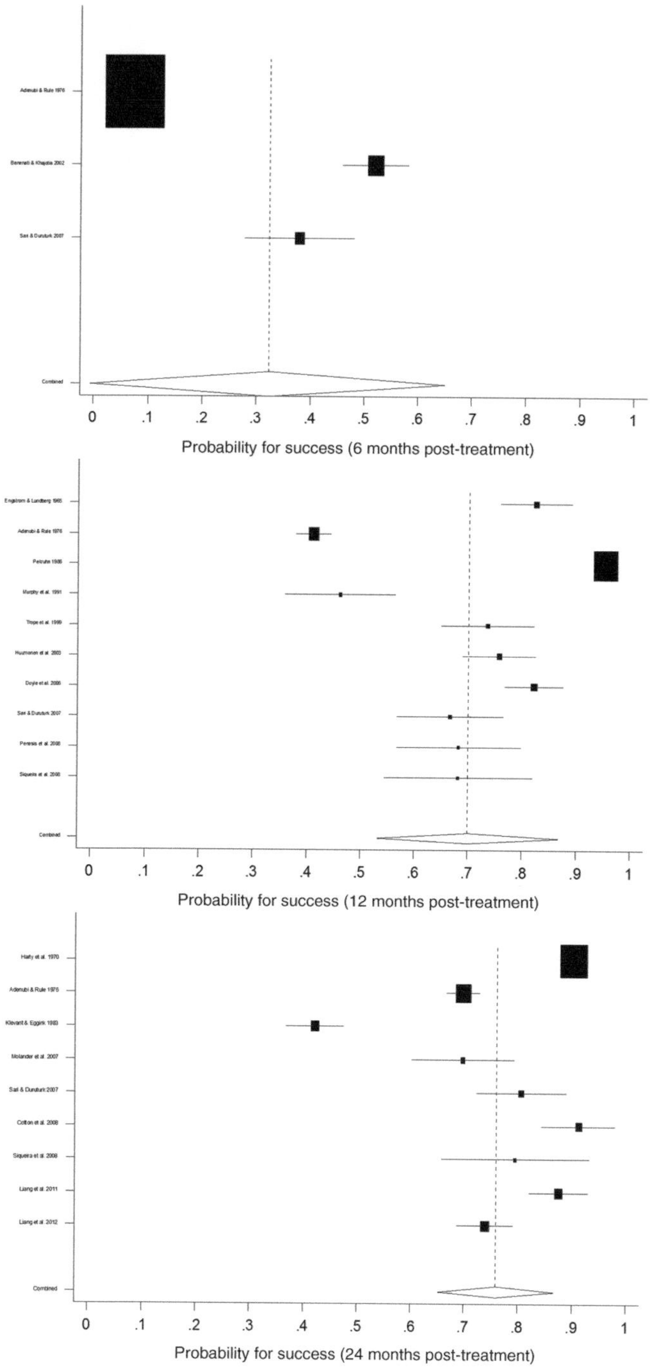

Fig. 18 (continued)

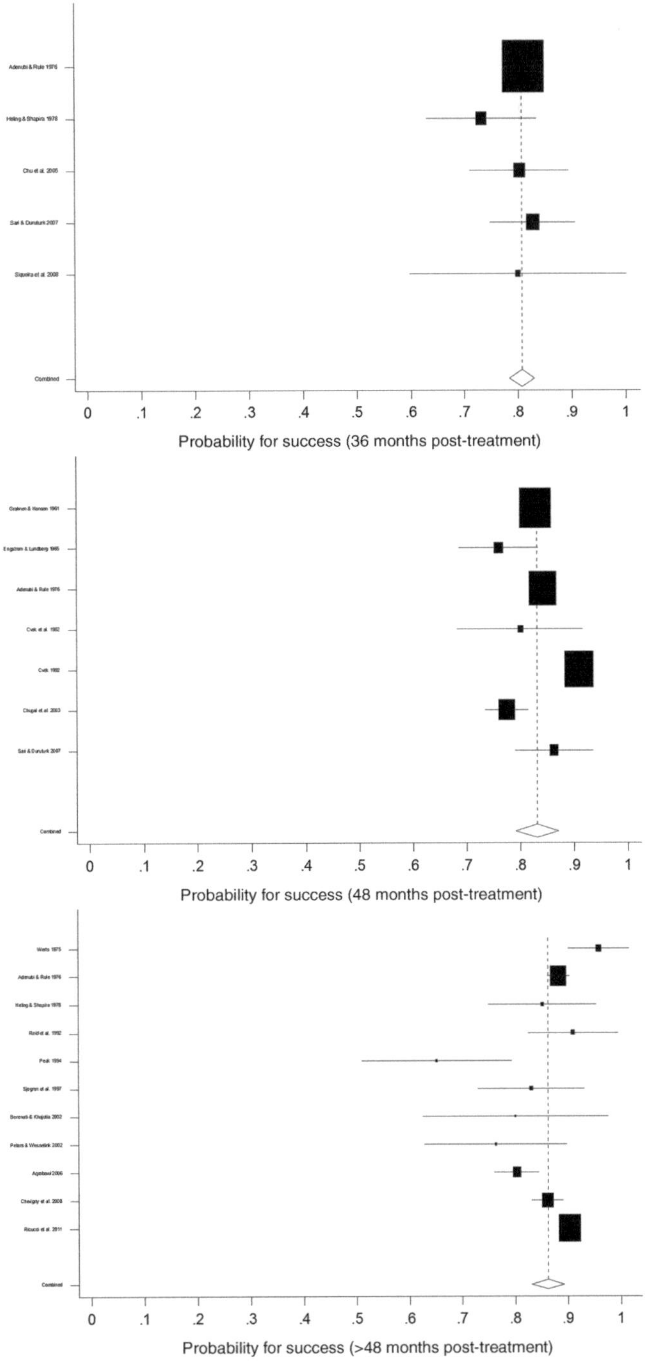

Fig. 18 Forest plots of outcomes of root canal treatment data partitioned by duration after treatment

References

Abdullah M, Ng Y-L, Gulabivala K, Moles DR, Spratt DA (2005) Susceptibilties of two Enterococcus faecalis phenotypes to root canal medications. J Endod 31(1):30–36
Abramson II (1961) A frank appraisal of the present status of the bacterial culture test as a positive endodontic procedure. Presented at the annual meeting of the American Association of Endodontists, Chicago, ILL
Adib V, Spratt D, Ng Y-L, Gulabivala K (2004) Cultivable microbial flora associated with persistent periapical disease and coronal leakage after root canal treatment: a preliminary study. Int Endod J 37(8):542–551
Akpata ES (1976) Effect of endodontic procedures on the population of viable microorganisms in the infected root canal. J Endod 2:369–373
Anderson AC, Al-Ahmad A, Elamin F et al (2013) Comparison of the bacterial composition and structure in symptomatic and asymptomatic endodontic infections associated with root-filled teeth using pyrosequencing. PLoS ONE 8(12), e84960
Anderson AC, Hellwig E, Vespermann R et al (2012) Comprehensive analysis of secondary dental root canal infections: a combination of culture and culture-independent approaches reveals new insights. PLoS ONE 7(11), e49576
Auerbach M (1953) Antibiotics vs. instrumentation in endodontics. NY State Dent J 19:225–228
Bence R, Madonia JV, Weine FS, Smulson MH (1973) A microbiologic evaluation of endodontic instrumentation in pulpless teeth. Oral Surg Oral Med Oral Pathol 35:676–683
Bender IB, Seltzer S, Turkenkopf S (1964) To culture or not to culture? Oral Surg Oral Med Oral Pathol 18:527–540
Bryce G, O'Donnell D, Ready D, Ng Y-L, Pratten J, Gulabivala K (2009) Contemporary root canal irrigants are able to disrupt and eradicate single- and dual-species biofilms. J Endod 35 (9):1243–1248
Buchbinder M (1941) A statistical comparison of cultured and non-cultured root canal cases. J Dent Res 20(2):93–96
Byers MR, Taylor PE, Khayat BG, Kimberly CL (1990) Effects of injury and inflammation on pulpal and periapical nerves. J Endod 16(2):78–84
Bystrom A, Happonen RP, Sjogren U, Sundqvist G (1987) Healing of periapical lesions of pulpless teeth after endodontic treatment with controlled asepsis. Endod Dent Traumatol 3(2):58–63
Bystrom A, Sundqvist G (1981) Bacteriologic evaluation of the efficacy of mechanical root canal instrumentation in endodontic therapy. Scand J Dent Res 89(4):321–328
Bystrom A, Sundqvist G (1983) Bacteriologic evaluation of the effect of 0.5 percent sodium hypochlorite in endodontic therapy. Oral Surg Oral Med Oral Pathol 55(3):307–312
Bystrom A, Sundqvist G (1985) The antibacterial action of sodium hypochlorite and EDTA in 60 cases of endodontic therapy. Int Endod J 18:35–40
Byström A, Claesson R, Sundqvist G (1985) The antibacterial effect of camphorated paramonochlorophenol, camphorated phenol and calcium hydroxide in the treatment of infected root canals. Dent Traumatol 1(5):170–175
Caliskan MK (1995) Pulpotomy of carious vital teeth with periapical involvement. Int Endod J 28 (3):172–176
Canadian Academy of Endodontics (2006) Standards of practices. http://www.caendo.ca/about_cae/standards/standards_english.pdf. Accessed 31 Jul 2008
Cheung GS, Ho MW (2001) Microbial flora of root canal-treated teeth associated with asymptomatic periapical radiolucent lesions. Oral Microbiol Immunol 16(6):332–337
Card SJ, Sigurdsson A, Orstavik D, Trope M (2002) The effectiveness of increased apical enlargement in reducing intracanal bacteria. J Endod 28:779–783
Cavalleri G, Cuzzolin L, Urbani G, Benoni G (1989) Root canal microflora: qualitative changes after endodontic instrumentation. J Chemother 1(2):101–102

Chávez de Paz LE, Dahlén G, Molander A, Möller A, Bergenholtz G (2003) Bacteria recovered from teeth with apical periodontitis after antimicrobial endodontic treatment. Int Endod J 36 (7):500–508

Costerton W, Veeh R, Shirtliff M, Pasmore M, Post C, Ehrlich G (2003) The application of biofilm science to the study and control of chronic bacterial infections. J Clin Invest 112 (10):1466–1477

Chu FC, Leung WK, Tsang PC, Chow TW, Samaranayake LP (2006) Identification of cultivable microorganisms from root canals with apical periodontitis following two-visit endodontic treatment with antibiotics/steroid or calcium hydroxide dressings. J Endod 32:17–23

Cvek M, Nord CE, Hollender L (1976) Antimicrobial effect of root canal debridement in teeth with immature root. A clinical and microbiologic study. Odontol Revy 27(1):1–10

Daeschlein G (2013) Antimicrobial and antiseptic strategies in wound management. Int Wound J 10(Suppl 1):9–14

Dahlen G, Möller Å (1992) Microbiology of endodontic infections. Oral microbiology with basic microbiology and immunology, 4th edn. Mosby, St Louis

Dalton BC, Orstavik D, Phillips C, Pettiette M, Trope M (1998) Bacterial reduction with nickel-titanium rotary instrumentation. J Endod 24:763–767

Egan MW, Spratt DA, Ng Y-L, Lam JM, Moles DR, Gulabivala K (2002) Prevalence of yeasts in saliva and root canals of teeth associated with apical periodontitis. Int Endod J 35(4):321–329

Eggink CO (1982) The value of the bacteriological culture in endodontics. I. The influence of infection during and after treatment. Int Endod J 15(2):79–86

Engstrom B (1964) Correlation of positive culture with the prognosis for root canal treatment. Odontol Revy 15:257–270

Engström B (1964) The significance of enterococci in root canal treatment. Odontol Revy 15:87–106

Engström B, Frostell G (1964) Experiences of bacteriological root canal control. Acta Odontol Scand 22:43–69

Engstrom B, Lundberg M (1965) The correlation between positive culture and the prognosis of root canal therapy after pulpectomy. Odontol Revy 16(3):193–203

European Society of Endodontology (2006) Quality guidelines for endodontic treatment: consensus report of the European Society of Endodontology. Int Endod J 39(12):921–930

Fabricius L, Dahlen G, Ohman AE, Moller AJ (1982) Predominant indigenous oral bacteria isolated from infected root canals after varied times of closure. Scand J Dent Res 90 (2):134–144

Fabricius L, Dahlen G, Sundqvist G, Happonen RP, Moller AJ (2006) Influence of residual bacteria on periapical tissue healing after chemomechanical treatment and root filling of experimentally infected monkey teeth. Eur J Oral Sci 114(4):278–285

Figdor D, Ggulabivala K (2008) Survival against the odds: microbiology of root canals associated with post-treatment disease. Endod Top 18(1):62–77

Fish E (1939) Bone infection. J Am Dent Assoc 26:691

Frostell G (1963) Clinical significance of the root canal culture. Transactions of the Third International Conference on Endodontics. University of Pensylvania, PA, pp 112–122

Fukushima H, Yamamoto K, Hirohata K, Sagawa H, Leung KP, Walker CB (1990) Localization and identification of root canal bacteria in clinically asymptomatic periapical pathosis. J Endod 16(11):534–538

Goldman M, Pearson AH (1969) Postdébridement bacterial flora and antibiotic sensitivity. Oral Surg Oral Med Oral Pathol 28(6):897–905

Gomes BP, Lilley JD, Drucker DB (1996) Clinical significance of dental root canal microflora. J Dent 24(1–2):47–55

Grahnén H, Krasse B (1963) The effect of instrumentation and flushing of non-vital teeth in endodontic therapy. Odontol Rev 14:167–177

Gulabivala K (2004) Species richness of gram-positive coccoid morphotypes isolated from untreated and treated root canals of teeth associated with periapical disease, University of London

Gulabivala K, Aung TH, Alavi A, Ng Y-L (2001) Root and canal morphology of Burmese mandibular molars. Int Endod J 34(5):359–370

Gulabivala K, Ng Y-L (2009) Endodontology. In: Wilson NHF (ed) Clinical dental medicine 2020. Quintessence, New Malden, pp 147–182

Gulabivala K, Ng Y-L, Gilbertson M, Eames I (2010) The fluid mechanics of root canal irrigation. Physiol Meas 31(12):R49–R84

Gulabivala K, Stock C (1995) Preparation of the root canal. Color atlas and text of endodontics, 2nd edn. Mosby-Wolfe, London, pp 97–144

Hall E (1928) Pulpless teeth. Dental Cosmo 19:145–150

Hancock HH 3rd, Sigurdsson A, Trope M, Moiseiwitsch J (2001) Bacteria isolated after unsuccessful endodontic treatment in a North American population. Oral Surg Oral Med Oral Pathol Oral Radiol Endod 91(5):579–586

Hashimura T, Sato M, Hoshino E (2001) Detection of Slackia exigua, Mogibacterium timidum and Eubacterium saphenum from pulpal and periradicular samples using the Polymerase Chain Reaction (PCR) method. Int Endod J 34(6):463–470

Heling B, Shapira J (1978) Roentgenologic and clinical evaluation of endodontically treated teeth, with or without negative culture. Quintessence Int Dent Dig 9(11):79–84

Ingle J, Zeldow B (1958) An evaluation of mechanical instrumentation and the negative culture in endodontic therapy. J Am Dent Assoc 57:471–476

Jones CE, Kennedy JP (2012) Treatment options to manage wound biofilm. Adv Wound Care (New Rochelle) 1(3):120–126

Jordan RE, Suzuki M, Skinner DH (1978) Indirect pulp-capping of carious teeth with periapical lesions. J Am Dent Assoc 97(1):37–43

Kalfas S, Figdor D, Sundqvist G (2001) A new bacterial species associated with failed endodontic treatment: identification and description of Actinomyces radicidentis. Oral Surg Oral Med Oral Pathol Oral Radiol Endod 92(2):208–214

Kapalas A, Spratt DA, Ng Y-L, Gulabivala K (2011) An investigation of a potential confounder in ex vivo microbiological studies–the bulk flow of fluid through apical foramina during tooth extraction. Int Endod J 44(6):534–542

Katebzadeh N, Hupp J, Trope M (1999) Histological periapical repair after obturation of infected root canals in dogs. J Endod 25(5):364–368

Katebzadeh N, Sigurdsson A, Trope M (2000) Radiographic evaluation of periapical healing after obturation of infected root canals: an in vivo study. Int Endod J 33(1):60–66

Klevant FJ, Eggink CO (1983) The effect of canal preparation on periapical disease. Int Endod J 16 (2):68–75

Koontongkaew S, Silapichit R, Thaweboon B (1988) Clinical and laboratory assessments of camphorated monochlorophenol in endodontic therapy. Oral Surg Oral Med Oral Pathol 65:757–762

Kvist T, Molander A, Dahlen G, Reit C (2004) Microbiological evaluation of one- and two-visit endodontic treatment of teeth with apical periodontitis: a randomized, clinical trial. J Endod 30:572–576

Lana MA, Ribeiro-Sobrinho AP, Stehling R et al (2001) Microorganisms isolated from root canals presenting necrotic pulp and their drug susceptibility in vitro. Oral Microbiol Immunol 16:100–105

Lussi A, Messerli L, Hotz P, Grosrey J (1995) A new non-instrumental technique for cleaning and filling root canals. Int Endod J 28(1):1–6

Markvart M, Dahlén G, Reit CE, Bjørndal L (2013) The antimicrobial effect of apical box versus apical cone preparation using iodine potassium iodide as root canal dressing: a pilot study. Acta Odontol Scand 71(3–4):786–791

Matsumiya S, Kitamura M (1960) Histo-pathological and histo-bacteriological studies of the relation between the condition of sterilization of the interior of the root canal and the healing process of periapical tissues in experimentally infected root canal treatment. Bull Tokyo Dent Coll 1(1):1–19

Mejàre B (1975) Characteristics of Streptococcus milleri and Streptococcus mitior from infected dental root canals. Odontol Rev 26(4):291–308

Melville TH, Slack GL (1961) Bacteria isolated from root canals during endodontic treatment. Br Dent J 110:127–130

Molander A, Reit C, Dahlen G (1990) Microbiological evaluation of clindamycin as a root canal dressing in teeth with apical periodontitis. Int Endod J 23:113–118

Molander A, Reit C, Dahlen G, Kvist T (1998) Microbiological status of root-filled teeth with apical periodontitis. Int Endod J 31(1):1–7

Molander A, Warfvinge J, Reit C, Kvist T (2007) Clinical and radiographic evaluation of one- and two-visit endodontic treatment of asymptomatic necrotic teeth with apical periodontitis: a randomized clinical trial. J Endod 33(10):1145–1148

Möller AJ (1966) Microbiological examination of root canals and periapical tissues of human teeth. Methodological studies. Odontol Tidskr 74(5 Suppl):1–380

Möller AJ, Fabricius L, Dahlen G, Ohman AE, Heyden G (1981) Influence on periapical tissues of indigenous oral bacteria and necrotic pulp tissue in monkeys. Scand J Dent Res 89(6):475–484

Möller AJ, Fabricius L, Dahlen G, Sundqvist G, Happonen RP (2004) Apical periodontitis development and bacterial response to endodontic treatment. Experimental root canal infections in monkeys with selected bacterial strains. Eur J Oral Sci 112(3):207–215

Moore DL (1967) Conservative treatment of teeth with vital pulps and periapical lesions: a preliminary report. J Prosthet Dent 18(5):476–481

Morse F, Yates M (1941) Follow up studies of root-filled teeth in relation to bacteriologic findings. J Am Dent Assoc 28:956–971

Myers JW, Marshall FJ, Rosen S (1969) The incidence and identity of microorganisms present in root canals at filling following culture reversals. Oral Surg Oral Med Oral Pathol 28(6):889–896

Nair PN (2004) Pathogenesis of apical periodontitis and the causes of endodontic failures. Crit Rev Oral Biol Med 15(6):348–381

Nair PN, Henry S, Cano V, Vera J (2005) Microbial status of apical root canal system of human mandibular first molars with primary apical periodontitis after "one-visit" endodontic treatment. Oral Surg Oral Med Oral Pathol Oral Radiol Endod 99(2):231–252

Nair PNR, Schroeder HE (1984) Periapical actinomycosis. J Endod 10:567–570

Nair PN, Sjögren U, Krey G, Kahnberg KE, Sundqvist G (1990a) Intra-radicular bacteria and fungi in root-filled, asymptomatic human teeth with therapy-resistant periapical lesions: a long-term light and electron microscopic follow-up study. J Endod 16:580–588

Nair PN, Sjögren U, Krey G, Sundqvist G (1990b) Therapy-resistant foreign body giant cell granuloma at the periapex of a root-filled human tooth. J Endod 16:589–595

Ng Y-L, Mann V, Gulabivala K (2011) A prospective study of the factors affecting outcomes of nonsurgical root canal treatment: part 1: periapical health. Int Endod J 44(7):583–609

Ng Y-L, Mann V, Rahbaran S, Lewsey J, Gulabivala K (2007) Outcome of primary root canal treatment: systematic review of the literature - part 1. Effects of study characteristics on probability of success. Int Endod J 40(12):921–939

Ng Y-L, Mann V, Rahbaran S, Lewsey J, Gulabivala K (2008) Outcome of primary root canal treatment: systematic review of the literature – Part 2. Influence of clinical factors. Int Endod J 41(1):6–31

Nicholls E (1962) The efficacy of cleansing of the root canal. Br Dent J 112(4):167–170

Olgart LG (1969) Bacteriological sampling from root canals directly after chemomechanical treatment. A clinical and bacteriological study. Acta Odontol Scand 27(1):91–103

Oliet S (1962) Evaluation of culturing in endodontic therapy. A preliminary clinical report. Oral Surg Oral Med Oral Pathol 15:727–730

Oliet S, Sorin SM (1969) Evaluation of clinical results based upon culturing root canals. Int Endod J 3(1):3–6

Ørstavik D, Kerekes K, Molven O (1991) Effects of extensive apical reaming and calcium hydroxide dressing on bacterial infection during treatment of apical periodontitis: a pilot study. Int Endod J 24(1):1–7

Ozok AR, Persoon IF, Huse SM et al (2012) Ecology of the microbiome of the infected root canal system: a comparison between apical and coronal root segments. Int Endod J 45(6):530–541

Paiva SS, Siqueira JF Jr, Rôças IN, Carmo FL, Leite DC, Ferreira DC, Rachid CT, Rosado AS (2013) Clinical antimicrobial efficacy of NiTi rotary instrumentation with NaOCl irrigation, final rinse with chlorhexidine and interappointment medication: a molecular study. Int Endod J 46(3):225–233

Paquette L, Legner M, Fillery ED, Friedman S (2007) Antibacterial efficacy of chlorhexidine gluconate intracanal medication in vivo. J Endod 33:788–795

Peciuliene V, Balciuniene I, Eriksen HM, Haapasalo M (2000) Isolation of Enterococcus faecalis in previously root-filled canals in a Lithuanian population. J Endod 26(10):593–595

Peciuliene V, Reynaud AH, Balciuniene I, Haapasalo M (2001) Isolation of yeasts and enteric bacteria in root-filled teeth with chronic apical periodontitis. Int Endod J 34(6):429–434

Peters LB, Wesselink PR (2002) Periapical healing of endodontically treated teeth in one and two visits obturated in the presence or absence of detectable microorganisms. Int Endod J 35:660–667

Pinheiro ET, Gomes BP, Ferraz CC, Sousa EL, Teixeira FB, Souza-Filho FJ (2003) Microorganisms from canals of root-filled teeth with periapical lesions. Int Endod J 36(1):1–11

Provenzano JC, Siqueira JF Jr, Rôças IN, Domingues RR, Paes Leme AF, Silva MR (2013) Metaproteome analysis of endodontic infections in association with different clinical conditions. PLoS ONE 8(10), e76108

Reit C, Dahlén G (1988) Decision making analysis of endodontic treatment strategies in teeth with apical periodontitis. Int Endod J 21:291–299

Reit C, Molander A, Dahlen G (1999) The diagnostic accuracy of microbiologic root canal sampling and the influence of antimicrobial dressings. Endod Dent Traumatol 15:278–283

Rhein M, Krasnow F, Gies W (1926) A prolonged study of the electrolytic treatment of dental focal infection–a preliminary report. Dent Cosmos 17:971–981

Richardson N, Mordan NJ, Figueiredo JA, Ng Y-L, Gulabivala K (2009) Microflora in teeth associated with apical periodontitis: a methodological observational study comparing two protocols and three microscopy techniques. Int Endod J 42(10):908–921

Rocas IN, Siqueira JF Jr (2008) Root canal microbiota of teeth with chronic apical periodontitis. J Clin Microbiol 46(11):3599–3606

Rocas IN, Siqueira JF Jr (2011) In vivo antimicrobial effects of endodontic treatment procedures as assessed by molecular microbiologic techniques. J Endod 37(3):304–310

Rolph HJ, Lennon A, Riggio MP et al (2001) Molecular identification of microorganisms from endodontic infections. J Clin Microbiol 39(9):3282–3289

Ruddle C (2007) Hydrodynamic disinfection: tsunami endodontics. Dent Today 26(5):110–112, 114

Sabeti MA, Nekofar M, Motahhary P, Ghandi M, Simon JH (2006) Healing of apical periodontitis after endodontic treatment with and without obturation in dogs. J Endod 32(7):628–633

Sakamoto M, Siqueira JF Jr, Rôças IN, Benno Y (2007) Bacterial reduction and persistence after endodontic treatment procedures. Oral Microbiol Immunol 22(1):19–23

Sakamoto M, Siqueira JF Jr, Rocas IN, Benno Y (2008) Molecular analysis of the root canal microbiota associated with endodontic treatment failures. Oral Microbiol Immunol 23 (4):275–281

Seltzer S, Bender IB, Turkenkopf S (1963) Factors affecting successful repair after root canal therapy. J Am Dent Assoc 67:651–662

Shuping GB, Orstavik D, Sigurdsson A, Trope M (2000) Reduction of intracanal bacteria using nickel-titanium rotary instrumentation and various medications. J Endod 26:751–755

Siqueira JF Jr, Magalhães KM, Rôças IN (2007a) Bacterial reduction in infected root canals treated with 2.5% NaOCl as an irrigant and calcium hydroxide/camphorated paramonochlorophenol paste as an intracanal dressing. J Endod 33(6):667–672
Siqueira JF Jr, Guimarães-Pinto T, Rôças IN (2007b) Effects of chemomechanical preparation with 2.5% sodium hypochlorite and intracanal medication with calcium hydroxide on cultivable bacteria in infected root canals. J Endod 33(7):800–805
Siren EK, Haapasalo MP, Ranta K, Salmi P, Kerosuo EN (1997) Microbiological findings and clinical treatment procedures in endodontic cases selected for microbiological investigation. Int Endod J 30(2):91–95
Sjogren U, Figdor D, Persson S, Sundqvist G (1997) Influence of infection at the time of root filling on the outcome of endodontic treatment of teeth with apical periodontitis. Int Endod J 30 (5):297–306
Sjögren U, Figdor D, Spangberg L, Sundqvist G (1991) The antimicrobial effect of calcium hydroxide as a short-term intracanal dressing. Int Endod J 24:119–125
Sjögren U, Sundqvist G (1987) Bacteriologic evaluation of ultrasonic root canal instrumentation. Oral Surg Oral Med Oral Pathol 63:366–370
Stewart GG (1955) The importance of chemomechanical preparation of the root canal. Oral Surg Oral Med Oral Pathol 8(9):993–997
Stewart GG, Cobe H, Rappaport H (1961) Study of new medicament in chemomechanical preparation of infected root canals. J Am Dent Assoc 63:33
Sundqvist G, Figdor D, Persson S, Sjogren U (1998) Microbiologic analysis of teeth with failed endodontic treatment and the outcome of conservative re-treatment. Oral Surg Oral Med Oral Pathol Oral Radiol Endod 85(1):86–93
Sundqvist G, Reuterving CO (1980) Isolation of *Actinomyces israelii* from periapical lesion. J Endod 6(6):602–606
Tani-Ishii N, Wang CY, Tanner A, Stashenko P (1994) Changes in root canal microbiota during the development of rat periapical lesions. Oral Microbiol Immunol 9(3):129–135
Tervit C, Paquette L, Torneck CD, Basrani B, Friedman S (2009) Proportion of healed teeth with apical periodontitis medicated with two percent chlorhexidine gluconate liquid: a case-series study. J Endod 35(9):1182–1185
Tsatsas B, Tzamouranis A, Mitsis F (1974) A bacteriological examination of root canals before filling. J Br Endod Soc 7(2):78–80
Vera J, Siqueira JF Jr, Ricucci D et al (2012) One- versus two-visit endodontic treatment of teeth with apical periodontitis: a histobacteriologic study. J Endod 38(8):1040–1052
Vianna ME, Horz HP, Conrads G, Zaia AA, Souza-Filho FJ, Gomes BP (2007) Effect of root canal procedures on endotoxins and endodontic pathogens. Oral Microbiol Immunol 22:411–418
Waltimo TM, Sirén EK, Torkko HL, Olsen I, Haapasalo MP (1997) Fungi in therapy-resistant apical periodontitis. Int Endod J 30(2):96–101
Waltimo T, Trope M, Haapasalo M, Orstavik D (2005) Clinical efficacy of treatment procedures in endodontic infection control and one year follow-up of periapical healing. J Endod 31(12):863–866
Wang CS, Arnold RR, Trope M, Teixeira FB (2007) Clinical efficiency of 2% chlorhexidine gel in reducing intracanal bacteria. J Endod 33:1283–1289
Wang QQ, Zhang CF, Chu CH, Zhu XF (2012) Prevalence of *Enterococcus faecalis* in saliva and filled root canals of teeth associated with apical periodontitis. Int J Oral Sci 4(1):19–23
Wolcott RD, Gontcharova V, Sun Y, Dowd SE (2009) Evaluation of the bacterial diversity among and within individual venous leg ulcers using bacterial tag-encoded FLX and Titanium amplicon pyrosequencing and metagenomic approaches. BMC Microbiol 9:226
Xavier AC, Martinho FC, Chung A et al (2013) One-visit versus two-visit root canal treatment: effectiveness in the removal of endotoxins and cultivable bacteria. J Endod 39:959–964
Yared GM, Dagher FE (1994) Influence of apical enlargement on bacterial infection during treatment of apical periodontitis. J Endod 20:535–537
Zeldow BI, Ingle JI (1963) Correlation of the positive culture to the prognosis of endodontically treated teeth: a clinical study. J Am Dent Assoc 66:9–13

Root Canal Irrigation

Luc van der Sluis, Christos Boutsioukis, Lei-Meng Jiang, Ricardo Macedo, Bram Verhaagen, and Michel Versluis

Abstract The aims of root canal irrigation are the chemical dissolution or disruption and the mechanical detachment of pulp tissue, dentin debris and smear layer (instrumentation products), microorganisms (planktonic or biofilm), and their products from the root canal wall, their removal out of the root canal system. Each of the endodontic irrigation systems has its own irrigant flow characteristics, which should fulfill these aims. Without flow (convection), the irrigant would have to be distributed through diffusion. This process is slow and depends on temperature and concentration gradients. On the other hand, convection is a faster and more efficient transport mechanism. During irrigant flow, frictional forces will occur, for example, between the irrigant and the root canal wall (wall shear stress). In this chapter the irrigant flow and wall shear stress produced by different irrigation systems will be described. Furthermore, the effect of the flow on the biofilm and the chemical effect of irrigants on the biofilm will be discussed.

1 Introduction

Root canal irrigation can be defined as the procedure to deliver a liquid or *irrigant* in the root canal system before, during, and after instrumentation of the root canal. The aims of this procedure are the *chemical* dissolution or disruption and the *mechanical* detachment of pulp tissue, dentin debris and smear layer (instrumentation products), microorganisms (planktonic or biofilm), and their products

L. van der Sluis (✉)
Center for Dentistry and Oral Hygiene, University Medical Center Groningen, Groningen, The Netherlands
e-mail: l.w.m.van.der.sluis@umcg.nl

C. Boutsioukis • L.-M. Jiang • R. Macedo
Department of Endodontology, Academic Centre for Dentistry Amsterdam (ACTA), Amsterdam, The Netherlands

B. Verhaagen • M. Versluis
Physics of Fluids group, MIRA Institute for Biomedical Technology and Technical Medicine, MESA+ Institute for Nanotechnology, University of Twente, The Netherlands

L.E. Chávez de Paz et al. (eds.), *The Root Canal Biofilm*, Springer Series on Biofilms 9, DOI 10.1007/978-3-662-47415-0_9

(all together hereafter named *substrate*) from the root canal wall and their removal out of the root canal system. The objectives of irrigation are to induce a flow of irrigant:

- Which ensures an adequate delivery throughout the root canal system and a close contact with the substrate
- That can achieve frequent refreshment and mixing of the irrigant, in order to retain an effective concentration of the active chemical component(s) and compensate for its rapid inactivation
- That carries away the substrate and provides lubrication for the instruments
- That applies a force on the root canal wall (wall shear stress), in order to detach/disrupt the substrate
- Which is restricted within the constraints of the root canal, thus preventing irrigant extrusion toward the periapical tissues

Each of the irrigation systems has its own flow characteristics, which should fulfill these objectives. Without the flow (also termed *convection*), the irrigant would have to be delivered throughout the root canal system by *diffusion*, which is the result of the random movement of individual particles (molecules/ions) in the fluid. This process is slow and depends, for example, on temperature and concentration gradients. On the other hand, convection is a faster and more efficient transport mechanism in which molecules are transported by the motion of fluid (Incropera and de Witt 1990) which contributes to effective delivery, refreshment, and mixing of the irrigant. The flow of irrigants is also affected by their physical properties, which will be discussed in section 5.2.

Within the irrigation procedure, two phases can be distinguished: the flow phase, during which the irrigant is delivered and flows in and out of the root canal, and a rest phase, when the irrigant is at rest in the root canal. Irrigant activation systems introduce an additional activation phase, to enable an energy source to enhance the streaming of the irrigant. The flow and activation phase assists the mechanical and chemical activity through convection and diffusion of the molecules/ions of the irrigant and the rest phase mostly through diffusion.

Because both the mechanical and chemical aspects are determined by the irrigant flow, the main goal of this chapter is to describe in detail this flow provoked by the different irrigation systems (syringe, negative pressure, manual dynamic activation (MDA), (ultra)sonics, laser) (Fig. 1) combined with its chemical or mechanical effects. Recently, a new irrigation system, the Multisonic Ultracleaning System (Sonendo Inc., Laguna Hills, CA, USA), has been introduced onto the market (Haapasalo et al. 2014). Although the manufacturer claims the working mechanism relies on "a wide frequency range of acoustic energy" (oral communication), they were not willing to share the working principles with us. Therefore, it is not possible to critically analyze this system within this chapter.

Unfortunately, only one randomized controlled trial (RCT) is available on the effect of root canal irrigation on the endodontic outcome (Liang et al. 2013).

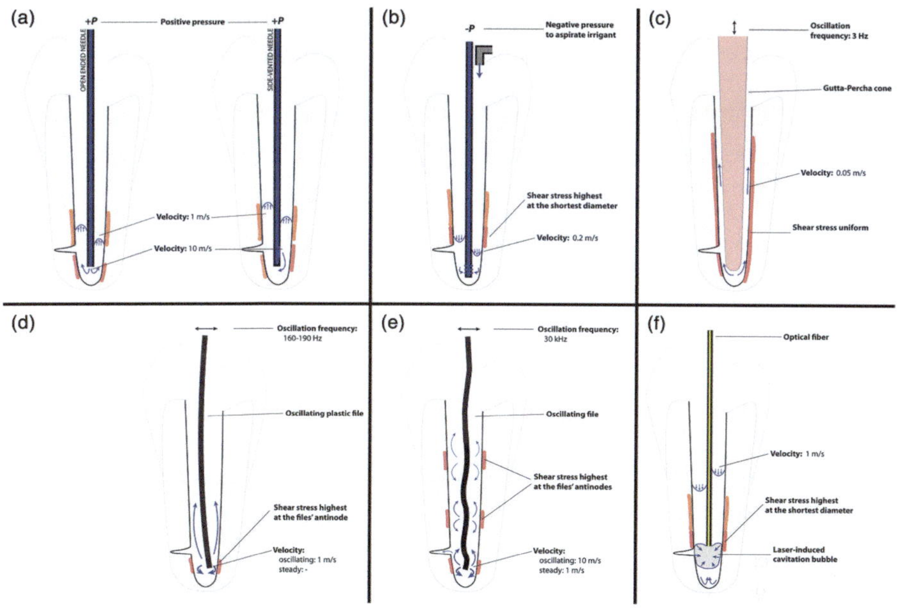

Fig. 1 Sketches of the flow induced by different irrigation systems: a) positive pressure, b) negative pressure, c) MDA, d) sonic activation, e) ultrasonic activation, and f) laser activation. The images are not to scale

Therefore, we have to rely on ex vivo and in vitro research, which mostly provides information on a certain predefined effect of different irrigation procedures, with the exception of some studies which provide data on the irrigant flow. Because often different methodologies are used, it is not always possible to compare the results. Furthermore, we excluded debris and smear layer studies conducted by SEM and studies with planktonic microorganisms or paper point analysis and only included biofilm studies. It is important to realize that the success of cleaning of the root canal system depends on many factors that are intertwined and interact with each other and that the exact contribution of each one of them is not known.

2 Irrigation Systems

Conventional irrigation with a syringe and needle remains widely accepted (Dutner et al. 2012; Savani et al. 2014; Willershausen et al. 2015). However, motivation to improve irrigation efficacy and safety led to the development of other systems (Gu et al. 2009). The mechanism of energy transfer determines the specific flow characteristics of the irrigation systems and consequently their efficacy. A non-instrumentation technique would still be the ideal method, preventing instrumentation of the root canal with its related disadvantages like smear layer and dentin debris production, iatrogenic errors, weakening of the root structure, and apical crack formation (Şen et al. 1995; Gorni and Gagliani 2004; Wu et al. 2004; Shemesh et al. 2010). Such a system has been proposed by Lussi et al. (1993) however, establishing an alternating negative and positive pressure which would enable an effective irrigation procedure and prevent extrusion of the irrigant without introducing instruments/devices in the apical part of the canal seems, for the moment, to be the major limitation. Combinations of syringe irrigation to deliver the irrigant in the root canal and various ways to activate/energize the irrigant have established a rich variety of irrigation techniques (Table 1).

3 Irrigant Flow (Table 2)

3.1 Irrigant Flow During Syringe Irrigation (Positive Pressure)

During syringe irrigation, the irrigant is normally delivered by a needle preferably positioned as close to WL as possible. For very low flow rates, in the order of 0.01 mL/s in combination with a 30G needle, a steady laminar flow is developed (Boutsioukis et al. 2009; Verhaagen et al. 2012b). For higher flow rates up to 0.26 mL/s, which is close to the clinical limits, the flow becomes unsteady, but remains laminar (Boutsioukis et al. 2009, 2010a; Verhaagen et al. 2012b), contrary to previous reports (Kahn et al. 1995). An unsteady flow changes over time, but is not necessarily turbulent. Turbulent flows are highly unsteady, random, unpredictable, and chaotic (Pope 2000). The *flow pattern* depends primarily on the type of needle used, while needle insertion depth, (apical) root canal size, and taper have a limited influence (Boutsioukis et al. 2009, 2010a, b, c, d, e; Verhaagen et al. 2012b) (Fig. 2). Based on the needle outlet and the resulting flow, needles can be categorized into two main groups, namely, the *open-ended* and *closed-ended* needles. All needles create a jet at their outlet, which can be defined as a high-velocity fluid stream forced out of a small-diameter opening or nozzle (Tilton 1999; White 1999); the shape of the outlet determines the orientation and, to some extent, the velocity and stability of the jet (Boutsioukis et al. 2010b; Verhaagen et al. 2012b).

When open-ended needles (flat, beveled, notched) are used, the velocity of the jet is relatively high, and the flow extends along the longitudinal axis of the root canal, apically to their tip. Within a certain distance, which depends on the geometry of the root canal and the insertion depth of the needle, the jet appears to break up gradually. In the case of the closed-ended needles (side vented, double side vented), the jet of irrigant is formed near the apical side of the outlet (the one proximal to the tip for the double side vented) and is directed toward the apex with a divergence of approximately 30° following a curved path around the tip (Boutsioukis et al. 2009, 2010b) (Fig. 2). A series of counterrotating vortices

Table 1 Different Irrigation Systems

Technique				Machine assisted		Simultaneous instrumentation	Flush method	Lit.
Delivery method		Activation method						
Syringe[P]		No		No		No	Cont	Boutsioukis et al. (2007)
		No		No	Brush needle	No	Cont	Al-Hadlaq et al. (2006)
		Gutta-percha		No		No	Int	Huang et al. (2008), Jiang et al. (2012)
		Brush		No		No	Int	Keir et al. (1990)
		Rotary brush		Yes		No	Int	Weise et al. (2007)
		Oscillating tip	Sonic	Yes		No	Int	Jiang et al. (2010a, b)
			Ultrasonic	Yes		No	Int	van der Sluis et al. (2006)
		Laser		Yes		No	Int	Blanken and Verdaasdonk (2007), de Groot et al. (2009)
Ultrasonic[P]	Needle	Oscillating needle		Yes		No	Cont	Castelo-Baz et al. (2012), Jiang et al. (2012)
	Handpiece	Oscillating tip		Yes		No	Cont	van der Sluis et al. (2006)
		File		Yes		Yes	Cont/int	Guerisoli et al. (1998)
NIT[P,N]		No		Yes		Yes	Cont	Lussi et al. (1993)
Quantec-E[P]		No		Yes		Yes	Cont	Walters et al. (2002)
SAF[P]		No		Yes		Yes	Cont	Metzger et al. (2010)
RinsEndo[P]		No		Yes		No	Cont	Hauser et al. (2007)
Safety irrigator[P]		No		Yes		No	Cont	Jiang et al. (2012)
IAC needle[N]		No		Yes		No	Cont	Fukumoto et al. (2006)
EndoVac[N]		No		Yes		No	Cont	Nielsen and Baumgartner (2007)

P = positive pressure; N = negative pressure

Table 2 Irrigation technique characteristics, assuming a root canal of size #35/.06 filled with water

Technique	Characteristics	Flow velocity	Pressure	Shear stress
Positive-pressure needle irrigation	Flow through a needle • Driving pressure: 150 kPa • Flow rate: 0.2 mL/s (Boutsioukis et al. 2007) • Needle size: 27–31G	Inside needle: 10 m/s (Boutsioukis et al. 2010d) Outside the needle: depends strongly on needle type and location in the root canal Flow is not likely to enter side canals	Apical pressure: Open-ended needle: 26 kPa Closed-ended needle: 18 kPa (Boutsioukis et al. 2010d) For a 45/.06 canal 3 mm away from WL	Up to 500 N/m^2 Highest shear stress locally, depending on needle type Away from needle shear stress much lower (Boutsioukis et al. 2010d)
Negative-pressure needle irrigation	Flow aspirated through a needle • Flow rate: 0.05 mL/s (Brunson et al. 2010) • Needle size: 30G	Inside needle: average 0.6 m/s, max 1.6 m/s Outside needle: 0.2 m/s between the needle and the walls; details in the apical area unknown Flow is not likely to enter side canals	Values for the apical pressure are not reported in literature, but expected to be small	Up to 100 N/m^2 Highest values near the tip of the micro-cannula
Manual dynamic irrigation	Gutta-percha cones that are moved up and down • Frequency: 3 Hz • Amplitude: 2.5 mm • Gutta-percha cone: fitting the canal (Jiang et al. 2012)	0.05 m/s, depending on how well the cone fits into the root canal	Apical pressures not reported in the literature	Around 1 N/m^2 Equally distributed on all walls of the root canal
Sonic irrigation	Oscillation of a plastic tip • Frequency: 160–190 Hz • Amplitude: 1 mm • File diameter: 200 µm (Data for EndoActivator) (Jiang et al. 2010a) Oscillating file has one node and one antinode	Steady: – Oscillating: 1 m/s Maximum velocities only at the free end of the plastic tip	Next to oscillating file: Steady: – Oscillating: 7 Pa Apical pressure: unknown, but expected to be small	Steady: – Oscillating: 8 N/m^2 (if it could oscillate unconstrained) Highest shear stress near the tip

Passive ultra-sonic irrigation	Oscillation of a file • Frequency: 30 kHz • Amplitude: 50 μm • File diameter: 200 μm Oscillating file has around six nodes and six antinodes, with a spacing of ca. 5 mm (Ahmad et al. 1993)	Steady: 2.5 m/s Oscillating: 10 m/s Maximum velocities at each anti-node; decreasing toward the nodes High directivity of the flow in the direction of oscillation	Next to oscillating file: Steady: 3 kPa Oscillating: 9 kPa Apical pressure: unknown, but expected to be small (Verhaagen et al. 2014b)	Steady: 2 N/m^2 Oscillatory: 3000 N/m^2 Highest shear stresses at each antinode (Verhaagen et al. 2014b)
Laser-activated irrigation	Transmission of laser energy into irrigant • Laser type: Er:YAG or Er,Cr: YSGG • Laser energy: < 250 mJ/pulse • Pulse repetition rate: 1–25 Hz (de Groot et al. 2009)	1–10 m/s (Blanken and Verdaasdonk 2007; de Groot et al. 2009)	Apical pressures not available in literature, but irrigant extrusion is reported (George and Walsh 2008)	1000 N/m^2 (de Groot et al. 2009)

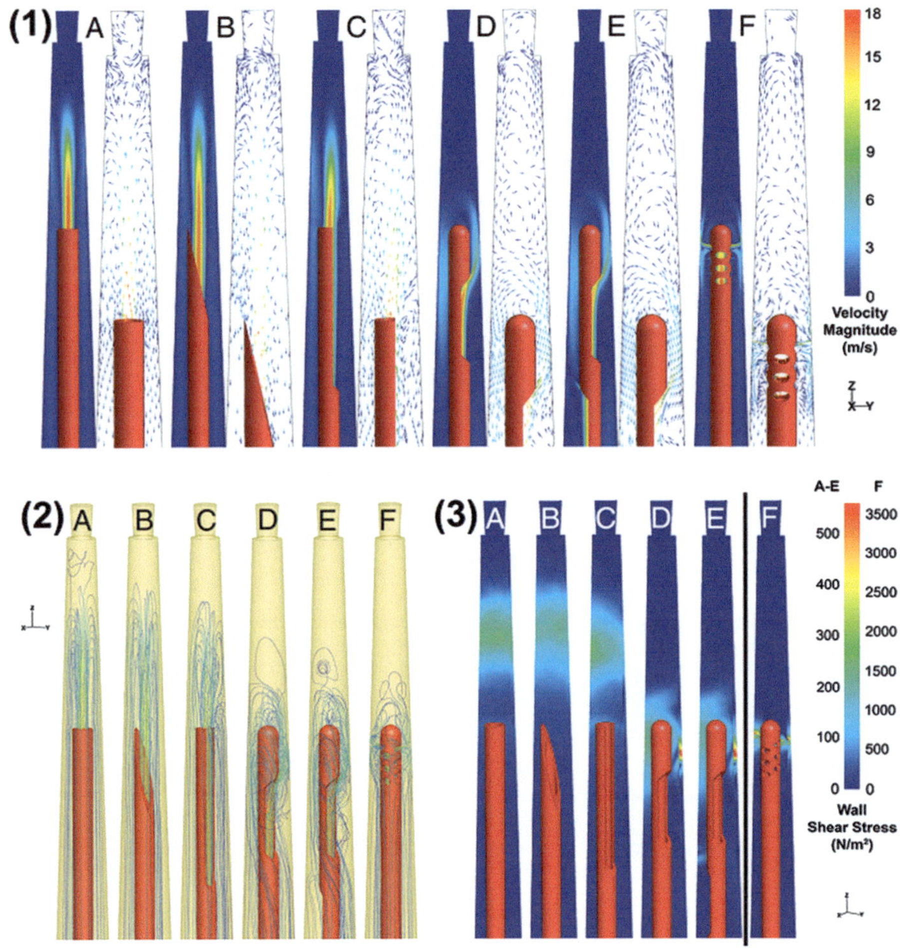

Fig. 2 (1) Time-averaged contours of velocity magnitude (left) and vectors (right) along the z-y plane in the apical part of the root canal for flat (A), beveled (B), and notched (C) open-ended needles and side-vented (D), double side-vented (E) and multi-vented (F) closed-ended needles. The reverse flow toward the canal orifice was noted mainly near the canal wall. (2) Streamlines indicating the route of massless particles released downstream from the needle inlet and colored according to time-averaged velocity magnitude. Particle trajectories provide visualization of the fresh irrigant main flow in three dimensions. Both presence and density of the streamlines are important to indicate the degree of irrigant penetration. (3) Time-averaged distribution of shear stress on the root canal wall. Only half of the root canal wall is presented to allow simultaneous evaluation of the needle position. Needles are colored in red. Reprinted from Boutsioukis et al. (2010b) with permission from Elsevier.

(flow structures where the fluid is rotating) extending to the apical terminus of the canal are formed apically to the tip. Their size, position, and number may differ according to needle insertion depth, root canal size and taper, and flow rate. The velocity of the irrigant inside each vortex decreases significantly toward the apex. When a double side-vented needle is used, 93.5 % of the irrigant flows out of the outlet proximal to the tip (Boutsioukis et al. 2010b). The apical irrigant refreshment/mixing of the closed-ended needles is limited and does not exceed 1–1.5 mm beyond the needle tip for flow rates up to 0.26 mL/s, even in large root canals (Boutsioukis et al. 2009; Boutsioukis et al. 2010b, c, d, e; Verhaagen et al. 2012b), in contrast to what is reported in earlier studies (Kahn et al. 1995), probably depending on the limitations of the study design. The jet formation and the differences between open-ended and closed-ended needles have also been identified in moderately curved root canals (Šnjarić et al. 2012).

A reverse flow toward the canal orifice occurs near the canal wall for both needle types. Regardless of type, small-diameter needles seem to be more effective even when positioned at the same depth (Chow 1983). Keeping the variables constant, the use of a larger needle would directly result in a decrease of the space available for the reverse flow between the needle and the root canal wall and consequently increase apical pressure for open-ended needles and decrease irrigant refreshment/mixing apically to the tip for closed-ended needles (Boutsioukis et al. 2010d, e). As expected, an increase in the preparation size or taper leads to a more efficient flow in the apical root canal (Chow 1983; Falk and Sedgley 2005; Hsieh et al. 2007; Huang et al. 2008; Bronnec et al. 2010a; Boutsioukis et al. 2010d, e). Effective irrigant refreshment/mixing starts from apical size 30, 0.06 taper, allowing refreshment/mixing 2 mm apically to an open-ended needle and 1 mm apically to the tip of a closed-ended needle (Boutsioukis et al. 2010e). The increase of size or taper results in a significant increase in irrigant refreshment/mixing for open-ended needles but not for closed-ended needles (Hockett et al. 2008; Boutsioukis et al. 2010d, e). Surprisingly, a minimally tapered root canal preparation (size 60, 0.02 taper) may present a significant advantage over the tapered ones in terms of irrigant refreshment/mixing (Boutsioukis et al. 2010e); however, a decision on the optimal instrumentation strategy should also take into account the possibility of iatrogenic errors, weakening of the root structure, and obturation technique requirements. A larger volume of irrigant has been associated with improved canal cleanliness (Baker et al. 1975) and irrigant refreshment/mixing (Sedgley et al. 2004, 2005; Bronnec et al. 2010b). An increase in the volume of irrigant can be translated in increased duration of irrigation, assuming a constant flow rate, which allows for improved irrigant refreshment/mixing and chemical effect. Root canal curvature doesn't seem to create additional obstacles for irrigant flow, provided that the needle is placed within 1 mm of working length (WL) (Nguy and Sedgley 2006). Small-size (30G) flexible irrigation needles available nowadays can facilitate irrigant delivery near WL even in severely curved canals, provided that the canal is enlarged at least to size 30 or 35. Tooth orientation (mandibular, maxillary, horizontal) has only a minor influence on the resulting irrigant flow (Boutsioukis et al. 2009, 2010a; Boutsioukis 2010). In a single-phase system, such as a root canal completely filled with irrigant, gravity affects the flow through hydrostatic pressure. The latter is very low compared to the dynamic pressure developed due to the flow of the irrigant. When considering root canal irrigation, it should be emphasized that the root canal is closed at the apex (Hockett et al. 2008; Boutsioukis et al. 2009; Parente et al. 2010; Tay et al. 2010). The apex being closed results in a more complicated flow pattern compared to a canal open from both sides.

3.2 Irrigant Flow During Negative-Pressure Irrigation

Negative-pressure irrigation systems are based on a flow generated by aspiration of irrigant through a needle, placed in the middle part of the root canal or close to or at WL. Fresh irrigant is delivered in the pulp chamber by a larger needle, so there is no forced ejection of fluid near WL and consequently no risk of irrigant extrusion through the apical foramen (Mitchell et al. 2010, 2011). Different negative-pressure systems have been described in the literature (Fukumoto et al. 2004, 2006), but because most of the information has been published on the EndoVac system, this system is used for the estimation of the flow characteristics. The maximum possible flow rate through the EndoVac micro-cannula (30G), and thus its cleaning efficiency, is determined by the aspiration pressure of the system, which is difficult to assess on the dental unit; under ideal conditions, the maximum flow rate is 0.05 mL/s for an aspiration pressure of −25.4 kPa (Brunson et al. 2010). The results of in vitro studies are therefore difficult to relate to clinical practice. Assuming a root canal of size 40, 0.04 taper, which is recommended for optimal flow rate under clinical conditions (Brunson et al. 2010), an average irrigant velocity of 1.1 m/s can be calculated for the root canal near the tip of the micro-cannula. Therefore, due to the low velocity, laminar flow is expected for the negative-pressure systems (Rothfus et al. 1950), which has been confirmed in a recent study (Chen et al. 2014).

3.3 Irrigant Flow During Manual Dynamic Activation

In manual dynamic activation (MDA), a tightly fitting gutta-percha (GP) cone is being moved in and out of the root canal in order to activate the irrigant. As the GP cone is moved toward WL, the change in the space between the cone and the root canal wall, the *reflux space* (Bronnec et al. 2010b), forces the fluid to be displaced coronally and apically. Closer adaptation of the tapered GP cone to the root canal wall could result in a higher irrigant velocity. An estimation of the average flow velocity during MDA can be obtained by considering the change in volume that is occupied by the irrigant. Assuming that the frequency of the up-and-down movements is 30 strokes per 10 s (3 Hz) (Jiang et al. 2012), the displacement is within the apical 5 mm of the root canal (Huang et al. 2008; Jiang et al. 2012), a 15-mm root canal prepared to size 35, .06 taper and a matching GP cone, the maximum velocities 5 mm from WL are on the order of 0.05 m/s. In reality the root canal geometry is irregular, with not only a non-smooth wall but also side canals, oval extensions, and isthmuses into which the fluid can be pushed. This could enhance the removal of debris from canal irregularities, but also increases the effective area and decreases the velocity in the main canal.

3.4 Irrigant Flow During (Ultra)Sonically Activated Irrigation (UAI, SAI)

The term *passive ultrasonic irrigation (PUI)* has been used in the past to describe ultrasonic activation of irrigants (Jensen et al. 1999; van der Sluis et al. 2007); however, recent findings suggest that this term may be inaccurate and confusing (Boutsioukis et al. 2013b).

During sonically and ultrasonically activated irrigation, instruments like files (cutting or non-cutting), plastic, or NiTi tips or needles oscillate while being driven at one end with a specific frequency, which is below 20 kHz for sonic devices and above 20 kHz for ultrasonic devices. For the relatively new sonic devices, like EndoActivator or Vibringe, the frequency was found to be in the range of 160–190 Hz (Jiang et al. 2010a); for systems like SONICflex (KaVo), the frequency goes up to 6000 Hz; for most ultrasonic devices, the frequency is approximately 30 kHz. The oscillation patterns of sonically and ultrasonically driven files are very distinct (Lumley et al. 1991), with sonically driven files having only one node and one antinode (Lumley et al. 1991; Jiang et al. 2010a), while ultrasonic files have around six nodes with a spacing of approximately 5 mm (Fig. 1) (Ahmad et al. 1993; Verhaagen et al. 2012a). The oscillation amplitude, measured outside the root canal, was around 1 mm for the EndoActivator tip (Jiang et al. 2010a) and around 50 μm for ultrasonically activated instruments (Ahmad et al. 1987; Lea et al. 2010; Verhaagen et al. 2012a). Oscillating instruments act as a mixer producing microstreaming of the irrigant that enhances delivery of the irrigant throughout the root canal system and refreshment/mixing of the irrigant. The ultrasonically oscillating instruments can generate acoustic streaming, contrary to the sonically driven systems (Duck and Smith 1979; Lumley et al. 1991; Jiang et al. 2010a; Verhaagen et al. 2014b). The induced flow consists of an *oscillatory component* that follows the oscillation of the instrument and dominates the flow near it and a *steady component* that always moves away from the file; the latter dominates the flow further away from the instrument (Fig. 3). The characteristics of each flow component are determined by the oscillation characteristics of the instrument together with the presence of a wall or confinement which affects the formation of steady jets and increases the oscillatory velocities and associated pressure and shear stress (Verhaagen et al. 2014b) (Fig. 4). The lateral component of the flow is relatively strong in comparison to the apical or coronal component. This lack of the reverse flow should be compensated by syringe irrigation after the activation of the irrigant or the use of ultrasonically activated needles allowing a continuous flow in the apical root canal and an effective reverse flow of the irrigant. A continuous flow directed into the pulp chamber does not result in an effective refreshment/mixing of irrigant in the apical root canal (van der Sluis et al. 2009). The lateral flow component can enhance irrigant flow and tissue dissolution in side canals (De Gregorio et al. 2009; Al-Jadaa et al. 2009a), isthmuses (Burleson et al. 2007), or oval extensions (Jiang et al. 2010b). The oscillation perpendicular to the main oscillation direction is less than 25 % of the main oscillation direction

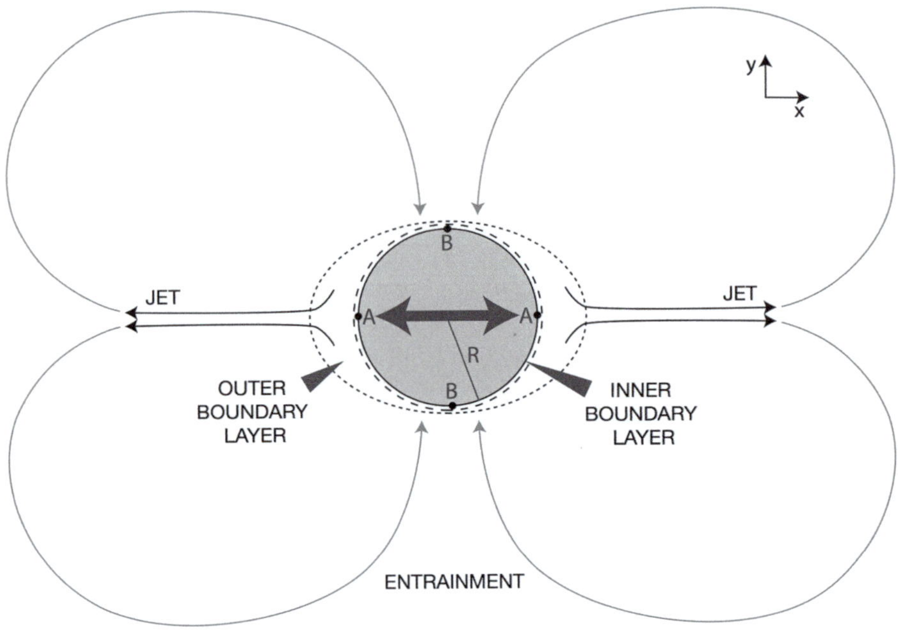

Fig. 3 Two-dimensional (cross section) view of the velocity vectors and pressure magnitudes induced by a file oscillating ultrasonically next to a wall (a) or inside a root canal (b). The radius of the file is 100 μm. The results are obtained with a computational fluid dynamics code and are shown at one instant of time (top row) and the time average over 5 cycles (bottom row). Reprinted with permission from Verhaagen et al. (2014b). Copyright 2014, Acoustic Society of America.

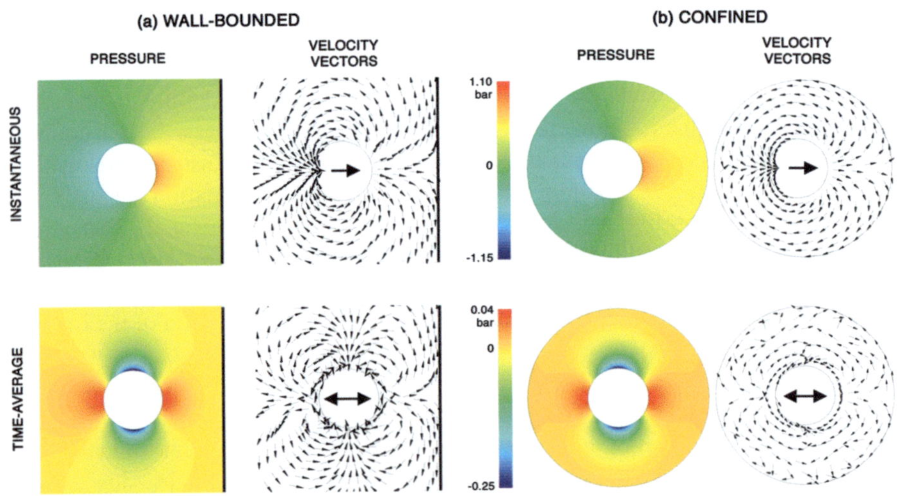

Fig. 4 Visualization of a laser-generated vapor bubble inside a glass root canal model. The laser energy was 60 mJ per pulse. Image sequence is from left to right; the interframe time is 140 Âμs. The first panel shows a sketch of the setup, with 1) the root canal model, 2) the laser fiber tip (outer diameter 280 lm), 3) the laser-induced cavitation bubble, and 4) a stable cavitation bubble at the apex. Reprinted with permission from Verhaagen et al. (2014b). Copyright 2014, Acoustic Society of America.

(Lea et al. 2010; Verhaagen et al. 2012a). For ultrasonic activation of the irrigant, an increase of the power setting results in an increase in the oscillation amplitude which enhances the irrigant flow and the cleaning efficacy mostly in the oscillation direction of the file (Lumley et al. 1993; Jiang et al. 2010b, c). The coronal-to-apical component of the flow extends beyond the tip of the instrument decreasing after 2 mm (Malki et al. 2012). It is assumed that the fluid dynamics is not affected by a curvature, as the typical streaming length scale is smaller than the curvature (Malki et al. 2012). However, it has been reported that in curved canals, the cleaning efficacy of sonic or ultrasonic files is affected by the curvature (Amato et al. 2011). Damping of the oscillatory movement by placing the ultrasonically activated instruments to WL (Boutsioukis et al. 2013b) could have been the reason for this effect. Pre-bending the instrument could bring it closer to WL and reduce contact and possible canal transportation (Al-Jadaa et al. 2009a); however, bending significantly changes the oscillation pattern (Ahmad et al. 1992; Lumley and Walmsley 1992; Verhaagen et al. 2012a). During ultrasonic activation of the irrigant, cavitation may be induced (Ahmad et al. 1988; Jiang et al. 2010c), which is the formation, behavior, and collapse of bubbles under an applied changing/oscillating/alternating pressure. The collapse of these bubbles close to a wall can generate a high-velocity jet directed toward the wall (a process called *transient cavitation*) (Brennen 1995), enhancing its cleaning (Ohl and Wolfum 2003). If present, soft material (like biofilm) might be pulled from the wall toward the bubble (Brujan et al. 2001). Gas-filled bubbles might be stable for a relatively long time and oscillate together with the oscillating pressure field induced by the oscillating file. This *stable cavitation* might enhance the streaming and consequently the cleaning locally, through unsteady oscillations of the bubble shape (Brennen 1995). Both transient and stable cavitations have been observed in a number of studies; the location of transient cavitation occurrence seemed to coincide with oscillation antinodes on the file (Ahmad et al. 1988; Lumley et al. 1988; Felver et al. 2009; Jiang et al. 2010c; Macedo et al. 2014a). Although stable cavitation was seen throughout the root canal, transient cavitation occurred closely to the file (Macedo et al. 2014a). Previous work has ruled out cavitation as a significant contribution to root canal cleaning (Ahmad et al. 1988; Lumley et al. 1988), but a recent study showed that it is not that clear-cut; its contribution depends on different factors (Macedo et al. 2014b). A non-tapered instrument will produce cavitation at all antinodes alongside the instrument, whereas tapered instruments only produce cavitation at the tip; in addition, a smaller confinement (root canal) leads to an increase in cavitation (Macedo et al. 2014a). More research is needed to clarify the role of cavitation in cleaning procedures.

3.5 Irrigant Flow During Laser-Activated Irrigation

Laser-activated irrigation (LAI) employs laser energy to activate the irrigant and relies primarily on cavitation. Normally, Er:YAG or Er,Cr:YSGG lasers are used with a wavelength in the infrared region (2796–2940 nm) which is well absorbed by water (Robertson and Williams 1971). The dynamics of LAI have been studied using high-speed imaging (Blanken and Verdaasdonk 2007; de Groot et al. 2009; Matsumoto et al. 2011). The working mechanism of LAI is believed to be related to the growth and collapse of a laser-induced cavitation bubble; its collapse causes shock waves and a fluid flow in the coronal, lateral, and mainly apical direction with velocities of a few meters per second (de Groot et al. 2009), resulting in a high cleaning efficacy (de Groot et al. 2009; de Moor et al. 2010). The effect of the implosion of the bubble seems to be directed to the apical root canal because of its smaller dimension. The size of the laser-induced bubble and consequently the collapse velocity depend on the energy output, pulse duration, and frequency, and its absorption by the irrigant; up to a certain extent, more energy means a larger bubble. However, the influence of the different parameters on the cleaning efficacy of the flow has not yet been clarified. Recently, good results have been published applying laser energy in the pulp chamber just above the canal orifice with specially developed tips (PIPS) (Peters et al. 2011; Lloyd et al. 2014; Ordinola-Zapata et al. 2014). It is believed that irrigant streaming in the complete root canal can be induced by shock waves. The mechanisms have not yet been clarified.

3.6 Apical Flow

It has been reported that negative-pressure irrigation, ultrasonic activation of the irrigant, patency files, or laser-activated irrigation can deliver irrigant solutions more efficiently into the apical third than syringe irrigation (de Gregorio et al. 2009, 2010; Vera et al. 2011, 2012; Munoz and Camacho-Cuadra 2012; Castelo-Baz et al. 2012; Peeters et al. 2014; Peeters and Gutknecht 2014). This has been related to bubble entrapment in the apical part of the root canal, also termed "apical vapor lock," during syringe irrigation, ex vivo (de Gregorio et al. 2009; Tay et al. 2010) and in vivo (Vera et al. 2011, 2012). This bubble could obstruct effective irrigant delivery specifically in the apical area of the root canal system. However, its formation and extent are dependent on the irrigant flow rate, needle type, and insertion depth and apical size of the root canal, and it has been shown in vitro that it can be removed or prevented by positioning the closed-ended needle within 1 mm away from WL or by using high irrigant flow rates in the order of 0.2 mL/s (Boutsioukis et al. 2014a). The abovementioned studies all used low flow rates and/or positioned the needle farther away from WL than 1 mm. A related argument is the so called "dead-water" or "stagnation" zone apically to the needle tip in areas of the root canal where no irrigant refreshment/mixing takes place, even if no bubble entrapment is assumed (Gao et al. 2009; Shen et al. 2010). However, high-speed imaging experiments and computer simulations have shown that in those areas, the irrigant flow is just very slow (Boutsioukis et al. 2010a, b, c, d, e, Verhaagen et al. 2012b).

3.7 *Irrigant Flow in Lateral Canals, Oval Extensions, and Isthmuses*

Irrigant flow into lateral canals (Al-Jadaa et al. 2009a; de Gregorio et al. 2010), isthmuses (Burleson et al. 2007), or oval extensions (Jiang et al. 2010a) is limited for irrigation systems producing a flow parallel to the root canal wall and for sonic activation of the irrigant. The flow in the main canal will drive the flow in the lateral canal, and penetration of the irrigant is limited to a few times the diameter of the canal entrance (Shankar and Deshpande 2000). It has been confirmed that syringe irrigation is not very effective in isthmuses and oval extensions (Burleson et al. 2007; Ricucci and Siqueira 2010) contrary to ultrasonic or laser-activated irrigation, which produces a lateral flow component, improving irrigant penetration and tissue dissolution in side canals (Al-Jadaa et al. 2009b; de Gregorio et al. 2009, 2010; Castelo-Baz et al. 2012), isthmuses (Burleson et al. 2007) or oval extensions (de Groot et al. 2009; de Moor et al. 2010; Jiang et al. 2010c). Recently, a modified apical negative-pressure technique was used to clean the isthmus between the mesial roots of mandibular molars. Irrigant was delivered near WL in one of the root canals by a syringe and needle, while a suction tip was positioned near WL in the second root canal, forcing the irrigant through the isthmus. The new technique showed improved removal of dentin debris from the isthmus, compared to syringe irrigation, a traditional apical negative-pressure system, and ultrasonic activation (Thomas et al. 2014).

3.8 *Irrigant Flow in Dentinal Tubules*

The geometry of the dentinal tubules (typical diameter 0.5–3.2 μm, length 1–2 mm) poses a serious difficulty for the irrigant to penetrate (Haapasalo and Orstavik 1987; Orstavik and Haapasalo 1990). Diffusion through the small opening in the root canal lumen is slow (Zou et al. 2010), whereas convection is limited to a few micrometers from the entrance, provided that the tubule is filled with liquid (Verhaagen et al. 2014a). Advanced treatment strategies specifically aimed at the tubules might be necessary to enhance the cleaning of the tubules (Shrestha et al. 2009).

3.9 *Irrigant Extrusion*

During root canal irrigation, part of the irrigant may be extruded toward the periapical tissues (Boutsioukis et al. 2013a). Irrigant extrusion does not seem to take place in vital cases, provided that root canal instrumentation has been limited within the root canal (Salzgeber and Brilliant 1977). The presence of a pulp stump and a healthy periodontal ligament seems to create a sufficient barrier against extrusion. However, in necrotic cases, extrusion seems more likely (Salzgeber and Brilliant 1977), which could be attributed to the absence of such a barrier. Several studies have investigated the irrigant pressure developed near the apical foramen during root canal irrigation, in an effort to compare the safety of different techniques (Boutsioukis et al. 2010b, c, d, e; Shen et al. 2010; Conard 2012; Verhaagen et al. 2012b; Park et al. 2013). Attempts to define a threshold for irrigant extrusion, such as the capillary pressure, have also been reported (Park et al. 2013). However, recently, it has been argued that apical irrigant pressure is not the only factor determining extrusion (Psimma et al. 2013a); therefore, extrusion cannot be predicted simply by comparing the irrigant pressure to a fixed threshold. Rather, the condition of the periapical tissues may have a significant effect on the pressure balance. A periapical lesion entirely surrounded by cortical bone could provide an effective barrier, thus reducing irrigant extrusion to a minimum. To the contrary, in cases of longstanding periapical lesions where the integrity of the cortical bone has been compromised and a direct pathway has been established from the apical foramen toward soft tissues, the oral cavity, or the maxillary sinus (e.g., in the form of a sinus tract or apical fenestration), the resistance (back pressure) of the lesion may be reduced. A series of ex vivo studies provided support for this hypothesis (Psimma et al. 2013a, b; Boutsioukis et al. 2013b). Extrusion during syringe irrigation depends on various technique-related parameters. Open-ended needles can extrude more irrigant than closed-ended needles under identical conditions. In both cases, extrusion decreases as the needles move away from WL or with increasing apical root canal size (Psimma et al. 2013a, b). Wedging of the needle inside the root canal causes a significant increase in extrusion, especially when open-ended needles are used (Psimma et al. 2013b). A higher irrigant flow rate has also been linked to increased extrusion (Boutsioukis et al. 2013b); however, contrary to common belief, flow rate should not be considered as the primary cause of irrigant extrusion in the absence of evidence (Boutsioukis et al. 2013a). Its effect seems comparable to the effect of other technique-related parameters discussed above and considerably inferior to the effect of the condition of the periapical tissues (Psimma et al. 2013a, b; Boutsioukis et al. 2013b). Anatomy-related factors such as the diameter of the apical constriction (range 0.15–0.35 mm) and the curvature of the root canal (range 0–30°) don't seem to affect irrigant extrusion during syringe irrigation (Psimma et al. 2013a, b; Boutsioukis et al. 2013b). Negative-pressure systems do not employ forced ejection of fluid near the WL, so the risk of irrigant extrusion through the apical foramen is very low (Mitchell et al. 2010, 2011), and these methods are generally considered safer than positive-pressure syringe irrigation. Similarly, sonic and ultrasonic agitation seems not to cause any measurable extrusion, irrespective of the power setting and the insertion depth of the file/tip. To the contrary, MDA has been verified to extrude significantly more irrigant compared to both sonic and ultrasonic agitation (Boutsioukis et al. 2014b). LAI also seems to extrude large volumes of irrigant ex vivo (George and Walsh 2008); however, the methodology used has major drawbacks, so it is difficult to extrapolate the results to the clinical situation.

An in vivo study demonstrated that when the laser tip is placed in the pulp chamber just above the root canal entrance, no extrusion of irrigant is to be expected (power setting 1 W and 35 Hz) (Peeters and Mooduto 2013). Furthermore, using PIPS, irrigant extrusion seems to be related with the power setting (Arslan et al. 2014).

It must be emphasized that irrigant extrusion of any amount of irrigant doesn't necessarily result into an extrusion accident with pronounced symptomatology. The minimum amount of extruded irrigant that can cause such accidents has not been elucidated (Boutsioukis et al. 2013a). For instance, passive extrusion of minute amounts of irrigant may continuously occur very slowly even during the rest phase (Chu 2010; Psimma et al. 2013a), but this has not been linked to an accident yet.

4 Fluid Wall Interaction (Wall Shear Stress) (Table 2)

4.1 Shear Stress

During irrigant flow, frictional forces occur within the irrigant and between the flowing irrigant and solid bodies at rest (e.g., root canal walls) or between a moving solid body (ultrasonically oscillating file) and irrigant at rest (Mott 1999; Tilton 1999; White 1999). The magnitude of such friction between the irrigant and the root canal wall is described by wall shear stress and is proportional to the difference of the velocity between adjacent "layers" of irrigant close to the wall. Irrigants with higher viscosity will develop higher wall shear stress; however, they will also resist flow and require more energy to be delivered or activated. Wall shear stress is of particular interest to irrigation because it determines the mechanical effect on the biofilm, tissue remnants, dentin debris, or smear layer attached to the root canal wall.

4.1.1 Wall Shear Stress Produced by Syringe Irrigation

Similar to the developed irrigant flow, two basic wall shear stress patterns can be distinguished for the open- and closed-ended needles. Regarding the open-ended needles, an area of increased shear stress (~200 N/m^2) with an approximately uniform circumferential distribution is developed apically to the needle tips, in the region where the unsteadiness, due to jet breakup, is strongest (Fig. 2). On the other hand, the closed-ended needles lead to higher maximum shear stress (~500 N/m^2); however, the areas are limited near the tip, on the wall facing the needle outlet (the proximal outlet for the double side vented) (Boutsioukis et al. 2010b; Chen et al. 2014). Therefore, optimum debridement is expected near the tip of the needle, as reported by Huang et al. (2008), and continuous displacement of the needle tip is necessary during syringe irrigation, to increase the area affected by high wall shear stress.

Needle depth, canal size, and taper do not seem to affect the *overall distribution* of wall shear stress (Fig. 2). The maximum shear stress decreases as needles move away from the working length or with increasing size or taper, because more space is available for the reverse flow of the irrigant. This reduces the irrigant velocity gradients, but the area affected by high shear stress becomes larger (Boutsioukis et al. 2010c, d, e). No data are available on the flow rate, but, as long as the flow characteristics are unchanged, a higher flow rate will lead to a higher shear stress.

4.1.2 Wall Shear Stress Produced by Negative-Pressure Systems

During negative-pressure systems, the flow that is aspirated toward the apex induces shear stress on the wall. The shear stress appears to have an approximately equal circumferential distribution along the main root canal walls, with highest values (50–100 N/m^2) occurring near the apex and decreasing magnitude toward the coronal part (Chen et al. 2014), as the flow velocity and shear stress are related to the increasing cross-section area between the micro-cannula and root canal wall. As irrigant velocities produced during positive-pressure irrigation (like syringe irrigation) are higher in certain areas (Boutsioukis et al. 2010a; Verhaagen et al. 2012b), negative-pressure systems seem to be inferior to syringe irrigation regarding the shear stress produced on the root canal wall (Table 2). This has been recently verified by computer simulations (Chen et al. 2014).

4.1.3 Wall Shear Stress Produced by MDA

The GP cone that is moved up and down in the root canal creates a flow and therefore a shear stress on the wall, whose magnitude depends on the reflux space available (Bronnec et al. 2010b; Jiang et al. 2012). With the help of a simple model based on two concentric cones, the shear stress can be estimated and seems to be on the order of 1 N/m^2, which is much lower than for other systems.

4.1.4 Wall Shear Stress Produced by (Ultra)Sonically Activated Irrigation

The oscillating flow induced by the sonic and ultrasonic devices gives rise to a nonuniform shear stress on the wall of the root canal and is highest where the oscillation amplitude is highest, i.e., near the antinodes of the file. This shear stress during ultrasonic activation has been estimated by computer simulation to be in the order of 500 N/m^2 (Chen et al. 2014), but the model lacked experimental validation. The influence of the oscillatory component of the flow on the pressure and shear stress is two or three orders of magnitude higher than the steady component, and the shear stress was calculated to be 3000 N/m^2 for the oscillatory and 2 N/m^2 for the steady component. The steady jets are known to exert a constant shear stress on the wall, which is highest at a location slightly off axis in the direction of oscillation (Deshpande and Vaishnav 1982; Phares et al. 2000) (Fig. 4). The oscillatory shear stress and pressure result in a loading cycle on the biofilm on the wall. Being a viscoelastic material, a biofilm will lose energy through a loading cycle (Guelon et al. 2011); therefore, the oscillation can lead to fatigue of the biofilm, which in turn leads to biofilm failure. In sonic irrigation, the frequency is much lower than in ultrasonic irrigation; therefore, biofilm fatigue is less likely to occur with sonic irrigation.

4.1.5 Wall Shear Stress Produced by Laser-Activated Irrigation

During LAI, the shear stress is expected to be induced by the flow produced by the formation and collapse of a laser-induced bubble. The observed flow is similar to negative-pressure systems, except that for LAI, the magnitude of the shear stress is one order of magnitude higher (1000 N/m^2).

4.2 *Biofilm Disruption by Wall Shear Stress*

Wall shear stress could mechanically disrupt biofilm. Biofilms are structured by a matrix of extracellular polymeric substance (EPS), which can reach up to 90 % of the biofilm. This matrix provides the biofilm its viscoelastic properties, facilitates nutrition, and acts as protection from chemical and mechanical attacks imposed by cleaning procedures and disinfectants (Stewart and Franklin 2008). Therefore, the viscoelastic properties can be seen as a virulence factor (He et al. 2013). Under small stresses, a biofilm can deform elastically, whereas under large stresses, it can flow viscously (Körstgens et al. 2001; Flemming and Wingerder 2010). Forces on the biofilm exerted by irrigant flow could cause absorption of energy into the biofilm leading to volumetric expansion (Busscher et al. 2003). Deformation beyond the yield point could disrupt the top layers of the biofilm, or its EPS matrix (cohesive failure), or could completely remove the biofilm (adhesive failure). If deformation is in the plastic range but below the yield point, the biofilm is expanded but not removed (Busscher et al. 2003). However, disruption of the top layers or EPS matrix or expansion of the biofilm facilitates irrigant penetration in the biofilm and could already enhance the chemical effect of irrigants (He et al. 2013). Furthermore, a disruption of the biofilm matrix could leave "footprints" in the remaining biofilm, facilitating adhesion of microorganisms, thereby influencing reorganization of the biofilm (Busscher et al. 2003). However, not much information is available in the literature on the effect of fluid flow on a biofilm, mainly because of a large variety of biofilm constituents and associated physical properties. Moreover, measurement of the mechanical properties should take place on a short time scale (within minutes), because the biofilm is a living organism and will adapt to its environment (Flemming et al. 2011). In addition, the time scale of irrigant activation systems should be considered, as the behavior of a biofilm under 30 kHz ultrasonic oscillations or laser pulsations may be different than when deforming it slowly.

Critical loads necessary to disrupt biofilms by a variety of different techniques have recently been reviewed (Böl et al. 2012). It was found that the sensitivity to certain loading modes, such as normal or shear stresses, varies extensively among biofilms. Furthermore, the reported values of adhesion strength depend greatly on the testing technique, which range from coarse macroscale measurements down to atomic force microscopy (AFM) operating on a nanoscale (Böl et al. 2012). Typical values found in the literature give an elastic modulus on the order of 10^{-1}–10^{2} Pa and a cohesive shear strength of 10^{1}–10^{3} Pa (Flemming et al. 2011; Böl et al. 2012). Pressures and shear stresses produced by different irrigation techniques show that some techniques should be able to remove biofilm. Unfortunately, the mechanical properties of an endodontic biofilm are not known; therefore, a prediction of the effect of fluidic stresses on biofilm removal in the root canal is not yet possible. A 3D numerical study on the effect of fluid flow on biofilm has shown that for high EPS matrix stability, only exposed structures at the surface of the biofilm are detached. Low EPS matrix stabilities may lead to the detachment of large portions from the top of the biofilm. Interestingly, it has been observed that a smooth basal biofilm surface structure remains after detachment in both cases (Böl et al. 2009). This is confirmed by another study where smooth base biofilms remain after the biofilm had been subjected to high shear stresses using the fluid dynamic gauging (FDG) technique (Möhle et al. 2007). These observations can be explained by the stratification of biofilms, which leaves older, stronger layers at the base of the biofilm, typically adhering strongly to the substrate (Möhle et al. 2007; Derlon et al. 2008). Therefore, complete removal of the biofilm from the root canal wall could be a difficult task, and a combination of mechanical *and* chemical stress on the biofilm remains crucial.

4.2.1 Interaction Between a Biofilm and the Flow Created by Sonic, Ultrasonic, or Laser Activation

Weak forces (low pressures and shear stresses) or a high EPS matrix stability causes only an elastic deformation of the biofilm that reverses as soon as the stress is removed. Repeated loading of the biofilm structure with a periodic stress, as is the case with sonic, ultrasonic, and laser activation, may result in cohesive failure or fatigue of the biofilm (Flemming et al. 2011); however, the threshold (force and number of loading cycles) for damage due to fatigue is unknown. At increased force (or lower EPS matrix stability), viscous deformation of the biofilm may occur. The biofilm deforms and is displaced in order to distribute and minimize the applied stresses (Klapper et al. 2002). When a steady force is applied, for example, in the case of a steady flow, the biofilm will attain a steady state, and no further deformation or removal will take place. Therefore, it may be advantageous to generate a nonsteady flow, for example, by unsteady oscillations of the ultrasonic file or by generating pulsations with ultrasound or laser. At stresses exceeding the cohesive or adhesive strength of the biofilm, parts of the biofilm may detach from the bulk biofilm or from the substrate, respectively.

4.2.2 Interaction with Cavitation Bubbles Created by Ultrasonic or Laser Activation

The gradients associated with the time scales and length scales of the exerted stress are important with regard to the behavior of a viscoelastic material. Cavitation bubbles, such as those created by fast irrigant vaporization with a laser device, typically exhibit large velocities and accelerations on a small time scale, making them efficient in plastic deformation of the biofilm. For transient cavitation, velocities of 100 m/s are feasible at micrometer scales (Brujan et al. 2001). To clean larger areas such as the entire root canal system, many cycles (laser pulses) may be required.

Biofilm models that allow for effectively screening the efficacy of irrigation systems are not available yet; therefore, it is difficult to determine the real impact of irrigation procedures on biofilms. To evaluate the effect of irrigant flow on the viscoelastic behavior of a biofilm, a standardized hydrogel of gelatine and hyaluronan has been used. When a hydrogel with a density of $1.07 \pm 0.07 \cdot 10^3$ kg/m^3 and a Young's modulus and adhesive strength of 10^{-1} Pa was used, the flow induced by an ultrasonically oscillating file at a distance of 100 μm from the hydrogel always removed the hydrogel, except at the lowest power setting. The visualizations confirmed that acoustic streaming consisted of a steady and oscillatory component of the flow and that the former is dominant for viscous deformation of the hydrogel and the latter induces an elastic deformation (Macedo et al. 2014b). More research is needed to study the interaction of a fluid flow with a biofilm in the typical endodontic environment.

5 Chemical Effect

5.1 Introduction

The chemical effect of irrigants will be significantly influenced by its concentration, the contact surface between the substrate and the irrigant, contact time, type of reactant, pH, temperature, interaction with other chemicals, and volume. Irrigant refreshment is a crucial requirement to ensure adequate chemical effect, as irrigants are (rapidly) inactivated when they come in contact with tissue remnants, dentin, microorganisms/biofilm (Moorer and Wesselink 1982; Druttman and Stock 1989; Haapasalo et al. 2005), or irrigants themselves (Rossi-Fedele et al. 2012). The efficacy of a certain flow in refreshing consumed irrigant can be characterized with the second Damköhler number, which is defined as the ratio of typical irrigant transport time to reaction time. The reaction time will be directly influenced by the reaction rate (RR) or speed of reaction, which is the change in concentration of the reactants or the change in concentration of the products of a given chemical reaction per unit time. The transport time can be determined from the local flow velocity U and vorticity Ω (measure of circulation of a fluid) scaled with a typical length L. For areas with low velocity, diffusion (with coefficient D) may also play a role.

The second Damköhler number (Da) is then given by

$$Da = \frac{\frac{L}{U_{\text{flow}} + \Omega_{\text{flow}} L + \frac{D}{L^2}}}{\tau_{\text{reaction}}}$$

At $Da \leq 1$, the mixing is optimal, meaning that the consumed irrigant is effectively replaced in time by fresh irrigant. At $Da > 1$ however, the reaction is slowed down because of a lack of fresh irrigant. Both the velocity U and vorticity Ω of the irrigant can be obtained from experiment; the diffusion coefficient D depends on the irrigant used. The length scale L is a typical length over which the reaction takes place; in a root canal, the L corresponds to the working length. Consequently, in order to predict the efficacy of our irrigation procedures, the RR of the irrigants is needed. Time is an important aspect and is directly related to the RR. However, because the properties of endodontic biofilm are unknown, it is impossible to make an estimation of the RR and consequently the required time of irrigant application. The chemical effects of the irrigants are regulated by convection or diffusion. During the rest phase, diffusion seems to be the main mechanism of molecular transport. To what extent and how fast this occurs will depend on the concentration gradients in the root canal and the contact surface between the irrigant and substrate (e.g., biofilm, pulp, dentin). However, the contact surface between the substrate and irrigant is quite limited in a human root canal, e.g., a root canal of size 30, .06 taper, with a length of 15 mm has a ratio of volume (V)/surface (S) of 0.2 mm ($V = 7.4\ mm^3$ and $S = 35.4\ mm^2$). Furthermore, the typical structure of the root canal system contains many areas which are difficult to reach, limiting both convection and diffusion of the irrigant. This will severely hamper the chemical effect of the irrigants (Rosenfeld et al. 1978; Moorer and Wesselink 1982). During the delivery and activation phase, convection is expected to be the dominant transport mechanism in the main canal, whereas in the confined areas of the root canal system, diffusion may still be important.

According to Zehnder (2006), the ideal irrigant should:

- Have a broad antimicrobial spectrum and high efficacy against anaerobic and facultative microorganisms organized in biofilms
- Dissolve pulp tissue remnants
- Inactivate endotoxin
- Prevent the formation of a smear layer during instrumentation or dissolve the latter once it has formed
- Be biocompatible

A high efficacy against biofilms is only possible when the irrigant has the ability to diffuse through or break down the EPS matrix of the biofilm; without this capacity, the irrigant cannot exert its antimicrobial action. This aspect is important when the biofilm is still intact but also when the top layers have been removed by instrumentation or irrigation because the structure of the remaining basal layer is characterized by high adhesion and cohesion forces. In this section we focus on the most frequently used irrigants in endodontics which are sodium hypochlorite (NaOCl), ethylenediaminetetraacetic acid (EDTA), and chlorhexidine (CHX) (Dutner et al. 2012; Willershausen et al. 2015).

NaOCl is widely used as the primary root canal irrigant (Dutner et al. 2012; Savani et al. 2014; Willershausen et al. 2015) due to its unparalleled action against microorganisms (McDonnell and Russell 1999), and biofilm (Bryce et al. 2009; Chávez de Paz et al. 2010; Stojicic et al. 2013), and its unique capacity to dissolve pulp tissue (Naenni et al. 2004; Sirtes et al. 2005) and organic components of the smear layer (Baumgartner and Mader 1987). In spite of its low clinical toxicity (Zehnder 2006), NaOCl has been shown to be extremely caustic when in contact with organic tissue in vitro (Pashley et al. 1985), even at concentrations lower than 0.1 % (Chang et al. 2001; Heling et al. 2001; Barnhart et al. 2005). The choice of the NaOCl concentration to be used is generally considered a trade-off between cleaning efficiency and tissue damage in the case of inadvertent extrusion (Spencer et al. 2007). Clinically, the use of concentrations between 0.5 % and 6 % is suggested; however, the optimal clinical concentration is still a subject of controversy (Zehnder 2006). A recent systematic review reported that higher concentrations appear to improve the outcome of root canal treatment, but the evidence was weak (Fedorowicz et al. 2012).

Once delivered inside the root canal, NaOCl reacts with its organic content such as pulp tissue, biofilm, or dentin (canal wall, smear layer, or debris) causing depletion of the free available chlorine (Baker 1947) and resulting in protein degradation, rise of temperature (Baker 1947), and changes in pH (Jungbluth et al. 2011; Macedo et al. 2014d). Even though this reaction occurs faster during the first minute, it evolves with time being limited by the amount of free available chlorine on the irrigant and/or the amount of organic substance available for the irrigant to react with. A stagnant (non-agitated) NaOCl solution can retain a considerable amount of available chlorine ($\geq$0.1 %) inside a root canal ex vivo even after 100 min (Ragnarsson et al. 2015), but this amount of chlorine may not actually be available near the substrate. A recent optical method to measure the NaOCl concentration in human root canals ex vivo was introduced in the endodontic literature (Rechenberg et al. 2014). The authors reported that NaOCl at clinically relevant initial concentrations ($\geq$1 %) remains chemically active (above 0.1 %) for more than 10 min (Ragnarsson et al. 2015). Nonetheless, such a method has a reduced accuracy to quantify concentrations higher than 0.1 %; therefore, such information has limited clinical value. The speed of the reaction between chlorine and substrate will be increased by the initial concentration of the irrigant, ultrasonic or laser activation (Macedo et al. 2010), frequent irrigant refreshment (Macedo et al. 2014c), and temperature (Sirtes et al. 2005). It should be noted that the temperature increase lasts only as long as the irrigant delivery (Macedo et al. 2014e). The pH of the NaOCl solution determines the equilibrium between the hypochlorite ion (OCl^-) and the hypochlorous acid (HOCl) (Baker 1947). The biological effect of NaOCl, which can be defined as its tissue-dissolving capacity and antimicrobial effect, will be influenced by this equilibrium. In alkaline solutions ($pH > 7.5$), OCl^- prevails, which has a powerful oxidative effect and therefore a higher tissue-dissolving capacity than HOCl (Baker 1947). On the other hand, HOCl prevails in acidic or neutral solutions ($3 < pH < 7.5$). It has a powerful bactericidal effect probably because it is a smaller and uncharged molecule, which can penetrate the bacterial membrane relatively easily. Differences in pH, which affects the free available chlorine form ($HOCl/OCl^-$), do not affect the reaction rate of 2 % NaOCl with dentin (Macedo et al. 2010). Therefore, the reported differences in the tissue dissolution capacity (Jungbluth et al. 2011) and/or antimicrobial efficacy (Bremer et al. 2002) can be better explained by chemical differences in the predominant chlorine form at $pH = 5$ (HOCl) and $pH = 12$ (OCl^-) than by the amount of molecules involved in the reaction. On the other hand, the reaction of the NaOCl with the organic content of the root canal buffers its pH (Baker 1947; Jungbluth et al. 2011; Macedo et al. 2014c); nonetheless, the observed buffer effect of dentin was too limited to change the form of free chlorine available ($HOCl/OCl^-$) in both $pH = 12$ and $pH = 5$ NaOCl solutions (Macedo et al. 2014d). Therefore, the biological effect (antimicrobial/tissue dissolution capacity) of the irrigant is not expected to change inside the root canal during final irrigation.

The effect of activation could be due to a sonochemical effect and/or agitation/mixing (Joyce Tiong and Price 2012; Macedo et al. 2014a, c). The influence of activation was also observed during the subsequent rest phase (Macedo et al. 2010). Nevertheless, it seems that frequent refreshment does not compensate for the lower concentration (Macedo et al. 2014c).

5.1.1 EDTA

Chelation is defined as a process whereby chemicals form soluble complexes with certain metal ions, binding the ions so that they cannot react with other molecules or ions. During root canal therapy, calcium-chelating agents such as 17 % EDTA (580 mM) or 30 % citric acid are used to dissolve inorganic components of the smear layer that NaOCl is unable to dissolve (Şen et al. 1995).

5.1.2 Chlorhexidine

CHX bis-guanide has a broad antimicrobial spectrum and is widely used in dentistry for chemical plaque control mostly in a 0.2 % concentration (Mohammadi and Abbott 2009). It is a positively charged hydrophobic molecule with high affinity to negatively charged molecules present in cell membranes (Rölla et al. 1970) and/or to the organic component of hard tissues (Carrilho et al. 2010). CHX penetrates the outer cell wall layers of the microorganisms and attacks the cytoplasmic membrane (Shaker et al. 1988; Russell and Day 1993). It has the ability to coagulate intracellular organelles of the microbial cells (McDonnell and Russell 1999). CHX does not have the capacity to dissolve organic tissues (Naenni et al. 2004), which impedes its use as a main root canal irrigant in endodontics (Zehnder 2006). As root canal irrigant, a high concentration (2 %) is mostly used. CHX presents a dose-dependent cytotoxic effect (Lee et al. 2010). It inhibits osteoblastic cells growth, proliferation, and collagen synthesis; therefore, its extrusion toward the periapical tissues should also be avoided.

5.2 *Physical Properties of Irrigants*

The flow of irrigants is affected by their physical properties, namely, density, viscosity, and surface tension (White 1999). These properties have been studied for commonly used endodontic irrigants, and they have been found very similar to those of distilled water (Guerisoli et al. 1998; Taşman et al. 2000; Estrela et al. 2005; Zehnder et al. 2005; Giardino et al. 2006; van der Sluis et al. 2010), which can be explained by the fact that irrigants are mainly sparse aqueous solutions. The surface tension of endodontic irrigants has received extra attention, under the assumption that it may have a significant effect on irrigant penetration in dentinal tubules and accessory root canals (Abou-Rass and Patonai 1982; Taşman et al. 2000) and on dissolution of pulp tissue (Stojicic et al. 2010). The addition of wetting agents (surfactants) to commonly used irrigation solutions has also been suggested to reduce their surface tension (Abou-Rass and Patonai 1982; Taşman et al. 2000), although their exact concentration in the solutions is rarely reported and mixing of NaOCl with other chemicals, like alcohol, in order to reduce the surface tension will reduce its effect (Cunningham et al. 1982). While density and viscosity affect the flow in all cases, the effect of surface tension is important only in cases where two immiscible fluids are present (e.g., irrigant and air) (White 1999; Kundu and Cohen 2004). However, dentin is hydrophilic, and the dentinal tubules are likely to contain dentinal fluid with fluidic properties similar to water (Berggren and Brännström 1965), which will probably soak the root canal wall directly limiting the effect of surfactants since the two fluids are miscible. Recent studies have also shown that surfactants did not enhance the ability of NaOCl to dissolve pulp tissue (Jungbluth et al. 2012; Clarkson et al. 2012; De-Deus et al. 2013) and the ability of common chelators to remove calcium from dentin (Zehnder et al. 2005) or to remove the smear layer (Lui et al. 2007; De-Deus et al. 2008). To the contrary, it appears that the use of irrigants with a reduced surface tension ex vivo may result in deeper penetration of the smear layer into the dentinal tubules (Aktener et al. 1989). The presence of an air bubble occupying the apical part of the root canal has been partially demonstrated ex vivo (Tay et al. 2010; de Gregorio et al. 2009; Vera et al. 2012) and in vivo (Vera et al. 2011, 2012) recently. In such a case, surface tension effects could be important for the irrigant flow. However, routine entrapment of air bubbles in the apical part or side canals of the root canal during endodontic treatment still remains a speculation (see also the section 3.6).

5.3 Reaction with the Biofilm

Because microorganisms in the root canal function in a biofilm state, they are consequently protected by the matrix structure (EPS). Chemical agents should diffuse or break through the EPS matrix before they can affect the microorganisms. Disruption or poration of this matrix will enhance an antimicrobial effect. Therefore, the potential of EPS penetration, disruption, and killing of the microorganisms are all inherently related. Disruption of the top layers or EPS matrix or expansion of the biofilm induced by shear stress on the biofilm during irrigation procedures facilitates irrigant penetration into the biofilm and could therefore enhance the chemical effect of irrigants (He et al. 2013). Furthermore, chemical agents can alter the mechanical properties of the EPS, which may be explained by an influence of these agents on the EPS network formation (Körstgens 2003). This alteration can directly influence the removal of the biofilm (Brindle et al. 2011). Mass transfer inside a biofilm occurs by convection and (mainly) diffusion processes (de Beer et al. 1994a, b). It has been described that the surface-averaged relative effective diffusion coefficient (D_{rs}) decreases from the top of the biofilm toward the bottom. The D_{rs} profiles differ for biofilms of different ages and generally decrease over time; furthermore, different biofilms tested showed similar D_{rs} profiles near the top of the biofilm but different D_{rs} profiles near the bottom of the biofilm (Renslow et al. 2010). This bottom layer also determines the attachment to the surface, in our case dentin, and is normally the most difficult to remove (Derlon et al. 2008). There are no data available on the diffusion through a biofilm for endodontic irrigants.

5.3.1 NaOCl

The EPS consists mainly of polysaccharides and different proteins (Flemming and Wingerder 2010). NaOCl reacts strongly with proteins (Baker 1947). Therefore, probably, NaOCl will react stronger with the proteins in the EPS than the polysaccharides which seems to be confirmed in the study of Agarwal and colleagues (Agarwal et al. 2012). If so, the distribution of proteins and polysaccharides in the EPS and the polysaccharide bonding structures will determine the penetration or disruption of the matrix and consequently its antimicrobial capacity. NaOCl reacts with proteins producing chloramines in the form of gas bubbles, which can be imprisoned in a polysaccharide matrix (unpublished data). Furthermore, the strong reactivity of NaOCl could hamper its diffusion in the biofilm. De Beer et al. (1994a, b) studied the penetration of chlorine through a *Pseudomonas aeruginosa*/*Klebsiella pneumoniae* biofilm and concluded that the diffusion into the biofilm is slow, the diffusion rate depends on the concentration, the chlorine is reduced in the matrix, there is a diffusion–reaction mechanism, and there is a large variation due to local differences (highly resistant spots). These highly resistant spots show a higher cell density with subpopulations with higher reducing capacity per cell. Furthermore, these spots have a higher density of EPSs with a higher reducing potential. Combined with the phenomenon of rapid regrowth after biocide treatment, these highly resistant spots are serious threats for our antimicrobial treatments. A NaOCl solution with a lower pH contains more HOCl which is more antimicrobial in relation to planktonic bacteria (Bremer et al. 2002). It is not completely clear if the same is true for biofilm (Bremer et al. 2002). However, it has been shown that the buffer capacity of dentin is such that a low pH solution of NaOCl will probably not result in a higher antimicrobial effect (Macedo et al. 2014d). More recent studies using biofilm models show an increasing antibiofilm effect when the concentration of the NaOCl solution increases (Arias-Moliz et al. 2009; Retamozo et al. 2010; Jiang et al. 2011; del Carpio-Perochena et al. 2011). We find contradictory results on the effect of lower concentrations of NaOCl on the biofilm. Some authors claim that a 1 % NaOCl solution can partially disrupt and decrease the viability of a biofilm (Chávez de Paz et al. 2010; Carpio-Perochena et al. 2011), whereas Ordinola-Zapata and coauthors (2012) show that 5-min application of 1 % NaOCl cannot disrupt or decrease the viability of a biofilm. On the other hand, it was still more effective than water. Retamozo and coauthors concluded that a 1.3 % NaOCl solution did not have any antibiofilm effect at all (Retamozo et al. 2010). It is reported that both 2.5 % and 5.25 % NaOCl solutions have an antibiofilm effect (disruption and decrease of viability) (Retamozo et al. 2010; del Carpio-Perochena et al. 2011). The antibiofilm effect seems to be time dependent; a 5.25 % solution was significantly more effective after 40 min compared to 15 min. Furthermore, a 1 % and 2.5 % NaOCl solution was almost as effective after 30 min as a 5.25 % NaOCl solution after 5 min (Retamozo et al. 2010; del Carpio-Perochena et al. 2011). In the study of del Carpio-Perochena, the irrigant was refreshed every 5 min in the 15- and 30-min group. Two different volumes of NaOCl were used, 500 μL and 1 mL, but there was no difference in the results indicating that under these circumstances, 500 μL already produced the maximum chemical effect for that typical concentration. The biofilm is more resistant against NaOCl when it is in the starvation phase (Liu et al. 2010). From these results we can conclude that the chemical antibiofilm effect of NaOCl increases with its concentration and the time of application provided an excess of available chlorine. Lower concentrations like the 1 % and 2 % are significantly less effective, and the reaction rate is lower, which could be a problem for the disruption for the apical biofilm, where fresh irrigant is only available when the root canal has enough space for the irrigant penetration. Furthermore, we have to realize that in all these studies, the biofilm was in direct contact with an excess of NaOCl. In the root canal, direct contact and excess of NaOCl will be difficult due to the morphology of the root canal system, as mentioned earlier.

5.3.2 EDTA

The generally anionic, negatively charged, nature of EPS encourages the interaction with positively charged, polyvalent ions, for instance, Ca^{2+} or Mg^{2+} (van der Waal and van der Sluis 2012). These calcium bonds strengthen the structure of the EPS. Calcium is present in the oral environment including the root canal and could therefore be incorporated in the EPS. Consequently, chelators could help disrupt the EPS by breaking the cationic bonds (van der Waal and van der Sluis 2012). In the biofilm literature, it has been reported that EDTA can decrease biofilm mass (Turakhia and Characklis 1989; Chen and Stewart 2000; Kite et al. 2004; Percival et al. 2005; Shanks et al. 2006; Devine et al. 2007), adhesion (Dunne and Burd 1992; Ramage et al. 2007), and formation (Rose and Turner 1998; Sherertz et al. 2006; Shanks et al. 2006). Additionally, a reduction of viscosity of the biofilm has been described (Chen and Stewart 2002). Also complete eradication of in vivo generated biofilms was observed after 24-h treatment with EDTA (Kite et al. 2004). The longer the application time, the better the chelating effect although it starts right from the first contact. A higher concentration results in higher chelation efficacy, and biofilm formation decreases when the EDTA concentration increases (Ramage et al. 2007; Ang et al. 2006). In the study of Ang and coworkers, the influence of pH, temperature, and flow velocity of the EDTA application on biofilm removal was evaluated. A higher pH and temperature were associated with improved cleaning (pH 4.9 and 11 and 20 °C and 40 °C); a higher flow velocity (volume, refreshment) did not make a difference. Most studies were performed without refreshment of the EDTA. It is important to realize that the contact time in the abovementioned studies varied from 1 to 24 h and the concentration from 10 to 100 mM which are both different from the endodontic literature. The application time is shorter and the concentration is higher (580 mM). The endodontic literature shows somewhat contrasting results on the biofilm disruption or antimicrobial effect of chelators on the biofilm and roughly varies from no effect to some effect, but always less effective than NaOCl (Ordinola-Zapata et al. 2012; Chávez de Paz et al. 2010; Arias-Moliz et al. 2009). This could be related to the short application time. A synergistic effect of EDTA with antimicrobials like antibiotics has been reported (Sherertz et al. 2006; Banin et al. 2006). Weakening the biofilm structure encourages diffusion of antimicrobials into the biofilm (Gordon et al. 1991). Therefore, a synergistic effect with NaOCl could be envisaged, which has already been reported in the endodontic literature (Soares et al. 2010; Ozdemir et al. 2010). However, more research is necessary to draw conclusions for the clinical application.

5.3.3 Chlorhexidine

Although CHX is very effective in killing planktonic microorganisms, it seems that it is not a powerful medicament to *disrupt* biofilm (Bryce et al. 2009). In the study of Ordinola-Zapata et al. (2012), neither disruption of the biofilm nor a decrease in the viability could be detected. Normally, an antimicrobial effect is seen there where CHX is able to penetrate the biofilm but without signs of biofilm disruption (Shen et al. 2010; Chávez de Paz et al. 2010). It should be taken into consideration that the use and function of CHX in biofilm control in the mouth are completely different from its use and possible function in the root canal. In the former, killing of the planktonic microorganisms could be more important than in the root canal where NaOCl is already a potent agent. Recently, a stiffening of a *S. epidermidis* biofilm was reported after the use of 0.2 % CHX (Brindle et al. 2011).

5.4 Reaction with Dentin

5.4.1 NaOCl

Dentin is a substrate with a complex organic and inorganic structure. Type I collagen dominates the organic matrix (around 30 % vol) and hydroxylapatite, the inorganic part. NaOCl has predominantly an impact on its organic matrix because it is a nonspecific proteolytic agent. It splits up the long peptide chains by chlorination of the final protein groups (Davies et al. 1993). Consequently, this will affect the mechanical properties of dentin (Saleh and Ettman 1999). It has been hypothesized that, in spite of the theory that inorganic components can largely protect the organic collagen matrix (Kronick and Cooke 1996) due to the low molecular size of hypochlorous acid (MW = 52.5) or the hypochlorite anion (MW = 51.5), NaOCl can penetrate and react with the apatite-encapsulated collagen matrix to a certain depth (Zhang et al. 2010a, b). Several studies reported that a 5.25 % solution of NaOCl degrades the physical properties of the dentin: the resistance to flexure, tension strength, the elastic modulus, and microhardness (Saleh and Ettman 1999; Sim et al. 2001; Marending et al. 2007; Grigoratos et al. 2001). However, none of these properties are directly related to fracture resistance of teeth (Kishen 2006). Hu and coworkers (2010) highlight that the concentration of NaOCl is the principal factor of dentin deproteination. Depending on the concentration of NaOCl, the application time will positively influence the effect (Hu et al. 2010; Zhang et al. 2010a, b). The concentration, activation, and time influence positively the reaction rate of NaOCl with dentin (Macedo et al. 2010).

5.4.2 EDTA Combined with NaOCl

In general we can say that NaOCl will remove mostly the organic component and chelators the inorganic component of dentin. Thus, after contact with EDTA, the inorganic component will be removed from the intertubular dentin to a depth of 1–5 μm and from the peritubular dentin up to 20 μm (Lottanti et al. 2009), leaving the organic matrix to be dissolved by NaOCl. The erosive effect is stronger when EDTA is used before NaOCl than the other way around (Moreira et al. 2009; Qian et al. 2011).

Sobhani et al. (2010) concluded that the alternate use of NaOCl and EDTA had less effect on the surface strength of a treated tooth in vitro than NaOCl alone (EDTA was replaced by NaOCl). They used a more clinical setup where the irrigation procedure was simulated. Zhang et al. (2010a, b) claim that the flexure strength significantly reduces after contact of a 5.25 % NaOCl solution with dentin for 1 h because of the collagen reduction of the dentin. The decline in flexure strength is probably attributed to the generation of a brittle layer of apatite crystallites that are not supported by a structurally intact collagen matrix. This process is time and concentration dependent. The effect is rather caused by NaOCl than by the application of EDTA during 2 min. In general, the studies quoted here make use of dentin powder or dentin disks, except for Sobhani et al. (2010). Therefore, the contact irrigant–dentin is different from the clinical setting where the surface contact irrigant–dentin is less than ideal. Moreover, the effect of irrigants on dentin is not only determined by the contact surface but also by the condition of the dentin surface. Furthermore, in the study of Sobhani et al. (2010), extracted teeth were used in an in vitro setup, so the tubules may not be filled with dentinal fluid. Therefore, the surface conditions were different from the clinical situation.

5.4.3 Chlorhexidine

Although CHX binds to dentin (Kim et al. 2010), the typical effects on the mineral or organic components of dentin are not well described except for the formation of a layer of debris (Perdigao et al. 1994). When CHX is mixed with NaOCl, a brown precipitate containing parachloroaniline (PCA) will form on dentin (Rossi-Fedele et al. 2012) leading to its discoloration. PCA has been suggested to be a toxic and carcinogenic substance; hence, its accumulation inside the canal or at the periapical tissues should be avoided.

5.5 Reaction with Pulp Tissue

5.5.1 NaOCl

Several studies reported a direct proportion between the concentration of NaOCl and its organic tissue dissolution capacity (Hand et al. 1978; Thé 1979; Cunningham and Balekjian 1980; Koskinen et al. 1980; Abou-Rass and Oglesby 1981; Moorer and Wesselink 1982; Sirtes et al. 2005; Christensen et al. 2008). For a concentration lower than 1 % at room temperature, the overall effect is comparable to water (Thé 1979), saline, and 3 % H_2O_2 (Hand et al. 1978). According to Sirtes et al. (2005), an increase of the solution's temperature will reduce the differences in the dissolution effect between a pre-heated 1 % and a non-heated 5.25 % solution, with a better result for the solutions with a low concentration. Furthermore, heated hypochlorite solutions remove organic debris more efficiently from dentin shavings than unheated counterparts (Sirtes et al. 2005). Recently, it has been shown that a significant decrease of the pH from 12 to 6 decreases the tissue dissolution capacity (Christensen et al. 2008). On the contrary, stabilization of the NaOCl solution at a higher pH of 13.5 improves the tissue dissolution and the effects on dentin properties (Jungbluth et al. 2011).

5.5.2 EDTA and Chlorhexidine

There is no available data to support that either EDTA or CHX dissolves pulp tissue.

6 Effect of Irrigation on Endodontic Outcome

Among the irrigant activation systems discussed in this chapter, ultrasonic and laser activation contributes positively to both the mechanical and the chemical aspects of the irrigation procedure. However, it is not exactly known to what extent this will contribute to the disinfection procedure in vivo and if this eventually will improve the outcome of the treatment. Both systems have the potential to disrupt or remove biofilm, but it is not known to what extent they can remove biofilm from the root canal wall and from more remote regions such as oval extensions, lateral canals, and tubules. No reliable endodontic biofilm models are currently available for research. Apical periodontitis is a multifactorial disease, and therefore, its healing also depends on a range of aspects, not only on the irrigation during the endodontic treatment. From clinical research, it is clear that the length and the quality of the root canal filling are risk factors. However, the influence of the irrigation procedure, complex canal anatomy (apical delta and dentinal tubules), structure of the biofilm, and extraradicular biofilm around the root apex on the endodontic outcome is not known. Recently, it was demonstrated in an RCT that an ultrasonic-assisted irrigation protocol did not improve significantly the endodontic outcome compared to syringe irrigation alone (Liang et al. 2013) although the improvement of the mechanical and chemical aspects of the irrigation procedures by ultrasound has convincingly been demonstrated in in vitro research. This could indicate that either we need a more significant improvement of the mechanical and chemical aspects of the irrigation procedure or that other influential factors are more important in determining the endodontic outcome.

7 Concluding Remarks

Although recently more information became available on flow characteristics of different irrigation systems, it is still difficult to establish evidence-based irrigation protocols. This is mainly due to the fact that the properties of endodontic biofilm are not known. Furthermore, our knowledge of the reaction rate of the irrigants used in endodontics is not complete. Another important aspect is that we do not know enough about the influential factors determining endodontic outcome. Therefore, to improve or establish our irrigation protocols, more knowledge of endodontic biofilm is crucial, and more RCTs are imperative to get to know more about endodontic procedures in vivo.

References

Abou-Rass M, Oglesby SW (1981) The effects of temperature, concentration, and tissue type on the solvent ability of sodium hypochlorite. J Endod 7:376–7

Abou-Rass M, Patonai FJ Jr (1982) The effects of decreasing surface tension on the flow of irrigating solutions in narrow root canals. Oral Surg Oral Med Oral Pathol 53:524–6

Agarwal A, Xu H, Ngab WJ, Liu Y (2012) Biofilm detachment by self-collapsing air microbubbles: a potential chemical-free cleaning technology for membrane biofouling. J Mater Chem 22:2203–7

Ahmad M, Pitt Ford TR, Crum LA (1987) Ultrasonic debridement of root canals: acoustic streaming and its possible role. J Endod 13:490–9

Ahmad M, Pitt Ford TR, Crum LA, Walton AJ (1988) Ultrasonic debridement of root canals: acoustic cavitation and its relevance. J Endod 14:486–93

Ahmad M, Roy RA, Kamarudin AG (1992) Observations of acoustic streaming fields around an oscillating ultrasonic file. Endod Dent Traumatol 8:189–94

Ahmad M, Roy RA, Kamarudin AG, Safar M (1993) The vibratory pattern of ultrasonic files driven piezoelectrically. Int Endod J 26:120–4

Aktener BO, Cengiz T, Pişkin B (1989) The penetration of smear material into dentinal tubules during instrumentation with surface-active reagents: a scanning electron microscopic study. J Endod 15:588–90

Al-Hadlaq SM, Al-Turaiki SA, Al-Sulami U, Saad AY (2006) Efficacy of a new brush-covered irrigation needle in removing root canal debris: a scanning electron microscopic study. J Endod 32:1181–4

Al-Jadaa A, Paqué F, Attin T, Zehnder M (2009a) Acoustic hypochlorite activation in simulated curved canals. J Endod 35:1408–11

Al-Jadaa A, Paqué F, Attin T, Zehnder M (2009b) Necrotic pulp tissue dissolution by passive ultrasonic irrigation in simulated accessory canals: impact of canal location and angulation. Int Endod J 42:59–65

Amato M, Vanoni-Heineken I, Hecker H, Weiger R (2011) Curved versus straight root canals: the benefit of activated irrigation techniques on dentin debris removal. Oral Surg Oral Med Oral Path Oral Rad Endod 111:529–34

Ang WS, Lee S, Elimelech M (2006) Chemical and physical aspects of cleaning of organic-fouled reverse osmosis membranes. J Membr Sci 272:198–210

Arias-Moliz MT, Ferrer-Luque CM, Espigares-Garcia M, Baca P (2009) Enterococcus faecalis biofilms eradication by root canal irrigants. J Endod 35:711–4

Arslan H, Akcay M, Ertas H, Capar ID, Saygili G, Meşe M (2014) Effect of PIPS technique at different power settings on irrigating solution extrusion. Lasers Med Sci In press DOI. doi:10.1007/s10103-014-1633-1

Baker RW (1947) Studies on the reaction between sodium hypochlorite and proteins: 1. Physico-chemical study of the course of the reaction. Biochem J 41:337–42

Baker NA, Eleazer PD, Averbach RE, Seltzer S (1975) Scanning electron microscopic study of the efficacy of various irrigating solutions. J Endod 1:127–35

Banin E, Brady KM, Greenberg EP (2006) Chelator-induced dispersal and killing of Pseudomonas aeruginosa cells in a biofilm. Appl Environm Microbiol 72:2064–2069

Barnhart BD, Chuang A, Lucca JJ, Roberts S, Liewehr F, Joyce AP (2005) An in vitro evaluation of the cytotoxicity of various endodontic irrigants on human gingival fibroblasts. J Endod 31:613–5

de Beer D, Stoodley P, Lewandowski Z (1994a) Liquid flow in heterogeneous biofilms. Biotechnol Bioeng 44:636–641

de Beer D, Srinivasan R, Stewart PS (1994b) Direct measurement of chlorine penetration into biofilms during disinfection. Appl Environ Microbiol 60:4339

Baumgartner JC, Mader CL (1987) A scanning electron microscopic evaluation of four root canal irrigation regimens. J Endod 13:147–57

Berggren G, Brännström M (1965) The rate of flow in dentinal tubules due to capillary attraction. J Dent Res 44:408–15

Blanken JW, Verdaasdonk RM (2007) Cavitation as a working mechanism of the Er, Cr:YSGG Laser in endodontics: a visualization study. J Oral Laser App 7:97–106

Böl M, Möhle RB, Haesner M, Neu TR, Horn H, Krull R (2009) 3D finite element model of biofilm detachment using real biofilm structures from CLSM data. Biotechnol Bioeng 103:177–86

Böl M, Ehret A, Bolea AA, Hellriegel J, Krul R (2012) Recent advances in mechanical characterisation of biofilm and their significance for material modeling. Crit Rev Biotechnol 33145–71

Boutsioukis C, Lambrianidis T, Kastrinakis E, Bekiaroglou P (2007) Measurement of pressure and flow rates during irrigation of a root canal ex vivo with three endodontic needles. Int Endod J 40:504–13

Boutsioukis C, Lambrianidis T, Kastrinakis E (2009) Irrigant flow within a prepared root canal using different flow rates: a Computational Fluid Dynamics study. Int Endod J 42:144–55

Boutsioukis C, Verhaagen B, Kastrinakis E, Versluis M, van der Sluis LWM (2010a) Irrigant flow in the root canal: experimental validation of a Computational Fluid Dynamics model using high-speed imaging. Int Endod J 43:393–407

Boutsioukis C, Verhaagen B, Versluis M, Kastrinakis E, Wesselink PR, van der Sluis LWM (2010b) Evaluation of irrigant flow in the root canal using different needle types by an unsteady computational fluid dynamics model. J Endod 36:87–9

Boutsioukis C, Verhaagen B, Kastrinakis E, Versluis M, van der Sluis LWM (2010c) The effect of needle penetration depth on the irrigant flow in the root canal: evaluation by an unsteady Computational Fluid Dynamics model. Int Endod J 36:1664–8

Boutsioukis C, Verhaagen B, Kastrinakis E, Versluis M, van der Sluis LWM (2010d) The effect of preparation size on the irrigant flow in the root canal: evaluation using an unsteady Computational Fluid Dynamics model. Int Endod J 43:874–81

Boutsioukis C, Verhaagen B, Kastrinakis E, Versluis M, van der Sluis LWM (2010e) The effect of root canal taper on the irrigant flow: evaluation using an unsteady Computational Fluid. Int Endod J 43:909–16

Boutsioukis C (2010) Application of a Computational Fluid Dynamics model to the study of root canal irrigation. Doctoral Dissertation. Thessaloniki, Greece: Aristotle University of Thessaloniki, 2010. pp.1-530.

Boutsioukis C, Psimma Z, van der Sluis LWM (2013a) Factors affecting irrigant extrusion during root canal irrigation: a systematic review. Int Endod J 46:599–618

Boutsioukis C, Verhaagen B, Walmsley AD, Versluis M, van der Sluis LWM (2013b) Measurement and visualization of file-to-wall contact during ultrasonically activated irrigation in simulated canals. Int Endod J 46:1046–55

Boutsioukis C, Kastrinakis E, Lambrianidis T, Verhaagen B, Versluis M, van der Sluis LWM (2014a) Formation and removal of apical vapor lock during syringe irrigation: a combined experimental and Computational Fluid Dynamics approach. Int Endod J 47:191–201

Boutsioukis C, Psimma Z, Kastrinakis E (2014b) The effect of flow rate and agitation technique on irrigant extrusion ex vivo. Int Endod J 47:487–96

Bremer P, Monk I, Butler R (2002) Inactivation of Listeria monocytogenes/Flavobacterium spp. biofilms using chlorine: impact of substrate, pH, time and concentration. Letters Appl Microbiol 35:321–325

Brennen CE (1995) Cavitation and bubble dynamics. Oxford University Press, New York (NY), USA. 1st ed.

Brindle ER, Miller DA, Stewart PS (2011) Hydrodynamic deformation and removal of Staphylococcus epidermidis biofilms treated with urea, chlorhexidine, iron chloride, or dispersinB. Biotechnol Bioeng 108:1–10

Bronnec F, Bouillaguet S, Machtou P (2010a) Ex vivo assessment of irrigant penetration and renewal during the cleaning and shaping of root canals: a digital subtraction radiographic study. Int Endod J 43:275–82

Bronnec F, Bouillaguet S, Machtou P (2010b) Ex vivo assessment of irrigant penetration and renewal during the final irrigation regimen. Int Endod J 43:663–72

Brujan E-A, Nahen K, Schmidt P, Volgel A (2001) Dynamics of laser-induced cavitation bubbles near an elastic boundary. J Fluid Mech 433:251–81

Brunson M, Heilborn C, Johnson JD, Cohenca N (2010) Effect of apical preparation size and preparation taper on irrigant volume delivered by using negative pressure irrigation system. J Endod 36:721–4

Bryce G, O'Donnell D, Ready D, Ng YL, Pratten J, Gulabivala K (2009) Contemporary root canal irrigants are able to disrupt and eradicate single- and dual-species biofilms. J Endod 35:1243–8

Burleson A, Nusstein J, Reader A, Beck M (2007) The in vivo evaluation of hand/rotary/ultrasound instrumentation in necrotic, human mandibular molars. J Endod 33:782–7

Busscher HJ, Rustema-Abbink M, Bruinsma GM, de Jager M, Gottenbos B, van der Mei HC (2003) Non-contact removal of coadhering and non-coadhering bacterial pairs from pellicle surfaces by sonic brushing and *de novo* adhesion. Eur J Oral Sci 111:459–64

del Carpio-Perochena AE, Bramante CM, Duarte MA, Cavenago BC, Villas-Boas MH, Graeff MS, Bernardineli N, de Andrade FB, Ordinola-Zapata R (2011) Biofilm dissolution and cleaning ability of different irrigant solutions on intraorally infected dentin. J Endod 37:1134–8

Carrilho MR, Carvalho RM, Sousa EN, Nicolau J, Breschi L, Mazzoni A, Tjäderhane L, Tay FR, Agee K, Pashley DH (2010) Substantivity of chlorhexidine to human dentin. Dent Mater 26:779–85

Castelo-Baz P, Martin-Biedma B, Cantatore G, Varela-Patino P (2012) In vitro comparison of passive and continuous ultrasonic irrigation in simulated lateral canals of extracted teeth. J Endod 38:688–91

Chang YC, Huang FM, Tai KW, Chou MY (2001) The effect of sodium hypochlorite and chlorhexidine on cultured human periodontal ligament cells. Oral Surg Oral Med Oral Pathol Oral Radiol Endod 92:446–50

Chávez de Paz LE, Bergenholtz G, Svensäter G (2010) The effects of antimicrobials on endodontic biofilm bacteria. J Endod 36:70–7

Chen JE, Nurbakhsh B, Layton G, Bussmann M, Kishen A (2014) Irrigation dynamics associated with positive pressure, apical negative pressure and passive ultrasonic irrigations: A computational fluid dynamics analysis. Aust Endod J 40:54–60

Chen X, Stewart PS (2000) Biofilm removal caused by chemical treatments. Wat Res 34:4229–33

Chen X, Stewart PS (2002) Role of electrostatic interactions in cohesion of bacterial biofilms. Appl Microbiol Biotechnol 59:718–720

Chow TW (1983) Mechanical effectiveness of root canal irrigation. J Endod 9:475–9
Christensen CE, McNeal SF, Eleazer P (2008) Effect of lowering the pH of sodium hypochlorite on dissolving tissue in vitro. J Endod 34:449–52
Chu A (2010) Penetration of sodium hypochlorite into periapical lesions in necrotic dog teeth. J Endod 2010;36:558 (abstract).
Clarkson RM, Kidd B, Evans GE, Moule AJ (2012) The effect of surfactant on the dissolution of porcine pulpal tissue by sodium hypochlorite solutions. J Endod 38:1257–60
Conard M (2012) A prospective study of fluid pressures of irrigation during root canal therapy. Ohio State University, Master Thesis, Columbus, OH, USA, pp 1–124
Cunningham WT, Balekjian AY (1980) Effect of temperature on collagen-dissolving ability of sodium hypochlorite as endodontic irrigant. Oral Surg Oral Med Oral Pathol 49:175–7
Cunningham WT, Cole JS, Balekjian AY (1982) Effect of alcohol on the spreading ability of sodium hypochlorite endodontic irrigant. Oral Surg Oral Med Oral Pathol 54:333–5
Davies JM, Hortwitz DA, Davies KJ (1993) Potential roles of hypochlorous acid and N-chloroamines in collagen breakdown by phagocytic cells in synovitis. Free Radical Biol & Med 15:637–643
De-Deus G, Namen F, Galan JJ, Zehnder M (2008) Soft chelating irrigation protocol optimizes bonding quality of Resilon/Epiphany root fillings. J Endod 34:703–705
De-Deus G, de Berredo Pinho MA, Reis C, Fidel S, Souza E, Zehnder M (2013). Sodium hypochlorite with reduced surface tension does not improve in situ pulp tissue dissolution. J Endod 39:1039–43
de Groot de SD, Verhaagen B, Versluis M, Wu M-K, Wesselink PR, van der Sluis LWM (2009) Laser-activated irrigation within root canals: Cleaning efficacy and flow visualization. Int Endod J 42:1077–83.
Derlon N, Massè A, Escudié R, Bernet N, Paul E (2008) Stratification in the cohesion of biofilms grown under various environmental conditions. Water Res 42:2102–10
Deshpande MD, Vaishnav RN (1982) Submerged laminar jet impingement on a plane. J Fluid Mech 114:213–36
Devine DA, Percival RS, Wood DJ et al (2007) Inhibition of biofilms associated with dentures and toothbrushes by tetrasodium EDTA. J Appl Microbiol 103:2516–24
Druttman ACS, Stock CJR (1989) An in vitro comparison of ultrasonic and conventional methods of irrigant replacement. Int Endod J 22:174–8
Duck PW, Smith FT (1979) Steady streaming induced between oscillating cylinders. J Fluid Mech 91:93–110
Dunne WM, Burd EM (1992) The effects of magnesium, calcium, EDTA, and pH on the in vitro adhesion of Staphylococcus epidermidis to plastic. Microbiol Immunol 36:1019–1027
Dutner J, Mines P, Anderson A (2012) Irrigation trends among American Association of Endodontists members: a web-based survey. J Endod 38:37–40
Estrela C, Estrela CRDA, Guimarães LF, Silva RS, Pécora JD (2005) Surface tension of calcium hydroxide associated with different substances. J Appl Oral Sci 13:152–6
Falk KW, Sedgley CM (2005) The influence of preparation size on the mechanical efficacy of root canal irrigation in vitro. J Endod 31:742–5
Fedorowicz Z, Nasser M, Sequeira-Byron P, de Souza RF, Carter B, Heft M (2012) Irrigants for non-surgical root canal treatment in mature permanent teeth. Cochrane Database Syst Rev 12;9:CD008948.
Felver B, King DC, Lea SC, Price GJ, Walsmley AD (2009) Cavitation occurrence around ultrasonic dental scalers. Ultrasonic Sonochem 16:692–7
Flemming HC, Wingerder J (2010) The biofilm matrix. Nature Rev Micro 8:623–33
Flemming HC, Wingender J, Szewzyk U (2011) Biofilm highlights, Springer Series on Biofilms, 1st edn. Springer, Berlin, Heidelberg, Germany
Fukumoto Y, Yoshioka T, Suda H (2004) Intra-canal aspiration technique for root canal irrigation. Part 1. Overflow of irrigating solution beyond the apical foramen. Japanese J Cons Dent 47:37–42

Fukumoto Y, Kikuchi I, Yoshioka T, Kobayashi C, Suda H (2006) An ex vivo evaluation of a new root canal irrigation technique with intracanal aspiration. Int Endod J 39:93–9

George R, Walsh LJ (2008) Apical extrusion of root canal irrigants when using Er:YAG and Er, CR:YSGG with optical fibers: an in vitro dye study. J Endod 34:706–8

Giardino L, Ambu E, Becce C, Rimondini L, Morra M (2006) Surface tension comparison of four common root canal irrigants and two new irrigants containing antibiotic. J Endod 32:1091–3

Gordon CA, Hodges NA, Marriott C (1991) Use of slime dispersants to promote antibiotic penetration through the extracellular polysaccharide of mucoid Pseudomonas aeruginosa. Antimicrob Agents Chemother 35:1258–60

Gorni FG, Gagliani MM (2004) The outcome of endodontic retreatment: a 2-yr follow-up. J Endod 30:1–4

de Gregorio C, Estevez R, Cisneros R, Heilborn C, Cohenca N (2009) Effect of EDTA, sonic, and ultrasonic activation on the penetration of sodium hypochlorite into simulated lateral canals: an in vitro study. J Endod 35:891–5

de Gregorio C, Estevez R, Cisneros R et al (2010) Efficacy of different irrigation and activation systems on the penetration of sodium hypochlorite into simulated lateral canals and up to working length: an in vitro study. J Endod 36:1216–21

Gao Y, Haapasalo M, Shen Y, Wu H, Li B, Ruse ND, Zhou X (2009) Development and validation of a three-dimensional Computational Fluid Dynamics model of root canal irrigation. J Endod 35:1282–7

Grigoratos D, Knowles J, Ng YL, Gulabivala K (2001) Effect of exposing dentin to sodium hypochlorite and calcium hydroxide on its flexural strength and elastic modulus. Inter Endod J 34:113–119

Gu L, Kim JR, Ling J, Choi KK, Pashley DH, Tay FR (2009) Review of contemporary irrigant agitation techniques and devices. J Endod 35:791–804

Guelon T, Mathias J-D, Stoodley P (2011) 'Advances in Biofilm Mechanics', In: Biofilm highlights, Flemming, H-C, Wingender, J and Szewzyk, U, eds. Springer. 1st ed.

Guerisoli DMZ, Silva RS, Pecora JD (1998) Evaluation of some physico-chemical properties of different concentrations of sodium hypochlorite solutions. Braz Endod J 3:21–3

Haapasalo M, Orstavik D (1987) In vitro infection and disinfection of dentinal tubules. J Dent Res 66:1375–9

Haapasalo M, Endal U, Zandi H, Coil JM (2005) Eradication of endodontic infection by instrumentation and irrigation solutions. Endo Top 10:77–102

Haapasalo M, Wang Z, Shen Y, Curtis A, Patel P (2014) Tissue dissolution by a novel multisonic ultracleaning system and sodium hypochlorite. J Endod 40:1178–81

Hand RE, Smith ML, Harrison JW (1978) Analysis of the effect of dilution on the necrotic tissue dissolution property of sodium hypochlorite. J Endod 4:60–4

Hauser V, Braun A, Frentzen M (2007) Penetration depth of a dye marker into dentine using a novel hydrodynamic system (RinsEndo). Int Endod J 40:644–52

He Y, Peterson BW, Jongsma MA, Ren Y, Sharma PK, Busscher HJ, van der Mei HC (2013) Stress relaxation analysis facilitates a quantitative approach towards antimicrobial penetration into biofilms. PLoS One 8, e63750. doi:10.1371/journal.pone.0063750

Heling I, Rotstein I, Dinur T, Szwec-Levine Y, Steinberg D (2001) Bactericidal and cytotoxic effects of sodium hypochlorite and sodium dichloroisocyanurate solutions in vitro. J Endod 27:278–80

Hockett JL, Dommisch JK, Johnson JD, Cohenca N (2008) Antimicrobial efficacy of two irrigation techniques in tapered and nontapered canal preparations: an *in vitro* study. J Endod 34:1374–7

Hsieh YD, Gau CH, Kung Wu SF, Shen EC, Hsu PW, Fu E (2007) Dynamic recording of irrigating fluid distribution in root canals using thermal image analysis. Int Endod J 40:11–7

Hu X, Peng Y, Sum CP, Ling J (2010) Effects of concentrations and exposure times of sodium hypochlorite on dentin deproteination: attenuated total reflection fourier transform infrared spectroscopy study. J Endod 36:2008–11

Huang TY, Gulabivala K, Ng YL (2008) A bio-molecular film ex-vivo model to evaluate the influence of canal dimensions and irrigation variables on the efficacy of irrigation. Int Endod J 41:60–71

Incropera FP, de Witt DP (1990) Fundamentals of heat and mass transfer, 3rd edn. John Wiley & Sons, NJ, USA

Jensen SA, Walker TL, Hutter JW, Nicoll BK (1999) Comparison of the cleaning efficacy of passive sonic activation and passive ultrasonic activation after hand instrumentation in molar root canals. J Endod 25:735–8

Jiang L-M, Verhaagen B, Versluis M, van der Sluis LWM (2010a) Evaluation of a sonic device designed to activate irrigant in the root canal. J Endod 36:143–6

Jiang L-M, Verhaagen B, Versluis M, van der Sluis LWM (2010b) Influence of the oscillation direction of an ultrasonic file on the cleaning efficacy of passive ultrasonic irrigation. J Endod 36:1372

Jiang LM, Verhaagen B, Versluis M, Zangrillo C, Cuckovic D, van der Sluis LWM (2010c) An evaluation of the effect of pulsed ultrasound on the cleaning efficacy of passive ultrasonic irrigation. J Endod 36:1887–91

Jiang LM, Hoogenkamp MA, van der Sluis LW, Wesselink PR, Crielaard W, Deng DM (2011) Resazurin metabolism assay for root canal disinfectant evaluation on dual-species biofilms. J Endod 37:31–5

Jiang L-M, Lak B, Eijsvogel E, Wesselink PR, van der Sluis LWM (2012) A cleaning efficacy comparison of different irrigation techniques as final irrigation. J Endod 38:838–41

Joyce Tiong T, Price GJ (2012) Ultrasound promoted reaction of Rhodamine B with sodium hypochlorite using sonochemical and dental ultrasonic instruments. Ultrasonic Sonochem 19:358–64

Jungbluth H, Marending M, De-Deus G, Sener B, Zehnder M (2011) Stabilizing sodium hypochlorite at high pH: effects on soft tissue and dentin. J Endod 37:693–6

Jungbluth H, Peters C, Peters O, Sener B, Zehnder M (2012) Physicochemical and pulp tissue dissolution properties of some household bleach brands compared with a dental sodium hypochlorite solution. J Endod 38:372–5

Kahn FH, Rosenberg PA, Gliksberg J (1995) An in vitro evaluation of the irrigating characteristics of ultrasonic and subsonic handpieces and irrigating needles and probes. J Endod 21:277–80

Keir DM, Senia ES, Montgomery S (1990) Effectiveness of a brush in removing postinstrumentation canal debris. J Endod 16:323–7

Kim J, Uchiyama T, Carillho M, Agee KA, Mazzoni A, Breschi L, Carvalho RM, Tjaderhane L, Looney S, Wimmer C, Tezvergil-Mutluay A, Tay FR, Pashley DH (2010) Chlorhexidine binding to mineralized versus demineralized dentin powder. Dent Materials 6:771–8

Kishen A (2006) Mechanisms and risk factors for fracture predilection in endodontically treated teeth. Endod Top 13:57–83

Kite P, Eatswood K, Sugden S et al (2004) Use on in vivo-generated biofilms from hemodialysis catheters to test the efficacy of a novel antimicrobial catheter lock for biofilm eradication in vitro. J Clin Microbiol 42:3073–76

Klapper I, Rupp CJ, Cargo R, Purvedorj B, Stoodley P (2002) Viscoelastic fluid description of bacterial biofilm material properties. Biotech Bioeng 80:289–96

Körstgens V, Flemming H-C, Wingender J, Borchard W (2001) Uniaxial compression measurement device for investigation of the mechanical stability of biofilms. J Microbiol Meth 46:9–17

Körstgens V (2003) Die mechanische stabilität bakterieller biofilme. Eine untersuchung verschiedener einflüsse auf biofilme von. Dissertation Universität Duisburg-Essen, Pseudomonas aeruginosa

Koskinen KP, Stenvall H, Uitto VJ (1980) Dissolution of bovine pulp tissue by endodontic solutions. Scand J Dent Res 88:406–11

Kronick PL, Cooke P (1996) Thermal stabilization of collagen fibers by calcification. Connect Tiss Res 33:275–82

Kundu PK, Cohen IM (2004) Fluid mechanics, 3rd edn. San Diego, CA, USA: Elsevier Academic Press 1–23:271–377

Lea SC, Walmsley AD, Lumley PJ (2010) Analyzing endosonic root canal file oscillations: an in vitro evaluation. J Endod 36:880–3

Lee TH, Hu CC, Lee SS, Chou MY, Chang YC (2010) Cytotoxicity of chlorhexidine on human osteoblastic cells is related to intracellular glutathione levels. Int Endod J 43:430–5

Liang Y-H, Jiang L-M, Jiang L, Chen X-B, Liu Y-Y, Tian F-C, Bao X-D, Gao X-J, Versluis M, Wu M-K, van der Sluis LWM (2013) Radiographic healing following root canal treatments performed in single-rooted teeth with and without ultrasonic activation of the irrigant: a randomized controlled trial. J Endod 39:1218–1225

Liu H, Wei X, Ling J, Wang L, Huang X (2010) Biofilm formation capability of Enterococcus faecalis cells in starvation phase and its susceptibility to sodium hypochlorite. J Endod 36:630–635

Lloyd A, Uhles JP, Clement DJ, Garcia-Godoy F (2014) Elimination of intracanal tissue and debris through a novel laser-activated system assessed using high-resolution micro-computed tomography: a pilot study. J Endod 40:584–7

Lottanti S, Gautschi H, Sener B, Zehnder M (2009) Effects of ethylenediaminetetraacetic, etidronic and peracetic acid irrigation on human root dentine and the smear layer. Int Endod J 42:335–43

Lui JN, Kuah HG, Chen NN (2007) Effect of EDTA with and without surfactants or ultrasonics on removal of smear layer. J Endod 33:472–5

Lumley PJ, Walmsley AD, Laird WRE (1988) An investigation into cavitational activity occurring in endosonic instrumentation. J Dent 16:120–2

Lumley PJ, Walmsley AD, Laird WRE (1991) Streaming patterns produced around endodontic files. Int Endod J 24:290–7

Lumley PJ, Walmsley AD (1992) Effect of precurving on the performance of endosonic K files. J Endod 18:232–6

Lumley PJ, Walmsley AD, Walton RE, Rippin JW (1993) Cleaning of oval canals using ultrasonic or sonic instrumention. J Endod 19:453–7

Lussi A, Nussbächer U, Grosrey J (1993) A novel noninstrumented technique for cleansing the root canal system. J Endod 19:549–53

Macedo RG, Wesselink PR, Zaccheo F, Fanali D, van Der Sluis LW (2010) Reaction rate of NaOCl in contact with bovine dentine: effect of activation, exposure time, concentration and pH. Int Endod J 43:1108–15

Macedo RG, Verhaagen B, Fernandez Rivas D, Gardeniers JGE, van der Sluis LWM, Wesselink PR, Versluis M (2014a) Sonochemical and high-speed optical characterization of cavitation generated by an ultrasonically oscillating dental file in root canal models. Ultrason Sonochem 21:324–35

Macedo RG, Robinson JP, Verhaagen B, Walmsley AD, Versluis M, Cooper PR, van der Sluis LW (2014b) A novel methodology providing insights into removal of biofilm mimicking hydrogel from lateral morphological features of the root canal during irrigation procedures. Int Endod J 47:1040–51

Macedo RG, Verhaagen B, Wesselink PR, Versluis M, van der Sluis LWM (2014c) Influence of refreshment/activation cycles and temperature rise on the reaction rate of sodium hypochlorite with bovine dentine during ultrasonic activated irrigation. Int Endod J 47:147–54

Macedo RG, Pascual Herrero N, Wesselink PR, Versluis M, van der Sluis LWM (2014d) Influence of the dentinal wall on the pH of NaOCl during root canal irrigation. J Endod 40:1005–8

Macedo RG, Verhaagen B, Versluis M, van der Sluis LWM (2014e) Temperature evolution of preheated irrigant injected into a root canal ex vivo. Int Endod J (submitted)

Malki M, Verhaagen B, Jiang L-M, Nehme W, Naaman A, Versluis M, Wesselink PR, van der Sluis LWM (2012) Irrigant flow beyond the insertion depth of an ultrasonically oscillating file in straight and curved root canals: visualization and cleaning efficacy. J Endod 38:657–61

Marending M, Luder HU, Brunner TJ, Knecht S, Stark WJ, Zehnder M (2007) Effect of sodium hypochlorite on human root dentin - mechanical, chemical and structural evaluation. Int Endod J 40:786–93

Matsumoto H, Yoshimine Y, Akamine A (2011) Visualization of irrigant flow and cavitation induced by Er:YAG laser within a root canal model. J Endod 37:839–43

McDonnell G, Russell AD (1999) Antiseptics and disinfectants: activity, action, and resistance. Clin Microbiol Rev 12:147–79

Metzger Z, Teperovich E, Cohen R, Zary R, Paqué F, Hülsmann M (2010) The self-adjusting file (SAF). Part 3: removal of debris and smear layer-A scanning electron microscope study. J Endod 36:697–702

Mitchell RP, Yang SE, Baumgartner JC (2010) Comparison of apical extrusion of NaOCl using the EndoVac or needle irrigation of root canals. J Endod 36:338–41

Mitchell RP, Baumgartner JC, Sedgley CM (2011) Apical extrusion of sodium hypochlorite using different root canal irrigation systems. J Endod 37:1677–1681

Mohammadi Z, Abbott PV (2009) Antimicrobial substantivity of root canal irrigants and medicaments: a review. Aust Endod J 35:131–9

Möhle RB, Langemann T, Haesner M, Augustin W, Scholl S, Neu TR, Hempel DC, Horn H (2007) Structure and shear strength of microbial biofilms as determined with confocal laser scanning microscopy and fluid dynamic gauging using a novel rotating disc biofilm reactor. Biotechnol Bioeng 98:747–55

de Moor RJ, Meire M, Goharkhay K, Moritz A, Vanobbergen J (2010) Efficacy of ultrasonic versus laser-activated irrigation to remove artificially placed dentin debris plugs. J Endod 36:1580–3

Moorer WR, Wesselink PR (1982) Factors promoting the tissue dissolving capability of sodium hypochlorite. Int Endod J 15:187–96

Moreira DM, Affonso Almeida JF, Ferraz CCR, Gomes BP, Line SR, Zaia AA (2009) Structural analysis of bovine root dentin after use of different endodontics auxiliary chemical substances. J Endod 35:1023–7

Mott RL (1999) Applied fluid mechanics, 5th edn. New Jersey, USA: Prentice Hall 1–188:221–38

Munoz HR, Camacho-Cuadra K (2012) In vivo efficacy of three different endodontic irrigation techniques for irrigant delivery to working length of mesial canals of mandibular molars. J Endod 38:445–8

Naenni N, Thoma K, Zehnder M (2004) Soft tissue dissolution capacity of currently used and potential endodontic irrigants. J Endod 30:785–7

Nguy D, Sedgley C (2006) The influence of canal curvature in the mechanical efficacy of root canal irrigation in vitro using real-time imaging of bioluminescent bacteria. J Endod 32:1077–80

Nielsen BA, Baumgartner CJ (2007) Comparison of the EndoVac system to needle irrigation of root canals. J Endod 33:611–5

Ohl C-D, Wolfum B (2003) Detachment and sonoporation of adherent HeLa-cells by shock wave-induced cavitation. Biochim Biophys Acta 1624:131–8

Ordinola-Zapata R, Bramante CM, Cavenago B, Graeff MS, Gomes de Moraes I, Marciano M, Duarte MA (2012) Antimicrobial effect of endodontic solutions used as final irrigants on a dentine biofilm model. Int Endod J 45:162–8

Ordinola-Zapata R, Bramante CM, Aprecio RM, Handysides R, Jaramillo DE (2014) Biofilm removal by 6% sodium hypochlorite activated by different irrigation techniques. Int Endod J 47: 659–666

Orstavik D, Haapasalo M (1990) Disinfection by endodontic irrigants and dressings of experimentally infected dentinal tubules. Endod Dent Traumatol 6:142–9

Ozdemir HO, Buzoglu HD, Calt S et al (2010) Effect of ethylenediaminetetraacetic acid and sodium hypochlorite irrigation on Enterococcus faecalis biofilm colonization in young and old human root canal dentin: in vitro study. J Endod 36:842–846

Parente JM, Loushine RJ, Susin L, Gu L, Looney SW, Weller RN, Pashley DH, Tay FR (2010) Root canal debridement using manual dynamic agitation or the EndoVac for final irrigation in a closed system and an open system. Int Endod J 43:1001–12

Park E, Shen Y, Khakpour M, Haapasalo M (2013) Apical pressure and extent of irrigant flow beyond the needle tip during positive-pressure irrigation in an in vitro root canal model. J Endod 39:511–5

Pashley EL, Birdsong NL, Bowman K, Pashley DH (1985) Cytotoxic effects of NaOCl on vital tissue. J Endod 11:525–8

Peeters HH, Mooduto L (2013) Radiographic examination of apical extrusion of root canal irrigants during cavitation induced by Er,Cr:YSGG laser irradiation: an in vivo study. Clin Oral Invest 17:2105–2112.

Peeters HH, Iskandar B, Suardita K, Suharto D (2014) Visualization of removal of trapped air from the apical region of the straight root canal models generating 2-phase intermittent counter flow during ultrasonically activated irrigation. J Endod 40:857–61

Peeters HH, Gutknecht N (2014) Efficacy of laser-driven irrigation versus ultrasonic in removing an airlock from the apical third of a narrow root canal. Aust Endod J 40:47–53

Percival SL, Kite P, Eastwood K et al (2005) Tetrasodium EDTA as a novel central venous catheter lock solution against biofilm. Infect Contr Hosp Epidem 26:515–19

Perdigao J, Denehy GE, Swift EJ (1994) Effects of chlorhexidine on dentin surfaces and shear bond strength. Am J Dent 7:81–4

Peters OA, Bardsley S, Fong J, Pandher G, Divito E (2011) Disinfection of root canals with photon-initiated photoacoustic streaming. J Endod 37:1008–12

Phares DJ, Smedley GT, Flagan RC (2000) The wall shear stress produced by the normal impingement of a jet on a flat surface. J Fluid Mech 418:351–75

Pope SB (2000) Turbulent flows. Cambridge, UK: Cambridge University Press:3–9.

Psimma Z, Boutsioukis C, Vasiliadis L, Kastrinakis E (2013a) A new method for real-time quantification of irrigant extrusion during root canal irrigation ex vivo. Int Endod J 46:619–31

Psimma Z, Boutsioukis C, Kastrinakis E, Vasiliadis L (2013b) The effect of needle insertion depth and root canal curvature on irrigant extrusion ex vivo. J Endod 39:521–4

Qian W, Shen Y, Haapasalo M (2011) Quantitative analysis of the effect of irrigant solution sequences on dentin erosion. J Endod 37:1437–1441

Ragnarsson KT, Rechenberg DK, Attin T, Zehnder M (2015) Available chlorine consumption from NaOCl solutions passively placed in instrumented human root canals. Int Endod J 48:435–40.

Ramage G, Wickes BL, López-Ribot JL (2007) Inhibition on Candida albicans biofilm formation using divalent cation chelators (EDTA). Mycopath 164:301–6

Rechenberg DK, Ragnarsson KT, Rüeger S, Held L, Mohn D, Zehnder M (2014) A new method to assess available chlorine in small volumes of liquid. J Endod 40:534–7

Renslow RS, Majors PD, McLean JS, Fredrickson JK, Ahmed B, Beyenal H (2010) In situ effective diffusion coefficient profiles in live biofilms using pulsed-field gradient nuclear magnetic resonance. Biotechnol Bioeng 106:928–37

Retamozo B, Shabahang S, Johnson N, Aprecio RM, Torabinejad M (2010) Minimum contact time and concentration of sodium hypochlorite required to eliminate Enterococcus faecalis. J Endod 36:520–3

Ricucci D, Siqueira JF Jr (2010) Fate of the tissue in lateral canals and apical ramifications in response to pathologic conditions and treatment procedures. J Endod 36:1–15

Robertson CW, Williams D (1971) Lambert absorption coefficients of water in the infrared. J Opt Soc Am 61:1316–20

Rölla G, Löe H, Rindom Schiøtt C (1970) The affinity of chlorhexidine for hydroxyapatite and salivary mucins. J Perio Res 5:90–5

Rose RK, Turner SJ (1998) Extracellular volume in streptococcal model biofilms: effects of pH, calcium and fluoride. Biochim Biophys Acta 1379:185–190

Rosenfeld EF, James GA, Burch BS (1978) Vital pulp tissue response to sodium hypochlorite. J Endod 4:140–6
Rossi-Fedele G, Doğramaci EJ, Guastalli AR, Steier L, Poli de Figueiredo JA (2012) Antagonistic interactions between sodium hypochlorite, chlorhexidine, EDTA, and citric acid. J Endod 38:426–31
Rothfus RR, Monrad CC, Senecal VE (1950) Velocity distribution and fluid friction in smooth concentric annuli. Ind Eng Chem 42:2511–20
Russell AD, Day MJ (1993) Antibacterial activity of chlorhexidine. J Hosp Infect 25:229–38
Saleh AA, Ettman WM (1999) Effect of endodontic irrigation solutions on microhardness of root canal dentine. J Dent 27:43–6
Salzgeber RM, Brilliant JD (1977) An in vivo evaluation of the penetration of an irrigating solution in root canals. J Endod 3:394–8
Savani GM, Sabbah W, Sedgley CM, Whitten B (2014) Current trends in endodontic treatment by general dental practitioners: report of a United States national survey. J Endod 40:618–24
Sedgley C, Applegate B, Nagel A, Hall D (2004) Real-time imaging and quantification of bioluminescent bacteria in root canals in vitro. J Endod 30:893–8
Sedgley CM, Nagel AC, Hall D, Applegate B (2005) Influence of irrigant needle depth in removing bacteria inoculated into instrumented root canals using real-time imaging in vitro. Int Endod J 38:97–104
Şen BH, Wesselink PR, Türkün M (1995) The smear layer: a phenomenon in root canal therapy. Int Endod J 28:141–8
Shaker LA, Furr JR, Russell AD (1988) Mechanism of resistance of Bacillus subtilis spores to chlorhexidine. J Appl Bacteriol 64:531–9
Shankar PN, Deshpande MD (2000) Fluid mechanics in the driven cavity. Ann Rev Fluid Mech 32:93–136
Shanks RMQ, Sargent JL, Martinez RM et al (2006) Catheter lock solutions influence staphylococcal biofilm formation on abiotic surfaces. Nephrol Dial Transplant 21:2247–55
Shemesh H, Wesselink PR, Wu MK. (2010) Incidence of dentinal defects after root canal filling procedures. Int Endod J.43:995–1000.
Shen Y, Gao Y, Qian W, Ruse ND, Zhou X, Wu H, Haapasalo M (2010) Three-dimensional numeric simulation of root canal irrigant flow with different irrigation needles. J Endod 36:884–9
Sherertz RJ, Boger MS, Collins CA et al (2006) Comparative in vitro efficacies of various catheter lock solutions. Antimicrob Agents Chemother 50:1865–8
Shrestha A, Fong S-W, Khoo B-C, Kishen A (2009) Delivery of antibacterial nanoparticles into dentinal tubules using high-intensity focused ultrasound. J Endod 35:1028–33
Sim TP, Knowles JC, Ng YL, Shelton J, Gulabiala K (2001) Effect of sodium hypochlorite on mechanical properties of dentin and tooth surface strain. Inter Endod J 34:120–132
Sirtes G, Waltimo T, Schaetzle M, Zehnder M (2005) The effects of temperature on sodium hypochlorite short-term stability, pulp dissolution capacity, and antimicrobial efficacy. J Endod 31:669–71
van der Sluis LW, Gambarini G, Wu MK, Wesselink PR (2006) The influence of volume, type of irrigant and flushing method on removing artificially placed dentine debris from the apical root canal during passive ultrasonic irrigation. Int Endod J 39:472–6
van der Sluis LW, Versluis M, Wu MK, Wesselink PR (2007) Passive ultrasonic irrigation of the root canal: a review of the literature. Int Endod J 40:415–26
van der Sluis L, Wu MK, Wesselink P (2009) Comparison of 2 flushing methods used during passive ultrasonic irrigation of the root canal. Quintessence Int 40:875–9
van der Sluis LWM, Vogels MP, Verhaagen B, Macedo R, Wesselink PR (2010) Study on the influence of refreshment/activation cycles and irrigants on mechanical cleaning efficiency during ultrasonic activation of the irrigant. J Endod 36:737–40
Šnjarić D, Čarija Z, Braut A, Halaji A, Kovačević M, Kuiš D (2012) Irrigation of human prepared root canal – ex vivo based computational fluid dynamics analysis. Croat Med J 53:470–9.

Soares JA, Roque de Carvalho MA, Cunha Santos SM et al (2010) Effectiveness of chemomechanical preparation with alternating use of sodium hypochlorite and EDTA in eliminating intracanal Enterococcus faecalis biofilm. J Endod 36:894–898

Sobhani OE, Gulabivala K, Knowles LC, Ng YL (2010) The effect of irrigation time, root morphology and dentin thickness on tooth surface strain when using 5% sodium hypochlorite and 17% EDTA. Inter Endod J 43:190–199

Spencer HR, Ike V, Brennan PA (2007) Review: the use of sodium hypochlorite in endodontics—potential complications and their management. Br Dent J 202:555–9

Stojicic S, Zivkovic S, Qian W, Zhang H, Haapasalo M (2010) Tissue dissolution by sodium hypochlorite: effect of concentration, temperature, agitation, and surfactant. J Endod 36:1558–62

Stojicic S, Shen Y, Haapasalo M (2013) Effect of the source of biofilm bacteria, level of biofilm maturation, and type of disinfecting agent on the susceptibility of biofilm bacteria to antibacterial agents. J Endod 39:473–7

Stewart PS, Franklin MJ (2008) Physiological heterogeneity in biofilms. Nature Rev Microbiol 6:199–210

Taşman F, Cehreli ZC, Oğan C, Etikan I (2000) Surface tension of root canal irrigants. J Endod 26:586–7

Tay FR, Gu LS, Schoeffel GJ, Wimmer C, Susin L, Zhang K, Arun SN, Kim J, Looney SW, Pashley DH (2010) Effect of vapor lock on root canal debridement by using a side-vented needle for positive-pressure irrigant delivery. J Endod 36:745–50

Thé SD (1979) The solvent action of sodium hypochlorite on fixed and unfixed necrotic tissue. Oral Surg Oral Med Oral Pathol 47:558–61

Thomas AR, Velmurugan N, Smita S, Jothilatha S (2014) Comparative evaluation of canal isthmus debridement efficacy of modified Endovac technique with different irrigation systems. J Endod 40:1676–80

Tilton JN (1999) Fluid and particle dynamics. In: Perry RH, Green DW, Maloney JO (eds) Perry's chemical engineer's handbook, 7th edn. McGraw-Hill, New York, USA, pp 6.1–6.50

Turakhia MH, Characklis WG (1989) Activity of Pseudomonas aeruginosa in biofilms: effect of calcium. Biotechnol Bioengin 33:406–414

Verhaagen B, Lea SC, De Bruin GJ, van der Sluis LWM, Walmsley AD, Versluis M (2012a) Oscillation characteristics of endodontic files: numerical model and its validation. IEEE - Trans Ultr Ferr Freq Contr 59:2448–59

Verhaagen B, Boutsioukis C, Heijnen GL, van der Sluis LWM, Versluis M (2012b) Role of the confinement of a root canal on jet impingement during endodontic irrigation. Exp Fluids 53:1841–53

Verhaagen B, Boutsioukis C, Sleutel CP, Kastrinakis E, Versluis M, van der Sluis LWM (2014a) Irrigant transport into dental microchannels. Microfluid Nanofluidics 16:1165–77

Verhaagen B, Boutsioukis C, van der Sluis LWM, Versluis M (2014b) Acoustic streaming induced by an ultrasonically oscillating endodontic file. J Acoust Soc Am 135:1717–30

Vera J, Arias A, Romero M (2011) Effect of maintaining apical patency on irrigant penetration into the apical third of root canals when using passive ultrasonic irrigation: an in vivo study. J Endod 37:1276–8

Vera J, Arias A, Romero M (2012) Dynamic movement of intracanal gas bubbles during cleaning and shaping procedures: the effect of maintaining apical patency on their presence in the middle and cervical thirds of human root canals-an in vivo study. J Endod 38:200–3

van der Waal SV, van der Sluis LWM (2012) Potential of calcium to scaffold an endodontic biofilm, thus protecting the micro-organisms from disinfection. Med Hypotheses 79:1–4

Walters MJ, Baumgartner JC, Marshall JG (2002) Efficacy of irrigation with rotary instrumentation. J Endod 28:837–9

Weise M, Roggendorf MJ, Ebert J, Petschelt A, Frankenberger R (2007) Four methods for cleaning simulated lateral extensions of curved root canals: a SEM evaluation. Int Endod J 40:991–2

White FM (1999) Fluid Mechanics, 4th edn. Boston, USA: McGraw-Hill: 1–56, 129–214, 541.

Willershausen I, Wolf TG, Schmidtmann I, Berger C, Ehlers V, Willershausen B, Briseño B (2015) Survey of root canal irrigating solutions used in dental practices within Germany. Int Endod J 48:654–60.

Wu MK, van der Sluis LW, Wesselink PR (2004) Comparison of mandibular premolars and canines with respect to their resistance to vertical root fracture. J Dent 32:265–8

Zhang K, Tay FR, Kyung Kim Y, Mitchell JK, Ryul KJ, Carrilhof M, Pashley DH, Ling J (2010a) The effect of initial irrigation with two different sodium hypochlorite concentrations on the erosion of instrumented radicular dentin. Dent Materials 26:514–523

Zhang K, Kyung Kim Y, Cadenaro M, Brian TE, Sidow SJ, Loushine RJ, Ling J, Pashley DH, Tay FR (2010b) Effect of different exposure times and concentrations of sodium hypochlorite/ethylenediaminetetraacetic acid on the structural integrity of mineralised dentin. J Endod 36:105–10

Zehnder M, Schmidlin P, Sener B, Waltimo T (2005) Chelation in root canal therapy reconsidered. J Endod 31:817–820

Zehnder M (2006) Root canal irrigants. J Endod 32:389–398

Zou L, Shen Y, Li W, Haapasalo M (2010) Penetration of sodium hypochlorite into dentin. J Endod 36:793–6

Inter-appointment Medication with Calcium Hydroxide in Routine Cases of Root Canal Therapy

Gunnar Bergenholtz, Calvin Torneck, and Anil Kishen

Abstract This chapter discusses the use of an inter-appointment application of calcium hydroxide in the endodontic treatment of teeth with apical periodontitis, from a historic and evidence-based perspective. Despite a long history of clinical use and the publication of numerous in vitro, in vivo, and treatment outcome studies, strong scientific evidence in support of its use in this capacity is still not available. Today's "best scientific evidence" is based primarily on a just a small number of studies that by today's standard are not considered "strong." However, given that elimination of bacteria from the root canal before placement of a root canal filling appears to enhance the treatment outcome, inter-appointment placement of calcium hydroxide could be an easy and appropriate way of promoting it. Any perceived advantage of its use, however, must be balanced by the risk of altering the physical properties of the root dentin due to its caustic action, particularly when it is present in the root canal for a long period of time. While clinical confirmation of an increased risk of root fracture in calcium hydroxide-treated fully developed teeth is still not verifiable, prudence dictates that caution should be exercised when calcium hydroxide is used in this manner in permanent teeth that are not completely developed.

1 Introduction

For over 100 years inter-appointment antiseptic dressings have been considered an important adjunct to root canal infection control. Over this period of time, many medicaments and compounds were used for this purpose, with ebbs and flows in the consensus as to which was considered best. In recent decades, a paste of calcium

G. Bergenholtz (✉)
Department of Endodontics, The Sahlgrenska Academy, University of Gothenburg, Gothenburg, Sweden
e-mail: gunnar.bergenholtz@odontologi.gu.se

C. Torneck • A. Kishen
Department of Endodontics, University of Toronto, Toronto, ON, Canada

L.E. Chávez de Paz et al. (eds.), *The Root Canal Biofilm*, Springer Series on Biofilms 9, DOI 10.1007/978-3-662-47415-0_10

hydroxide has been used as a treatment dressing and has been viewed as a valuable adjunct to control infection and subsequently improving treatment outcome. By virtue of its tissue-dissolving property, calcium hydroxide has also been used to necrotize pulp tissue remnants left on the canal walls after pulpectomy, remnants that can be more readily removed with sodium hypochlorite irrigation at a subsequent treatment session. Additionally, calcium hydroxide has been used in the long term in cases of teeth with a necrotic pulp and a large periapical lesion, symptomatic apical periodontitis, advanced root resorption, and root perforation (for reviews see, e.g., Fava and Saunders 1999; Mohammadi and Dummer 2011).

The lack of convincing scientific evidence in support of calcium hydroxide treatment, augmented by the recent introduction of advanced options for root canal irrigation, has challenged the use of calcium hydroxide and the concept of inter-appointment dressings in general. The use of rotary root canal instrumentation in combination with super-flexible Ni–Ti files currently allows the clinician to complete root canal preparation for most teeth within a relatively short time. Thus, to many endodontists, in such instances, it seems impractical to extend the treatment time unless there is compelling evidence to the contrary. In addition, patients often appreciate a single-visit treatment when they have to spend as little time in the chair as possible, if the same result can be achieved. However, concerns have been raised by clinicians and dental researchers alike as to whether it is possible to effectively inhibit persistence of bacterial aggregates and biofilms in an infected root canal in one treatment session and thereby minimize persistent or recurrent periapical disease (Trope and Bergenholtz 2002; Bergenholtz and Spångberg 2004). This philosophy also has been voiced in numerous clinical and in vitro observational studies, where the total removal of infected tissue from the root canal systems, whether by hand or rotary instrumentation, has been deemed to be an unachievable goal (Walton 1976; Wu and Wesselink 1995). All too often, tissue remnants, debris, and bacteria are left behind in the root canal, especially in oval canals and "c"-shaped canals, and in intercanal isthmuses of multirooted teeth. This provides the bacteria that survive the initial treatment with both substrate and an environment necessary to develop virulent characteristics that may lead to treatment failure. This concept has been supported by clinical studies and case reports, which have found that the intracanal biofilm remains intact, even after the conscientious completion of treatment (Nair et al. 2005; Carr et al. 2009; Vera et al. 2012).

This chapter will discuss the value of inter-appointment medication in the endodontic management of teeth with an infected root canal and will focus mainly on the use of calcium hydroxide and calcium hydroxide-based medicaments for that purpose. Their use today will be supported by historical perspectives and a narrative review of the studies that provide us with what has been termed "scientific evidence." It will be obvious to the reader that despite the publication of numerous clinical and laboratory reports, strong scientific evidence in support of inter-appointment dressings with calcium hydroxide is not yet available and that the current "best scientific evidence," which is drawn upon to formulate clinical protocol, is based upon only on a small number of relatively weak clinical studies.

2 Historical Perspectives

The need for an inter-appointment dressing in root canal therapy has been a topic of debate for more than a century. In the late nineteenth century, an ongoing debate between Drs. C. Edmund Kells Jr. and J. Smith Dodge Jr. regarding the appropriateness of immediate root canal filling was published in the Dental Cosmos (1887). Dr. J. Smith Dodge Jr. (Dodge 1887) argued that immediate root filling was suitable only for teeth with vital pulp tissue, while Dr. C. Edmund Kells Jr. stated (Kells 1887): "I believe that dentists all over the country are filling at one sitting, pulpless roots, roots in all stages of inflammation, as well as those in which the pulps are wholly or partly alive; and not only filling them, but doing so successfully. Any root may be cleaned as well in one sitting as in a dozen." It should be noted that this debate took at a time, when the role of microbes in the etiology of the pulp and periapical inflammation was poorly understood. Willoughby D Miller's hallmark textbook *Microorganisms of the Human Mouth*(1890) introduced the possibility of a causal relationship between microorganisms and oral disease. This information had a great impact on the profession and supported the view that effective elimination of bacteria from the infected root canals was a necessary component to its treatment, which required the use of bactericidal chemicals in addition to mechanical debridement and root filling. Onderdonk (1901) (cited by Coolidge 1960) went further and proposed the routine bacteriological culturing of the root canals to assess the efficacy of antimicrobial treatment. This step not only supported but also mandated the concept of multiple treatment appointments in root canal treatment. A virtual hunt was subsequently initiated in search for an effective antimicrobial for root canal treatment. In that context, numerous drugs and compounds were clinically tested in a more or less trial and error type of studies over the ensuing years with the patient as the test subject!

During the early 1900s, strong antiseptics were used for root canal disinfection, and little attention was given to their potential harmful local and systemic effects. Creosote, called oil of smoke, and phenol were in common use, having been used earlier to control oral malodor (Grossman 1976). Compounds based on paraformaldehyde including formalin, tricresolformalin and formocresol were also advocated. Paraformaldehyde, used earlier along with arsenic to devitalize pulps that had been exposed by caries (so-called mortal treatment), became a popular disinfection agent to avoid pain experienced during pulp extirpation, when profound local anesthetics were not as yet available. Because these were all highly toxic agents, concerns were expressed by several opinion leaders in the profession, such as G.V. Black (Grossman 1971) and Carl Grove (1913), that they had a potential to severely injure periapical tissue, when placed in a root canal. Formaldehyde and formaldehyde compounds nevertheless gained widespread popularity and remain in common use today as a root canal disinfectant. They are also an active component of some root filling materials such as N2, Endomethasone, and SPAD. It is noteworthy to mention that Strindberg, in his classic thesis on treatment outcomes (1956), used tricresolformalin as a root canal disinfectant in 35 % of the teeth included in his study. Similarly, in an often-cited treatment outcome study undertaken by undergraduate dental students in Oslo (Kerekes and Tronstad 1979), a solution of 3.7 % formaldehyde was used to disinfect the root canals of teeth with necrotic pulps.

Calcium hydroxide surfaced, at the beginning of the twentieth century, in the profession's search for an effective root canal disinfectant. That was at a time when endodontics was condemned as a health risk by those who believed in the focal infection theory. Their published and voiced disapproval of endodontics retarded its scientific advancement for many years (Grossman 1976). Although highly caustic in nature, calcium hydroxide was introduced as a potential replacement for the more tissue-toxic and potentially allergenic medicaments that were commonly used in treatment at that time. BW Hermann (1920) has been generally acknowledged as the individual who introduced calcium hydroxide into endodontics. However, according to Staehle (1990), calcium hydroxide had been used in root canal therapy long before then. Nevertheless, Hermann in his dissertation and his subsequent publications was admittedly the first to emphasize its antimicrobial effect and its tissue compatibility when placed in a root canal (Staehle 1990). He reported that a dressing of calcium hydroxide, when sealed into the root canal over a period of time, could produce bacteria-free root canal and induce apical root closure by hard tissue (cementum or a dentin-like tissue) (Hermann 1930). However, this occurred less frequently than when it was used as a capping agent for an exposed vital pulp (Hermann 1930). The above observations paved the way for the use of calcium hydroxide and calcium hydroxide-containing materials as a routine agent for capping dental pulps exposed by caries and trauma in the years that followed (for reviews, see by Staehle 1990 and Bergenholtz 2005).

In the wake of Hermann's observations, many other clinical applications for calcium hydroxide were identified (for a recent review, see Mohammadi and Dummer 2011). Calcium hydroxide, however, never gained universal acceptance as a root filling material or as an inter-appointment dressing until well into the second half of the twentieth century. The basis for the use of calcium hydroxide in the treatment of exposed vital pulp (Rohner 1940) and in the endodontic management of immature permanent teeth with pulp necrosis (Granath 1959; Frank 1966; Heithersay 1970) was nevertheless established. In 1960, Matsumiya and Kitamura (1960) demonstrated that a paste of calcium hydroxide could reduce or eliminate the microbial population present in the experimentally infected root canals of dog teeth. Engström and Spångberg (1967) described the benefits of calcium hydroxide use in the management of partially pulpectomized teeth. They observed in their study that when a portion of the pulp of young, human, healthy premolar teeth was not removed and calcium hydroxide was placed into the root canal, hard tissue was deposited at the interface of calcium hydroxide and residual pulp and at the orifices of accessory canals present along the main canal. A likely reason for the delay in the application of these findings to clinical practice was the strong belief held by clinicians of that time, that liquid antimicrobials were superior to calcium hydroxide use in root canal therapy. This philosophy prevailed in both North America and in Scandinavia.

Although proposed earlier, routine use of culture tests was not incorporated into clinical endodontic practice until the late 1930s, when follow-up studies highlighting the use of bacteriological controls to improve treatment outcomes were published (Coolidge 1960). Subsequently, culture tests and inter-appointment liquid medicaments became a standard component of endodontic practice and a protocol recommended in contemporary textbooks (e.g., Sommer et al. 1966). The protocol, however, had several shortcomings. One limitation was the volatility of most liquid antimicrobials that resulted in their antibacterial effectiveness being rapidly depleted in the root canal environment (Messer and Chen 1984). This meant that the root canals medicated with these agents remained virtually unmedicated between appointments, which allows residual microorganisms to repopulate the root canal prior to the placement of root canal filling. This was later confirmed in reports published by Byström and Sundqvist in 1983and 1985. Another reason was the need to schedule at least three sessions to complete the endodontic treatment to comply with the protocol (Molander 2000). Mechanical and chemical cleansing of the root canal and the placement of the liquid disinfectant into the root canal for a period of days was done at the first treatment session. The second treatment session was devoted to bacterial sampling of the root canal and the placement of another antimicrobial dressing, and filling of the root canal was done at the third treatment session, if the tooth was asymptomatic and no growth of bacteria was noted in the culture sample. A positive culture sample on the third visit would prompt repetition of the disinfection procedure necessitating even more appointments to be scheduled. Needless to say this approach was inconvenient to both practitioner and patient.

The lack of scientific validation for this protocol in subsequent clinical follow-up studies and the doubts that some researchers had about the reliability of the culture test finally led to the elimination of bacteriologic control in endodontics at many dental schools. Seltzer and Bender in a classic debate article (1965) pointed out the inherent risk of false-positive and false-negative results in the culture testing and stated: "the possibilities of obtaining false negative cultures are so numerous that the credibility of a negative culture is constantly in doubt. All we hope for is a reduction in the number of microorganisms." Their paper contributed not only to the abandonment of the culture concept in clinical practice but unfortunately helped de-emphasize the singular role of microorganisms in the development of primary periapical disease. Their philosophy was quickly adopted in North America and prompted a shift in the direction of endodontic research from infection control to endodontic technology. A widely read chapter in a textbook authored by Schilder (1974) supported this technical perspective and stated that only technical excellence, that is, proper root canal preparation and filling of the canal space in all its dimensions, was decisive in predicting a favorable outcome of treatment.

The belief that elimination of bacteria from the root canal prior to filling was critical to a favorable treatment outcome remained firm in the minds of clinicians in other parts of the world, particularly in Scandinavia. Investigators there lacked confidence in the ability of a mechanical approach to adequately eliminate bacteria from the root canal and the ability of contemporary root filling materials to provide an effective seal against their ingress and egress from the root canal. No medicament appeared to be available at that time, to ensure safe and effective infection control, until it was realized that calcium hydroxide might be an answer to that problem. This breakthrough occurred when it was noted that calcium hydroxide, placed in the root canal of teeth with a suppurating periapical lesion, could initiate remission in the suppurative process. An important contributing factor to this effect was the ability of the calcium hydroxide to effectively block the apical foramen. By eliminating seepage of inflammatory exudate into the root canal from the periapical tissue, a nutrient source for the proteolytic microorganisms, which was responsible for the infection, was thus eliminated. Subsequently bacterial activity was lessened and the apical inflammation reduced leading to an abatement of the clinical symptoms. It was reasoned that if this was an effective approach to curb active signs of root canal infection, why not place calcium hydroxide into all teeth receiving root canal therapy to act as a "space holder" after instrumentation, regardless of the presenting clinical symptoms. It was also speculated that the high pH of the material (>12) might suppress the growth of bacteria that remain in the root canal after it had been mechanically debrided. Hence a two-appointment approach to endodontic treatment was adopted by some clinicians. The concept of a two-appointment approach using calcium hydroxide as an inter-appointment dressing gained wider recognition after the publication of case series reports by Byström and Sundqvist in 1981, 1983, 1985 and Byström et al. (1985) that demonstrated that mechanical instrumentation, irrigation with different strengths of sodium hypochlorite, or inter-appointment medication with paramonochlorophenol or phenol camphor could not predictably eliminate cultivable bacteria from the infected root canal. This, they claimed, could only be achieved over time with an intracanal dressing of calcium hydroxide.

3 Scientific Evidence

3.1 Experimental Observations

Over the years numerous in vitro studies were carried out to evaluate the effect of calcium hydroxide on bacterial cells and bacterial byproducts. Various strains of *Enterococcus faecalis* were often employed as the test organism, due to their ability to survive in the presence of many antimicrobial drugs and the harsh nutritional conditions present in a mechanically prepared root canal. Stevens and Grossman (1983) concluded from a combined series of in vivo and in vitro experiments that calcium hydroxide, in a water-based slurry, had a limited ability to destroy this microbe. Using an agar diffusion test, they noted, that there were only small zones of inhibition around calcium hydroxide suggesting that the slurry needed close contact with the microbes in order to exert a killing effect. Haapasalo and Ørstavik (1987) confirmed this observation using a dentin model infected with *E. faecalis*.

Calcium hydroxide failed to eliminate the microorganism from dentinal tubules in 24 h, while camphorated p-monochlorophenol did. A subsequent study using the same model reported that with a 10-day application of calcium hydroxide, it was possible to significantly reduce their presence in the tubules (Ørstavik and Haapasalo 1990). Parmar et al. (2011) used a similar in vitro model but employed fluorescence image analysis to identify live/dead bacteria in the tubules. They too reported that calcium hydroxide significantly reduced the number of viable bacterial cells when compared to nonmedicated controls. Another study (Barnard et al. 1996) demonstrated that calcium hydroxide was effective against *Actinomyces israelii*. Estrela et al. (1999a, b), on the other hand, showed that calcium hydroxide had no significant antimicrobial activity against *E. faecalis*, *S. aureus*, *B. subtilis*, and *P. aeruginosa* after 2, 3, and 7 days of application.

Gomes et al. (2002) suggested that anaerobic Gram-negative bacteria were more susceptible to calcium hydroxide pastes than facultative Gram-positive microorganisms. Wei et al. (2003) showed that the antimicrobial action of calcium hydroxide could be enhanced when it is mixed with 2 % chlorhexidine rather than water. However, it was observed that the antibacterial efficacy of calcium hydroxide was inactivated by dentin powder, hydroxylapatite, and serum (Portenier et al. 2001, see also review by Haapasalo et al. 2007), while it cannot disrupt biofilm matrix in vitro (Upadya et al. 2011).

The resistance of *E. faecalis* to calcium hydroxide raised several questions on their resistance mechanisms. In an attempt to answer at least part of the above questions, Evans et al. (2002) studied the mechanism that permitted *E. faecalis* to survive the high environmental pH related to the presence of calcium hydroxide. It was observed that *E. faecalis* was resistant to calcium hydroxide at a pH of 11.1, but not at a pH of 11.5. It was also noted that there was no difference in *E. faecalis* survival when protein synthesis was blocked during stress induction. Interestingly, the addition of a proton pump inhibitor to the media resulted in a dramatic reduction in the survival of *E. faecalis* when it was reexposed to the calcium hydroxide. It was therefore concluded that the survival of *E. faecalis* in calcium hydroxide was a function of the proton pump, which was activated as the pH rose to a point, where it reached its maximum capacity. Distel et al. (2002) studied the ability of *E. faecalis* to form biofilms on calcium hydroxide-medicated root dentin in vitro and concluded that *E. faecalis* had the potential to form biofilms even after the dentin was previously medicated. Exposure of *E. faecalis* to collagen and EDTA was also shown to increase its resistance to calcium hydroxide disinfection (Kayaoglu et al. 2009; George and Kishen 2008). Chávez de Paz et al. (2007) demonstrated that a biofilm of *E. faecalis*, *Lactobacillus paracasei*, *Olsenella uli*, *Streptococcus anginosus*, *S. gordonii*, *S. oralis*, or *Fusobacterium nucleatum* was more resistant to alkaline stress than growth present in planktonic form.

The effect of calcium hydroxide treatment on bacterial endotoxin/lipopolysaccharide (LPS) and lipoteichoic acid (LTA) components of the Gram-negative and Gram-positive bacterial cell wall, respectively, has been studied because of the role they play in periapical inflammation (Hong et al. 2004; Costa Junior et al. 2003). Inactivation of LPS by calcium hydroxide was first reported by Safavi and Nichols in 1993 using LPS derived from the Gram-negative bacterium *Salmonella typhimurium*. The study was repeated in 1994 using LPS derived from the root canal bacterium *P. intermedia*. Calcium hydroxide's ability to eliminate LPS bioactivity in both of these studies was attributed to its ability to hydrolyze ester bonds and cause a release of fatty acids from the LPS molecule (Safavi and Nichols 1993, 1994). Like Safavi and Nichols, Buck et al. (2001) used *S. typhimurium* LPS to test the efficacy of calcium hydroxide, but did so by comparing its efficacy relative to other agents commonly used in endodontic treatment. While chlorhexidine (0.12 %), sodium hypochlorite (2.62 %), EDTA (15 %), and ethanol (95 %) all proved to be ineffective against LPS when used separately, they did prove to have the same effectiveness in 30 min as calcium hydroxide had in 1 day, if used in combination. This trend, however, was short-lived, and in days 2 and 5 of the study, calcium hydroxide proved to be the most effective. Several years later, Baik et al. (2011) reported that, while calcium hydroxide displayed limited efficacy against *E. faecalis*, it was able to inactivate *E. faecalis* LTA through a deacylation of its lipid moiety. These studies all used an in vitro model. Calcium hydroxide efficacy against LPS was also shown in an in vivo animal study. Tanomaru et al. (2003) demonstrated that only calcium hydroxide was effective in suppressing the pro-inflammatory activity of LPS when compared with 5 % sodium hypochlorite and 2 % chlorhexidine digluconate. An in vitro study by de Oliveira et al. (2007)) later confirmed those findings. Additional support for calcium hydroxide's ability to inactivate LPS appeared in a recent paper by Guo et al. (2014) that reported that calcium hydroxide could suppress *P. endodontalis* LPS from inducing osteoclastogenesis in a mouse calivarium model.

In vitro studies have yielded variable results regarding the antibacterial effect of calcium hydroxide on Gram-positive and Gram-negative microorganisms. Many studies have employed *E. faecalis* as the test organism and reported a limited ability of calcium hydroxide in an aqueous slurry to effectively kill this organism, especially when it was present in dentinal tubules or in a biofilm. It should be remembered that while commonly recovered from the root canals of endodontically treated teeth, this species has a low incidence in primary root canal infections. Further, investigations comparing the prevalence of *E. faecalis* in teeth with/without periapical lesions suggested that *E. faecalis*, though present, does not play a significant role in the posttreatment endodontic infections (Kaufman et al. 2005; Zoletti et al. 2006.

3.2 Clinical Observations

Systematic reviews have examined the strength of the scientific evidence for and against the use of a calcium hydroxide dressing in teeth with vital and non-vital pulps (Sathorn et al. 2005, 2007; Figini et al. 2008; SBU 2010). Primarily two clinical outcome parameters were used for the assessment: culture-negative samples (Sathorn et al. 2007) and an absence of clinical and/or radiographical signs of apical periodontitis following a variable postoperative period (Sathorn et al. 2005; Figini et al. 2008; SBU 2010).

A meta-analysis by Sathorn et al. (2007) (included the following eight studies Ørstavik et al. 1991; Sjögren et al. 1991; Yared and Dagher 1994; Shuping et al. 2000; Peters et al. 2002; Kvist et al. 2004; McGurkin-Smith et al. 2005; Waltimo et al. 2005) concluded that a calcium hydroxide root canal dressing has only limited efficacy in eliminating cultivable bacteria from root canals of human teeth. The study failed to confirm the results of the "classic" study published by Byström et al. (1985) several years earlier. Though no significant effect was reported, all the studies included in this systematic review with the exception of Peters et al. (2002) and Waltimo et al. (2005) reported a reduced number of culture-positive samples, when a calcium hydroxide dressing was used when compared to the control. However, the incidence of culture-positive samples in the calcium hydroxide group ranged from 0 % (Sjögren et al. 1991) to 71 % (Peters et al. 2002) and demonstrated a wide inconsistency that was worthy of further investigation.

There are several reasons for inconsistencies in the culture test results reported in these studies. In the paper by Sjögren et al. (1991) and the case series reports published by Byström and Sundqvist (1981–1985), the high incidence of culture-negative test results occurred only in single-rooted teeth with intact pulp chambers, that is, teeth with only minor or no restoration. The studies that reported a higher incidence of culture-positive test results, after calcium hydroxide use (Reit and Dahlén 1988; Peters et al. 2002; Kvist et al. 2004), included multirooted teeth (teeth with more complex canal anatomy), as well as teeth with less stringent restrictions on the preoperative status of the crown than those in the Byström and Sundqvist (1981–1985) reports. Hence, the root canal microbiota present in the treatment and control group teeth may have differed significantly in regard to their type, number, and antimicrobial sensitivity. In addition, teeth with more complex root canal anatomy of multirooted teeth have proven to be more difficult to clean and disinfect, particularly when biofilm is present in non-accessible areas. Calcium hydroxide is only slightly soluble and therefore has limited antimicrobial action at sites distant to where it is placed (Nerwich et al. 1993). This limits its antimicrobial effect not only in dentinal tubules but also in the irregularities and canal spaces distant from the main root canal. Another factor that may have contributed to the variable results seen in culture-dependent studies is the difference in the degree of calcium hydroxide carry-over from the root canal into the culture media, when a sample is taken. Significant amounts of carryover can impede bacterial growth and lead to a false-negative finding, when the results are tabulated. The era of noncultivation (molecular) identification of bacteria has also shown that many of the bacteria found in infected root canals are noncultivable even when present in large numbers (Siqueira and Rôças 2009; Ribeiro et al. 2011; Sakamoto et al. 2006, see also chapter "Bacterial Biofilms and Endodontic Disease: Histo-Bacteriological and Molecular Exploration" by Siqueira et al.). These nonculturable bacteria are not reported, when cultivation-based identification is used. While the role played by many of the noncultivable species found in the root canal has yet to be determined, others such as species in the phyla Synergistetes and Spirochaetes have been classified as putative endodontic pathogens (Foschi et al. 2006; Vianna et al. 2007; Montagner et al. 2010). Other factors such as variances in collection methods, specificity of the growth media used, and variances in sample handling could also skew results and lead to false conclusions.

The systematic review undertaken by Sathorn et al. (2007) was also weakened by the fact that only one of the studies they included, Kvist et al. (2004), was a randomized controlled study. The others had variable designs and small patient numbers. Therefore, it could be argued that even with a meta-analysis, these studies were too weak to scientifically support or refute the efficacy of intracanal calcium hydroxide in altering the outcome of treatment. Its impact was also weakened by the fact that only studies that used cultivation identification of bacteria were included in the systematic review. From what we have already noted, this would overestimate the incidence of a negative culture result (Siqueira and Rôcas 2005). It is nevertheless interesting to note that in several studies that used the cultivation method to identify bacteria in the root canal, a positive relationship was found between the incidence of apical and the absence of bacteria in the culture test (Sjögren et al. 1997; Waltimo et al. 2005; Fabricius et al. 2006). In a case series that included 55 single-rooted teeth with pulp necrosis and apical infection, Sjögren et al. (1997) reported complete healing in the radiograph 5-year after treatment in 94 % of teeth with a negative culture test prior to root canal filling, while the healing rate for those with a positive culture test was only 68 %. The difference proved to be significantly lower. In a randomized controlled study based on the cases included in the Kvist et al. (2004) study, Molander et al. (2007) found no difference in the treatment outcomes for teeth medicated between appointments with calcium hydroxide or 5 % iodine potassium iodide, but did find there was a significantly more favorable healing outcome in teeth where prefill culture was negative as compared to those that were positive.

There have been only three published studies that meet the requirements necessary for a proper systematic review (Sathorn et al. 2005). These were Trope et al. (1999), Weiger et al. (2000), and Peters and Wesselink (2002). The Cochrane Institute rated only two of these studies, Trope et al. and Weiger et al. low risk for bias, but neither was classified as a strong study (Figini et al. 2008). Another study concerned with the outcome of calcium hydroxide use after pulpectomy was also rated as low risk for bias (Gesi et al. 2006), but this study was not concerned with non-vital, infected root canals. In summary then, it is not possible to take a scientifically defensible position on the positive use of calcium hydroxide dressings in the endodontic treatment of teeth with a necrotic pulp and apical disease. The evidence as we have seen is controversial and weak by Cochrane standards, and we will have to wait until stronger evidence is available to conclusively answer that question.

In a recent survey conducted by SBU (Swedish Council on Health Technology Assessment, a Swedish Governmental Agency for the critical evaluation of methods for preventing, diagnosing, and treating health problems), it was reported that only two studies satisfied their criteria for a moderate level of evidence (Weiger et al. 2000; Molander et al. 2007). Both studies reported similar incidences of periapical healing regardless of whether the teeth received an inter-appointment dressing of calcium hydroxide or were prepared and root filled in one session. SBU also concluded that there is a lack of scientific evidence to verify that root canal medication with calcium hydroxide has a positive effect on the outcome of treatment.

The significance of a negative prefilling culture test on apical healing has been investigated in an animal model (Fabricius et al. (2006). In that study, the healing of lesions created by exposure to specific groups of bacteria placed into the root canals of monkeys was correlated with the absence of the very same bacteria at the time the root canal filling was placed. An assessment of healing of lesions induced was made 2–2.5 years after the completion of treatment. They found that when bacteria were recovered from the root canal prior to filling, 79 % of the apical lesions failed to heal. By contrast, when bacteria were not recovered, 28 % of the lesions failed to heal. It was a reverse way of showing that the presence of cultivable bacteria at the time of root canal filling had a negative impact on the treatment outcome. An interesting finding in that study was the relationship of the technical quality of the root canal filling to the presence of recoverable bacteria. It appeared that quality of the root canal filling only was a significant factor in treatment outcome, when bacteria were present at the time of the filling. It was not a factor when the filling of the canal met what they stated to be "an acceptable standard."

To date, clinical outcome studies appear to be the best and most practical way to assess the clinical impact of residual bacteria in the root canal. A radiographic and clinical assessment of endodontically treated teeth over time can determine whether a new periapical lesion has emerged and whether an existing lesion has persisted, enlarged, or resolved, either in total or in part. They can also determine whether symptoms of apical disease have appeared, abated, or persisted. While the assessment of the clinical status is relatively straightforward, the accuracy of an intraoral radiograph to assess the periapical status is more challenging. In a large cohort of 888 consecutive patients, Estrela et al. (2008) compared the ability of intraoral radiography to cone beam computed tomography (CBCT) in their ability to depict periapical bone loss. They found that many more areas of bone loss could be detected when CBCT was used. In view of CBCT's greater sensitivity, Wu et al. (2009) suggested that future outcome studies of endodontic treatment should utilize CBCT, rather than intraoral radiography, which had been used in the past. However, it was pointed out that, to date, CBCT results have not been measured against a proper reference standard and may, by virtue of the greater ability to show minor bone tissue changes, lead to false conclusions when healing assessments are made (Petersson et al. 2012; Pope et al. 2014). Furthermore, the increased radiation necessary to generate a CBCT image cannot be justified, especially when dealing with a young patient, when a standard intraoral radiograph can in most cases supply the same information (European Society of Endodontology 2014).

To conclude, there is as yet no clinical study that gives convincing and unequivocal support to the premise that an inter-appointment dressing with calcium hydroxide is important to the outcome of root canal therapy. Currently published reports are often based upon relatively small numbers of patient, and only, but in a few instances, have they followed a rigid, well-controlled, experimental protocol. There is also a lack of convincing scientific evidence to show that the use of a calcium hydroxide intracanal dressing impacts upon the short-term incidence of postoperative symptoms and flare-ups during treatment. This implies that despite the publication of numerous studies concerned with calcium hydroxide use in the management of root canal infection, the clinician has relatively weak evidence to support the choice of protocol. Some evidence does exist to indicate that a reduced bacterial presence at the time of root canal filling may lead to a better treatment outcome. Reducing the presence of bacteria at a site of injury is in keeping with the general philosophy of wound management and seems a reasonable course to follow for root canal therapy. However, we will have to wait until better scientific evidence is presented before a valid answer to the question "to dress or not to dress" is available.

3.3 Effects of Calcium Hydroxide on the Mechanical Characteristics of Dentin

Dentin is a composite connective tissue consisting of organic (30 %) and inorganic (50 %) components as well as water (20 %) by weight (Linde and Goldberg 1993). The mineral fraction of dentin is mainly carbonated nanocrystalline apatite, deposited on a tight mesh of randomly oriented type I collagen fibrils (Butler and Ritchie 1995). Type I collagen is a heterotrimer, composed of two α1 (I) and one α2 (I) chains. Hydrogen bonding between closely packed trimers stabilizes the secondary and tertiary structure of collagen (Brodsky et al. 2008). Other non-collagenous proteins and enzymes crucial in the formation of dentin are bound up in the mineralized dentin matrix. These include phosphoproteins, Gla proteins, proteoglycans, acidic glycoproteins, etc. and constitute about 10 % of the organic matrix (Linde 1989). The inorganic, organic, and water fractions of dentin play a distinctive role in the mechanical characteristics of dentin. The inorganic phase contributes to its elastic modulus and compressive strength, while the organic phase contributes to its toughness and tensile strength. The water phase contributes to its fracture resistance and viscoelasticity (Kishen et al. 2006). Biased removal or depletion of any phase or combination of phases, from the dentin during root canal treatment, can, over time, alter its physical and mechanical characteristics. Concerns have been raised that calcium hydroxide medication of the root canal can weaken the root by denaturing its collagen content (Whitbeck et al. 2011). This, in the presence of other attending factors, may predispose a root treated with calcium hydroxide to fracture. While clinical confirmation of this potential risk is lacking for treated teeth with fully developed roots, caution is advisable when treating incompletely developed, permanent teeth (Andreasen et al. 2006; Rosenberg et al. 2007). Cvek (1992) in a retrospective study that included 885 luxated non-vital incisors with immature roots, dressed with calcium hydroxide for 3–54 months, reported that 19 % of the teeth sustained a cervical root fracture during the 3.5–5 years follow-up period. He reported a relationship between the potential for root fracture and the stage of root development.

Calcium hydroxide, also known as slaked lime, is made by adding water to calcium oxide. It has a molecular mass of 74.093 g/mol and a low solubility index of approximately 0.185 g/100 cm^3 at room temperature (25 °C). Unlike many substances, calcium hydroxide solubility is exothermic, meaning that it becomes less soluble as the temperature rises. It has a pH of 12–12.5. In vitro experiments have shown that the hydroxyl ions released, when calcium hydroxide is placed into the root canal, diffuse throughout the root dentin. Nerwich et al. (1993) using microelectrodes buried in small cavities in the apical and coronal parts of the root, as well as the lumen of the root canal, and the exterior surface of the root, found that when calcium hydroxide was placed into the root canal, the pH of dentin rose to 10.8 in the cervical portion of the tooth and 9.7 near the apex. The pH of the dentin near the root surface rose rapidly from day 1 to day 7 and dropped to 9.3 at the cervix and 9 near the apex 2–3 weeks later. They noted that the carrier medium used with the calcium hydroxide and the duration of calcium hydroxide played an important role in the degree of pH change that occurred. Hosoya et al. (2001) suggested that water was an important carrier medium, while Alaçam et al. (1998) suggested that water in combination with glycerin facilitated ionic diffusion and therefore was better than water alone. It is highly probable, that the degree of pH change observed in dentin exposed to calcium hydroxide is sufficient to alter the organic content of its matrix. In addition, since calcium hydroxide can also denature and hydrolyze the other extracellular matrix proteins, it can also disrupt interactions between collagen fibrils and the hydroxyapatite crystals. Collectively the effect can undermine the mechanical integrity of the dentin (Andreasen et al. 2002; Grigoratos et al. 2001).

The Grigoratos et al. (2001) study evaluated the effect of sodium hypochlorite solutions (3 % and 5 %) and saturated calcium hydroxide solution, individually and consecutively, on the flexural strength and elastic modulus of a standardized dentin specimen. The flexural strength represents the highest stress experienced within a material at the moment of fracture, and elastic modulus represents the stiffness of a material. The higher its stiffness, the higher its elastic modulus. Their investigation revealed that 3 % and 5 % sodium hypochlorite significantly decreased the modulus of elasticity and the flexural strength of dentin. Calcium hydroxide, on the other hand, significantly reduced flexural strength but had no effect on its elastic modulus. Dentin treated with sodium hypochlorite, followed by calcium hydroxide, had a significantly different elastic modulus and flexural strength than those treated with sodium hypochlorite alone.

In light of the decrease in mechanical strength of dentin noted when calcium hydroxide is placed into the root canal, it is not surprising to find that a similar effect occurs in dentin when it is exposed to mineral trioxide aggregate (MTA) (White et al. 2002). After 5 weeks of exposure the decrease was similar and reached 32% for calcium hydroxide and 33% for MTA. A most notable decrease in mechanical strength, 59%, was noted with sodium hypochlorite (White et al. 2002). In a later study, Doyon et al. (2005), demonstrated a significant decrease in peak load at fracture in human dentin exposed to calcium hydroxide for 180 days as compared with dentin exposed for 30 days. He reported that a 10–20 % decrease in dentin strength was sufficient to significantly increase the likelihood of fracture in teeth that were already structurally compromised.

Changes in the fracture resistance of human root dentin treated with calcium hydroxide over different time intervals were also the topic of investigation by Zarei et al. (2013). In this study, the root canals of single-rooted decoronated human premolars were endodontically prepared to a standard size. Half of the roots were treated with aqueous calcium hydroxide, the remainder was left untreated, and both groups were incubated at body temperature for periods of 1 week, 1 month, 3 months, or 6 months before testing. A significant reduction in compressive strength to fracture was found to be present in roots treated with calcium hydroxide after 1, 3, and 6 months, leading them to conclude that the extended use of a calcium hydroxide intracanal dressing significantly reduces the physical and mechanical characteristics of the dentin and therefore should be avoided when possible.

4 Clinical Perspectives

Current scientific evidence does not support the choice of single or multiple visits as an approach to the endodontic management of apical periodontitis. Furthermore, controlled clinical trials do not confirm an enhanced treatment outcome with the use of calcium hydroxide or, for that matter, any other intracanal medicament. However, while reviewing the details of many of the studies used as a basis, one must bear in mind that these studies, especially those advocating single-visit treatment, did not take into account the presence of situations, other than those that were bacteria related, that might preclude the use of a one-visit approach. Examples of such situations might be (a) the extended time required to adequately find and prepare the root canals of teeth with complicated root canal anatomy and (b) cases with suppuration or bleeding from the apical lesion that cannot be arrested within in the time period available to the clinician. In addition, since properly controlled clinical trials require proper case selection to avoid bias between test and control groups, teeth that presented extraordinary challenges were most likely excluded from some or all of the outcome studies. In face of the realities of endodontic practice, decisions on the number of appointments necessary to complete the treatment must be based on factors other than scientific evidence, even when the strength of that evidence is even higher. Hence, they have to take into account all the features specific to a case and its clinical presentation. This means that some situations will lend themselves to completion in one visit, while others will require multiple sessions, if all the criteria for a favorable treatment outcome are to be met.

What are the reasonable clinical considerations that lend some teeth to be treated in a single visit while others require more? The health status of the pulp is certainly an important criterion. Teeth with inflamed and infected pulps and no evidence of apical periodontitis generally can be viewed as good candidates for preparation and filling in one appointment. During the early phases of pulp infection, much of the pulp is still intact, and the inflammation present is likely to be caused only by bacterial byproducts from infected dentin and superficially located planktonic bacteria. It is highly likely that all bacteria will be removed from the root canal in these cases. Immediate filling and sealing of the access opening will reduce the chances of new bacterial ingress. Certainly from a clinical perspective, such teeth are more easily managed than teeth with advanced infection, where the entire pulp tissue is necrotic and the root canals is replete with dense bacterial biofilm.

On the other hand, calcium hydroxide medication has long been seen as a valuable adjunct in treatment after pulpectomy. One advantage of calcium hydroxide is its ability to necrotize pulp tissue remnants that are inadvertently left on the root canal walls after the instrumentation procedure (Hasselgren et al. 1988; Andersen et al. 1992). These necrotized tissue remnants are removed more easily at a second appointment by sodium hypochlorite irrigation in that state. The use of calcium hydroxide in such instances theoretically renders the root canal cleaner and thereby optimizes the potential of filling the root canal space more completely in the final phase of treatment (Tûrkün and Cengiz 1997). It also removes a potential source of nutrients in the event that bacteria remain in the root canal after instrumentation and irrigations or gain access to the root canal after treatment has been completed. The efficacy of this regimen was not substantiated in a randomized clinical trial, where instrumentation and filling of the root canal of pulpectomized teeth completed in one session was compared to a two-session treatment in which calcium hydroxide was employed as an inter-appointment dressing (Gesi et al. 2006). Both treatment regimens resulted in similar and highly favorable outcomes clinically and radiographically, 2–3 years after treatment was completed. It should be stressed that in this study a rigid aseptic approach to treatment was followed leading the authors to also conclude that when bacterial contamination of the root canal system is effectively prevented during pulpectomy, predictable, uneventful wound healing is likely to take place, if the material used to fill the root canal is biocompatible as demonstrated by Engström and Spångberg (1967).

Teeth with non-vital pulps and no obvious clinical signs of root canal infection, viz., painful symptoms, and displaying radiographic evidence of apical periodontitis are also candidates for one-visit treatment. Teeth in this category include those with ischemic pulp necrosis caused by trauma and previously root-filled teeth with technical deficiencies that require retreatment for restorative reasons. In both instances, the endodontic treatment could be considered prophylactic in nature, aimed at preventing infection, due to a compromised clinical state, than in treating it. Despite a high probability for a successful treatment outcome, some teeth in this category may offer a high risk for misadventure and special care, and extra time may be required to minimize the risk. Teeth that have a prior root filling or teeth that have been previously post restored present a risk of root fracture. Extrusion of material into the apical tissue and perforation of the root canal wall during retreatment are examples of some of the misfortunes that might occur. These events can establish a pathway for the ingress of bacteria from the oral cavity or provide nutrients for dormant microorganisms present in the root canal system, leading to a persistence or initiation of apical disease. Clinical follow-up studies of teeth retreated due to a technically poor root canal filling have shown that overinstrumentation of the root canal leads to an increased risk of posttreatment disease (Bergenholtz et al. 1979). While some clinicians have recommended the use of a calcium hydroxide dressing, when the root canal is overinstrumented, to allow time to clinically assess the effect of a complication over time, currently there is no convincing scientific proof to support its efficacy.

The approach to teeth with necrotic infected pulps or previously treated teeth with an apical lesion may vary. When there is seepage of inflammatory exudate from an apical lesion into the root canal that cannot be stopped, attempting to complete the treatment in a single visit is fraught with risk. A wet root canal implies wet dentin, and this is likely to undermine the ability of the clinician to properly "seal" the root canal system. As previously cited, clinical case reports have indicated that an inter-appointment calcium hydroxide dressing or a series of dressings can be helpful to suppress the exudate production in these cases and improve the chances for a proper seal and decrease the risk of a poor treatment outcome (Heithersay 1975; Calişkan 2004).

In teeth with non-suppurating apical lesions, the complexity of the root canal anatomy and the time necessary to achieve all the treatment goals often dictate the strategy used in treatment. It is generally understood that the more difficult it is to access the root canal and the more complex the root canal anatomy, the longer it will take to complete its preparation and disinfection. It may be more convenient and more practical for both patient and clinician to schedule more than one treatment session. In fact, it is practical to view the treatment in such teeth as consisting of two critical and separate steps, (1) the preparation and antimicrobial treatment of the root canal system and (2) placement of a filling of a root canal, where every attempt has been made to reduce its content of necrotic and bacterial contents. In teeth that present no anatomical or technical complications and canal preparation, in all aspects, has been uneventful, completion of the treatment in one session can most likely be completed in one treatment session without an impact on the treatment outcome (Weiger et al. 2000; Molander et al. 2007).

In instances where the clinician chooses to postpone the placement of a root filling for another session, a question arises as to whether it is necessary to place a root canal dressing and if so which one. While randomized clinical trials have not supported either choice, there are compelling theoretical reasons to support the use of an interim dressing rather than allowing canal(s) to remain "empty" between patient visits. Several studies that have used bacteriological controls during the various stages of treatment have demonstrated that elimination of all cultivable bacteria from an infected root canal is rarely possible (Byström and Sundqvist 1983; 1985; Byström et al. 1985; Sjögren et al. 1997; Dalton et al. 1998; Shuping et al. 2000; Sundqvist et al. 1998; Siqueira et al. 2007). This means that without a long-acting antimicrobial in the instrumented root canal space between treatment visits, bacteria, although greatly reduced in number at the completion of the preparation phase, continue to grow and can achieve amounts comparable to those that were originally present. It is also possible that some of the surviving strains can emerge as a more virulent and resistant strain and thereby make their subsequent removal more difficult (Chávez de Paz et al. 2003).

Calcium hydroxide, particularly in a paste-like form, possesses several important properties that make it particularly useful as an intracanal dressing between treatment visits. Its alkaline capacity not only helps destroy bacteria in the main portions of the root canal but also serves as an important deterrent to the growth of bacteria in the noninstrumented sites of the root canal and in dentinal tubules adjacent to the main canal. It produces contact killing of most planktonic bacteria species, wherever it is placed, and its physical presence occupies space, limiting that available for bacterial regrowth. As pointed out previously, the dressing will also deprive surviving organisms of a nutritional supply by separating them from residual tissue within the root canal and limiting the seepage of apically derived exudate into the root canal. Hence, when the entire instrumented root canal space is properly filled with a calcium hydroxide dressing, microbial–host interaction in the periapex is reduced to a minimum. While we have advocated caution with regard to the deleterious effect on dentin with long-term use, it is less likely to do so when used in the short term.

Convenient scheduling to allow assessment of the effects of infection control on the signs and symptoms of periapical disease is another clinical reason for the inter-appointment use of a calcium hydroxide dressing. Not all cases are straightforward, and a period of time is at times required to determine whether additional and/or different approach to treatment is necessary. Examples of this are teeth with large periapical lesions, teeth with apical–marginal communications, teeth with root perforations, teeth with open or grossly overinstrumented apices, and teeth with long-standing clinical symptoms. Disappearance of painful symptoms and sinus tracts and a reduction in the size of a periapical lesion are positive signs of successful infection control, and it offers encouragement that with completion of the treatment, that is, after filling the root canal, a successful outcome can be expected. This insight is of practical value, when the patient's treatment plan includes complex and costly restorations. In such cases the increased risk of adversely changing the physical properties of the dentin also has to be considered.

Support for a calcium hydroxide inter-appointment dressing's ability to suppress growth of bacteria in the endodontically prepared root canal appeared in a recently published study by Vera et al. (2012) and is worthy of mention. They examined the postoperative microbiological status of the endodontically treated mesial roots of mandibular molars with primary apical periodontitis. Teeth were treated in either one or two visits. A dressing of calcium hydroxide mixed in saline was placed into the prepared root canals for 1 week prior to the completion of treatment in the teeth that were treated in two visits. This represented the only difference in treatment between the teeth in the two groups. After treatment, teeth were extracted and prepared histologically using H&E and a modified Brown and Brenn stain, to identify the presence of bacteria. Microscopic examination revealed a reduced bacterial presence in the root canals of teeth in the medicated (2 visit) group, especially in areas where calcium hydroxide had been placed. The root canals in 2 of the teeth in the two-visit group were found to be free of bacteria. In the root canals of the remaining 5 teeth in this group, bacterial concentrations appeared markedly reduced in the main portion of the root canal and in the adjacent dentinal tubules. The root canals in the nonmedicated teeth (1 visit) all had bacteria present in the main portion of the root canal and in their ramifications and isthmuses as well as the dentinal tubules adjacent to it. Since there are limitations associated with the identification of bacteria by Brown and Brenn stain and since this was only a short-term study, it is difficult to predict whether the differences found between the two groups would impact on the future expression of disease. It was, however, a visual endorsement of the value of a calcium hydroxide dressing in endodontic treatment. It was also in agreement with the general principles of wound management, the elimination of bacteria from the injured site to the extent that is possible.

5 Concluding Remarks

Over the years, total disinfection of the root canal has been, and still remains, an important goal to achieve in endodontic treatment. Disinfecting agents are still used for root canal irrigation and as an intracanal dressing for short and long periods of time to reduce its microbial concentrations. In recent years, calcium hydroxide has been accepted by many clinicians as a preferred inter-appointment dressing. While subject to limitations, its antimicrobial efficacy has been demonstrated in laboratory and clinical studies. There is, to date, no convincing scientific evidence to indicate that its use enhances the eventual outcome of treatment. Clinical, as well as experimental evidence, nevertheless suggests that the absence of demonstrable bacteria in the root canal prior to root canal filling is a strategic goal in treatment and that the intracanal placement of calcium hydroxide between appointments is an easy and appropriate adjunct in achieving it. As with all the choices made by the clinician, the advantages of its use must be balanced by the potential risk in its use, and like all wise choices, it should always be "what is best for the patient" that becomes the deciding factor.

References

Alaçam T, Yoldaş HO, Gülen O (1998) Dentin penetration of two calcium hydroxide combinations. Oral Surg Oral Med Oral Pathol Oral Radiol Endod 86:469–472

Andersen M, Lund A, Andreasen JO, Andreasen FM (1992) In vitro solubility of human pulp tissue in calcium hydroxide and sodium hypochlorite. Endod Dent Traumatol 8:104–108

Andreasen JO, Farik B, Munksgaard EC (2002) Long-term calcium hydroxide as a root canal dressing may increase risk of root fracture. Dent Traumatol 18:134–137

Andreasen JO, Munksgaard EC, Bakland LK (2006) Comparison of fracture resistance in root canals of immature sheep teeth after filling with calcium hydroxide or MTA. Dent Traumatol 22:154–156

Baik JE, Jang KS, Kang SS, Yun CH, Lee K, Kim BG, Kum KY, Han SH (2011) Calcium hydroxide inactivates lipoteichoic acid from Enterococcus faecalis through deacylation of the lipid moiety. J Endod 37:191–196
Barnard D, Davies J, Figdor D (1996) Susceptibility of Actinomyces israelii to antibiotics, sodium hypochlorite and calcium hydroxide. Int Endod J 29:320–326
Bergenholtz G (2005) Advances since the paper by Zander and Glass (1949) on the pursuit of healing methods for pulpal exposures – historical perspectives. Oral Surg Oral Med Oral Pathol Oral Radiol Endod 100(suppl):S102–S108
Bergenholtz G, Spångberg L (2004) Controversies in endodontics. Crit Rev Oral Biol Med 15:99–114
Bergenholtz G, Lekholm U, Milthon R, Engström B (1979) Influence of apical overinstrumentation and overfilling on re-treated root canals. J Endod 5:310–314
Brodsky B, Thiagarajan G, Madhan B, Kar K (2008) Triple-helical peptides: an approach to collagen conformation, stability, and self-association. Biopolymers 89:345–353
Buck RA, Cai J, Eleazer PD, Staat RH, Hurst HE (2001) Detoxification of endotoxin by endodontic irrigants and calcium hydroxide. J Endod 27:325–327
Butler WT, Ritchie H (1995) The nature and functional significance of dentin extracellular matrix proteins. Int J Dev Biol 39:169–179
Byström A, Sundqvist G (1981) Bacteriologic evaluation of the efficacy of mechanical root canal instrumentation in endodontic therapy. Scand J Dent Res 89:321–328
Byström A, Sundqvist G (1983) Bacteriologic evaluation of the effect of 0.5 percent sodium hypochlorite in endodontic therapy. Oral Surg Oral Med Oral Pathol 55:307–312
Byström A, Sundqvist G (1985) The antibacterial action of sodium hypochlorite and EDTA in 60 cases of endodontic therapy. Int Endod J 18:35–40
Byström A, Claesson R, Sundqvist G (1985) The antibacterial effect of camphorated paramonochlorophenol, camphorated phenol and calcium hydroxide in the treatment of infected root canals. Endod Dent Traumatol 1:170–175
Calişkan MK (2004) Prognosis of large cyst-like periapical lesions following nonsurgical root canal treatment: a clinical review. Int Endod J 37:408–416
Carr GB, Schwartz RS, Schaudinn C, Gorur A, Costerton JW (2009) Ultrastructural examination of failed molar retreatment with secondary apical periodontitis: an examination of endodontic biofilms in an endodontic retreatment failure. J Endod 35:1303–1309
Chávez de Paz LE, Dahlén G, Molander A, Möller Å, Bergenholtz G (2003) Bacteria recovered from teeth with apical periodontitis after antimicrobial endodontic treatment. Int Endod J 36:500–508
Chávez de Paz LE, Bergenholtz G, Dahlén G, Svensäter G (2007) Viability and mechanisms of response to alkaline stress by selected root canal bacteria. Int Endod J 40:344–355
Coolidge ED (1960) Past and present concepts in endodontics. J Am Dent Assoc 61:676–688
Costa Junior SV, deSouza-Filho FJ, Barbosa SV (2003) Tissue reactions to a component of root canal system bacteria: lipoteichoic acid. Braz Dent J 14:95–98
Cvek M (1992) Prognosis of luxated non-vital maxillary incisors treated with calcium hydroxide and filled with gutta-percha. A retrospective clinical study. Endod Dent Traumatol 8:45–55
Dalton BC, Ørstavik D, Phillips C, Pettiette M, Trope M (1998) Bacterial reduction with nickel-titanium rotary instrumentation. J Endod 24:763–767
de Oliveira LD, Jorge AO, Carvalho CA, Koga-Ito CY, Valera MC (2007) In vitro effects of endodontic irrigants on endotoxins in root canals. Oral Surg Oral Med Oral Pathol Oral Radiol Endod 104:135–142
Distel JW, Hatton JF, Gillespie MJ (2002) Biofilm formation in medicated root canals. J Endod 28:689–693
Dodge JS Jr (1887) Immediate root-filling. Dent Cosmos 29:234–235
Doyon GE, Dumsha T, von Fraunhofer JA (2005) Fracture resistance of human root dentin exposed to intracanal calcium hydroxide. J Endod 31:895–897
Engström B, Spångberg L (1967) Wound healing after partial pulpectomy. A histological study performed on contralateral tooth pairs. Odontol Tidskr 75:5–18

Estrela C, Pécora JD, Souza-Neto MD, Estrela CR, Bammann LL (1999a) Effect of vehicle on antimicrobial properties of calcium hydroxide pastes. Braz Dent J 10:63–72
Estrela C, Pimenta FC, Ito IY, Bammann LL (1999b) Antimicrobial evaluation of calcium hydroxide in infected dentinal tubules. J Endod 25:416–418
Estrela C, Bueno MR, Leles CR, Azevedo B, Azevedo JR (2008) Accuracy of cone beam computed tomography and panoramic and periapical radiography for detection of apical periodontitis. J Endod 34:273–279
European Society of Endodontology, Patel S, Durack C, Abella F, Roig M, Shemesh H, Lambrechts P, Lemberg K (2014) European society of endodontology position statement: the use of CBCT in endodontics. Int Endod J 47:502–504
Evans M, Davies JK, Sundqvist G, Figdor D (2002) Mechanisms involved in the resistance of Enterococcus faecalis to calcium hydroxide. Int Endod J 35:221–228
Fabricius L, Dahlén G, Sundqvist G, Happonen R-P, Möller ÅJ (2006) Influence of residual bacteria on periapical tissue healing after chemomechanical treatment and root filling of experimentally infected monkey teeth. Eur J Oral Sci 114:278–285
Fava LRG, Saunders WP (1999) Calcium hydroxide pastes: classification and clinical indications. Int Endod J 32:257–282
Figini L, Lodi G, Gorni F, Gagliani M (2008) Single versus multiple visits for endodontic treatment of permanent teeth: a Cochrane systematic review. J Endod 34:1041–1047
Foschi F, Izard J, Sasaki H, Sambri V, Prati C, Muller R, Stashenko P (2006) Treponema in disseminating endodontic infections. J Dent Res 85:761–765
Frank AL (1966) Therapy for the divergent pulpless tooth by continuing apical formation. J Am Dent Assoc 72:87–93
George S, Kishen A (2008) Effect of sub-lethal doses of EDTA and calcium hydroxide on the physiochemical surface properties of *Enterococcus faecalis* and its adherence to collagen-in vitro study. Biofilm. Cambridge University Press, pp 1–10
Gesi A, Hakeberg M, Warfvinge J, Bergenholtz G (2006) Incidence of osteolytic lesions and clinical symptoms after pulpectomy – a clinical evaluation of one versus two-session treatment. Oral Surg Oral Med Oral Pathol Oral Radiol Endod 101:379–388
Gomes BP, Ferraz CC, Garrido FD, Rosalen PL, Zaia AA, Teixeira FB, de Souza-Filho FJ (2002) Microbial susceptibility to calcium hydroxide pastes and their vehicles. J Endod 28:758–761
Granath L-E (1959) Några synpukter på behandlingen av traumatiserade incisiver på barn. Odontol Revy 10:272–286
Grigoratos D, Knowles J, Ng YL, Gulabivala K (2001) Effect of exposing dentine to sodium hypochlorite and calcium hydroxide on its flexural strength and elastic modulus. Int Endod J 34:113–119
Grossman LI (1971) Endodontics: then and now. Oral Surg Oral Med Oral Pathol 32:254–259
Grossman LI (1976) Endodontics 1776–1996: a bicentennial history against the background of general dentistry. J Am Dent Assoc 93:78–87
Grove CJ (1913) A warning against the indiscriminate use of formaldehyde preparations. Dent Cosmos 50:147–150
Guo J, Yang D, Okamura H, Teramachi J, Ochiai K, Qiu L, Haneji T (2014) Calcium hydroxide suppresses Porphyromonas endodontalis lipopolysaccharide-induced bone destruction. J Dent Res 93:508–513
Haapasalo M, Ørstavik D (1987) In vitro infection of dentinal tubules. J Dent Res 66:1375–1379
Haapasalo M, Qian W, Portenier I, Waltimo T (2007) Effects of dentin on the antimicrobial properties of endodontic medicaments. J Endod 33:917–925
Hasselgren G, Olsson B, Cvek M (1988) Effects of calcium hydroxide and sodium hypochlorite on the dissolution of necrotic porcine muscle tissue. J Endod 14:125–127
Heithersay GS (1970) Stimulation of root formation in incompletely developed pulpless teeth. Oral Surg Oral Med Oral Pathol 29:620–630
Heithersay GS (1975) Calcium hydroxide in the treatment of pulpless teeth with associated pathology. J Br Endod Soc 8:74–98
Hermann BW (1920) Calciumhydroxyd als Mittels zum Behandel und Füllen von Zahnwurzelkanälen. Würzburg Medical Dissertation 29

Hermann BW (1930) Dentinobliteration der Wurzelkanälen nach Behandlung mit Kalzium. Zahnärztl Rundschau 21:888–899

Hong C-Y, Lin S-K, Kok S-H, Cheng S-J, Lee M-S, Wang T-M (2004) The role of lipopolysaccharide in infectious bone resorption of periapical disease. J Oral Path Med 33:162–169

Hosoya N, Takahashi G, Arai T, Nakamura J (2001) Calcium concentration and pH of the periapical environment after applying calcium hydroxide into root canals in vitro. J Endod 27:343–346

Kaufman B, Spångberg L, Barry J, Fouad AF (2005) Enterococcus spp. in endodontically treated teeth with and without periradicular lesions. J Endod 31:851–856

Kayaoglu G, Erten H, Bodrumlu E, Ørstavik D (2009) The resistance of collagen-associated, planktonic cells of Enterococcus faecalis to calcium hydroxide. J Endod 35:46–49

Kells CE Jr (1887) Immediate root-filling. Dent Cosmos 29:366–367

Kerekes K, Tronstad L (1979) Long-term results of endodontic treatment performed with a standardized technique. J Endod 5:83–90

Kishen A, George S, Kumar R (2006) Enterococcus faecalis-mediated biomineralized biofilm formation on root canal dentine in vitro. J Biomed Mater Res A 77:406–415

Kvist T, Molander A, Dahlén G, Reit C (2004) Microbiological evaluation of one- and two-visit endodontic treatment of teeth with apical periodontitis: a randomized, clinical trial. J Endod 30:572–576

Linde A (1989) Dentin matrix proteins: composition and possible functions in calcification. Anat Rec 224:154–166

Linde A, Goldberg M (1993) Dentinogenesis. Crit Rev Oral Biol Med 4:679–728

Matsumiya S, Kitamura M (1960) Histopathological and histobacteriological studies of the relation between the condition of sterilization of the interior of the root canal and the healing process of periapical tissues in experimentally infected root canal treatment. Bull Tokyo Dent College 1:1–19

McGurkin-Smith R, Trope M, Caplan D, Sigurdsson A (2005) Reduction of intracanal bacteria using GT rotary instrumentation, 5.25 % NaOCl, EDTA, and $Ca(OH)_2$. J Endod 31:359–363

Messer HH, Chen RS (1984) The duration of effectiveness of root canal medicaments. J Endod 10:240–245

Miller WD (1890) Microorganisms of the human mouth. The local and general diseases which are caused by them. Philadelphia, pp 1–364

Mohammadi Z, Dummer P (2011) Properties and applications of calcium hydroxide in endodontics and dental traumatology. Int Endod J 44:697–730

Molander A (2000) On testing microbial presence in the root canal. Thesis, University of Gothenburg

Molander A, Warfvinge J, Reit C, Kvist T (2007) Clinical and radiographic evaluation of one- and two-visit endodontic treatment of asymptomatic necrotic teeth with apical periodontitis: a randomized clinical trial. J Endod 33:1145–1148

Montagner F, Jacinto RC, Signoretti FG, Gomes BP (2010) Treponema species detected in infected root canals and acute apical abscess exudates. J Endod 36:1796–1799

Nair PN, Henry S, Cano V, Vera J (2005) Microbial status of apical root canal system of human mandibular first molars with primary apical periodontitis after "one visit" endodontic treatment. Oral Surg Oral Med Oral Pathol Oral Radiol Endod 99:231–252

Nerwich A, Fidgor D, Messer HH (1993) pH changes in root dentine over a 4-week period following root canal dressing with calcium hydroxide. J Endod 19:302–306

Ørstavik D, Haapasalo M (1990) Disinfection by endodontic irrigants and dressings of experimentally infected dentinal tubules. Endod Dent Traumatol 6:142–149

Ørstavik D, Kerekes K, Molven O (1991) Effects of extensive apical reaming and calcium hydroxide dressing on bacterial infection during treatment of apical periodontitis: a pilot study. Int Endod J 24:1–7

Parmar D, Hauman CH, Leichter JW, McNaughton A, Tompkins GR (2011) Bacterial localization and viability assessment in human ex vivo dentinal tubules by fluorescence confocal laser scanning microscopy. Int Endod J 44:644–651

Peters L, Wesselink PR (2002) Periapical healing of endodontically treated teeth in one and two visits obturated in the presence or absence of detectable microorganisms. Int Endod J 35:660–667

Peters LB, van Winkelhoff AJ, Buijs JF, Wesselink PR (2002) Effects of instrumentation, irrigation and dressing with calcium hydroxide on infection in pulpless teeth with periapical bone lesions. Int Endod J 35:13–21

Petersson A, Axelsson S, Davidson T, Frisk F, Hakeberg M, Kvist T, Norlund A, Mejàre I, Portenier I, Sandberg H, Tranæus S, Bergenholtz G (2012) Radiological diagnosis of periapical bone tissue lesions in endodontics. A systematic review. Int Endod J 45:783–801

Pope O, Sathorn C, Parashos P (2014) A comparative investigation of cone-beam computed tomography and periapical radiography in the diagnosis of a healthy periapex. J Endod 40:360–365

Portenier I, Haapasalo H, Rye A, Waltimo T, Ørstavik D, Haapasalo M (2001) Inactivation of root canal medicaments by dentine, hydroxylapatite and bovine serum albumin. Int Endod J 34:184–188

Reit C, Dahlén G (1988) Decision making analysis of endodontic treatment strategies in teeth with apical periodontitis. Int Endod J 21:291–299

Ribeiro AC, Matarazzo F, Faveri M, Zezell DM, Mayer MP (2011) Exploring bacterial diversity of endodontic microbiota by cloning and sequencing 16S rRNA. J Endod 37:922–926

Rohner W (1940) Calxyl als Wurzelfüllungs Material nach Pulpa Extirpation. Schweiz Monatssch Zahnmed 50:903–948

Rosenberg B, Murray PE, Namerow K (2007) The effect of calcium hydroxide root filling on dentin fracture strength. Dent Traumatol 23:26–29

Safavi KE, Nichols FC (1993) Effect of calcium hydroxide on bacterial lipopolysaccharide. J Endod 19:76–78

Safavi KE, Nichols FC (1994) Alteration of biological properties of bacterial lipopolysaccharide by calcium hydroxide treatment. J Endod 20:127–129

Sakamoto M, Rôcas IN, Siqueira JF Jr, Benno Y (2006) Molecular analysis of bacteria in asymptomatic and symptomatic endodontic infections. Oral Microbiol Immunol 21:112–122

Sathorn C, Parashos P, Messer HH (2005) Effectiveness of single- versus multiple-visit endodontic treatment of teeth with apical periodontitis: a systematic review and meta-analysis. Int Endod J 38:347–355

Sathorn C, Parashos P, Messer HH (2007) How useful is root canal culturing in predicting treatment outcome? J Endod 33:220–225

Schilder H (1974) Cleaning and shaping the root canal. Dent Clin North Am 18:269–296

Seltzer S, Bender IB (1965) Cognitive dissonances in endodontics. Oral Surg 20:505–515

Shuping GB, Ørstavik D, Sigurdsson A, Trope M (2000) Reduction of intracanal bacteria using nickel-titanium rotary instrumentation and various medications. J Endod 26:751–755

Siqueira JF Jr, Rôcas IN (2005) Exploiting molecular methods to explore endodontic infections: part 2 – Redefining the endodontic microbiota. J Endod 31:488–498

Siqueira JF Jr, Rôças IN (2009) Diversity of endodontic microbiota revisited. J Dent Res 88:969–981

Siqueira JF Jr, Magalhães KM, Rôças IN (2007) Bacterial reduction in infected root canals treated with 2.5 % NaOCl as an irrigant and calcium hydroxide/camphorated paramonochlorophenol paste as an intracanal dressing. J Endod 33:667–672

Sjögren U, Figdor D, Spångberg L, Sundqvist G (1991) The antimicrobial effect of calcium hydroxide as a short-term intracanal dressing. Int Endod J 24:119–125

Sjögren U, Figdor D, Persson S, Sundqvist G (1997) Influence of infection at the time of root filling on the outcome of endodontic treatment of teeth with apical periodontitis. Int Endod J 30:297–306

Sommer RF, Ostrander FD, Crowley MC (1966) Clinical endodontics. A manual of scientific endodontics, 3rd edn. WB Saunders, Philadelphia/London

Staehle HJ (1990) Calciumhydroxid in der Zahnheilkunde. Carl Hanser Verlag, München, Wien

Stevens RH, Grossman LI (1983) Evaluation of the antimicrobial potential of calcium hydroxide as an intracanal medicament. J Endod 9:372–374

Strindberg LZ (1956) The dependence of the results of pulp therapy on certain factors. An analytical study based on radiographic and clinical follow-up examinations. Acta Odontol Scand 14 (suppl 21)

Sundqvist G, Figdor D, Persson S, Sjögren U (1998) Microbiologic analysis of teeth with failed endodontic treatment and the outcome of conservative re-treatment. Oral Surg Oral Med Oral Pathol Oral Radiol Endod 85:86–93

Swedish Council on Health Technology Assessment (SBU) (2010) Methods of diagnosis and treatment in endodontics. A systematic review. Report no. 203. Available at: http://www.sbu.se

Tanomaru JHG, Leonardo MR, Filho M, Silva LAB (2003) Effect of different irrigation solutions and calcium hydroxide on bacterial LPS. Int Endod J 36:733–739

Trope M, Bergenholtz G (2002) Microbiological basis for endodontic treatment: can a maximal outcome be achieved in one visit. Endod Topics 1:40–53

Trope M, Delano EO, Ørstavik D (1999) Endodontic treatment of teeth with apical periodontitis: single vs. multivisit treatment. J Endod 25:345–350

Tûrkün M, Cengiz T (1997) The effect of sodium hypoclorite and calcium hydroxide on tissue dissolution and root canal cleanliness. Int Endod J 30:335–342

Upadya M, Shrestha A, Kishen A (2011) Role of efflux pump inhibitors on the antibiofilm efficacy of calcium hydroxide, chitosan nanoparticles, and light-activated disinfection. J Endod 237:1422–1426

Vera J, Siqueira JF Jr, Ricucci D, Loghin S, Fernández N, Flores B, Cruz AG (2012) One- versus two-visit endodontic treatment of teeth with apical periodontitis: a histobacteriologic study. J Endod 38:1040–1052

Vianna ME, Conrads G, Gomes BPFA (2007) Quantification and characterization of synergistes in endodontic infections. Oral Microbiol Immunol 22:260–265

Waltimo T, Trope M, Haapasalo M, Ørstavik D (2005) Clinical efficacy of treatment procedures in endodontic infection control and one year follow-up of periapical healing. J Endod 31:863–866

Walton RE (1976) Histologic evaluation of different methods of enlarging the pulp canal space. J Endod 2:304–311

Wei M, Evans JH, Bostrom T, Grøndahl L (2003) Synthesis and characterization of hydroxyapatite, fluoride-substituted hydroxyapatite and fluorapatite. J Mater Sci Mater Med 14:311–320

Weiger R, Rosendahl R, Löst C (2000) Influence of calcium hydroxide intracanal dressings on the prognosis of teeth with endodontically induced periapical lesions. Int Endod J 33:219–226

Whitbeck ER, Quinn GD, Quinn JB (2011) Effect of calcium hydroxide on the fracture resistance of dentin. J Res Natl Inst Stand Technol 116:743–749

White JD, Lacefield WR, Chavers LS, Eleazer PD (2002) The effect of three commonly used endodontic materials on the strength and hardness of root dentin. J Endod 28:828–830

Wu MK, Wesselink PR (1995) Efficacy of three techniques in cleaning the apical portion of curved root canals. Oral Surg Oral Med Oral Pathol Oral Radiol Endod 79:492–496

Wu MK, Shemesh H, Wesselink PR (2009) Limitations of previously published systematic reviews evaluating the outcome of endodontic treatment. Int Endod J 42:656–666

Yared GM, Dagher FE (1994) Influence of apical enlargement on bacterial infection during treatment of apical periodontitis. J Endod 20:535–537

Zarei M, Afkhami F, Malek Poor Z (2013) Fracture resistance of human root dentin exposed to calcium hydroxide intervisit medication at various time periods: an in vitro study. Dent Traumatol 29:56–60

Zoletti GO, Siqueira JF Jr, Santos KR (2006) Identification of Enterococcus faecalis in root-filled teeth with or without periradicular lesions by culture-dependent and-independent approaches. J Endod 32:722–726

Advanced Therapeutic Options to Disinfect Root Canals

Anil Kishen

Abstract Bacterial biofilms in the root canals are challenging targets for conventional antimicrobial irrigants and medicaments. Advanced endodontic disinfection strategies are tested to enhance the antibiofilm effectiveness during root canal treatment. The primary goal of different advanced theraputic options is to eliminate intracanal biofilms from the anatomical complexities and uninstrumented portions of the root canal system, without producing deleterious effects on the host tissues. This chapter describes the challenges offered by bacterial biofilm as a therapeutic target and discusses the current antibiofilm options in root canal disinfection.

1 Bacterial Biofilm as a Therapeutic Target in Root Canal Disinfection

The infected root canal harbors a diverse population of aerobic, anaerobic, gram-positive, gram-negative, spiral, and filamentous bacteria often existing as a structured community, known as a biofilm (Nair et al. 1990; Nair 1987, 2006; Sundqvist and Figdor 2003; Baumgartner et al. 2008). Gram-positive and gram-negative bacteria manifest significant differences in the structure of their cell wall (Denyer and Maillard 2002). The susceptibility of bacteria to antimicrobial agents depends upon the nature of their cell wall and the capsule that at times surrounds it. Bacteria present in a biofilm demonstrates a considerably higher degree of resistance to antimicrobial agents than those that are planktonic (free floating) (Costerton et al. 1994). The increased antimicrobial resistance of biofilm bacteria can generally be attributed to (1) *the presence of extracellular polymeric substance (EPS), a polysaccharide and protein matrix that surrounds and embeds the bacteria,* (2) *an increased rate of growth and nutrient availability,* and (3) *the evolution of resistant phenotypes due to horizontal transfer of resistance genes.* In general, biofilm bacteria do not rely on a single resistance mechanism, but utilize different

A. Kishen (✉)
Department of Endodontics, University of Toronto, Toronto, ON, Canada
e-mail: Anil.Kishen@dentistry.utoronto.ca

L.E. Chávez de Paz et al. (eds.), *The Root Canal Biofilm*, Springer Series on Biofilms 9, DOI 10.1007/978-3-662-47415-0_11

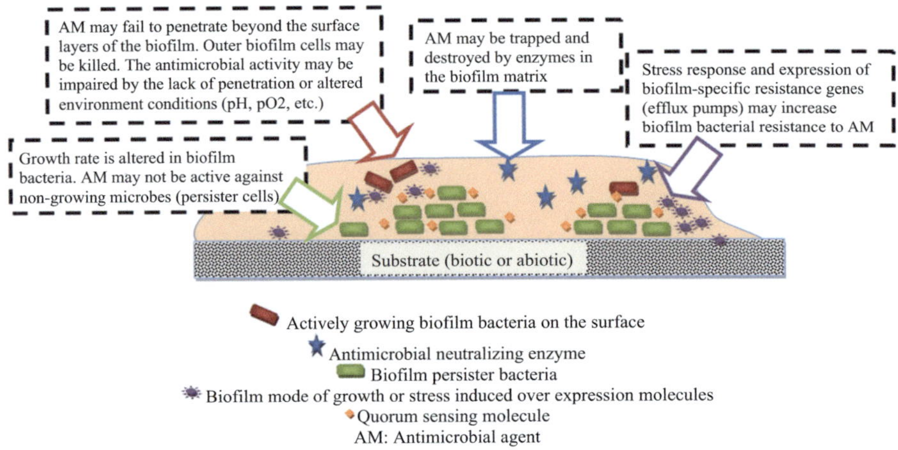

Fig. 1 Schematic diagram showing different methods by which bacteria in a biofilm gain resistance against antimicrobials (Kishen 2010)

mechanisms that act in concert, to evade destruction (Dunne et al. 1993; Gilbert et al. 2002) (Fig. 1).

In the root canal, mature biofilm consists of multiple layers of bacteria embedded in EPS, which often accounts for over 50 % of the biofilm content. EPS has the potential to alleviate the response of the resident bacteria to antimicrobial agents by acting as a diffusion shield and reaction neutralizer. This is because of its highly charged and interwoven structure (Anderl et al. 2000; Vrany et al. 1997; Prince 2002). The chemical components in a biofilm matrix further enhance its barrier effect since certain constituents in the matrix posses a potential to react chemically, and subsequently neutralize, some antimicrobials, such as iodine, iodine-polyvinylpyrrolidone complexes, chlorine, and peroxygens (Nichols et al. 1989; del Pozo and Patel 2007). The antimicrobial resistance associated with biofilm bacteria is also linked to the slow growth and starvation of the bacteria that exist in a biofilm. The spatial arrangement of bacterial species within a biofilm, allows those that inhabit the deeper layers to receive lesser nutrients and have a lower redox potential than those closer to the surface. This makes antimicrobial agents that use penetration, as a vector, less effective. Since nutrient and gas gradients increase with the thickness and maturity of a biofilm, the resistance of deeply located bacteria becomes even more marked in a matured or aged biofilm (del Pozo and Patel 2007; Al Cunningham et al. 2011).

Certain bacteria when exposed to environmental stress or low-level antimicrobial develop as survivor cells called *persister cells* (Lewis 2005). These persister cells are nongrowing phenotypic variants of their general cell population. Following the removal of environmental and chemical stresses, persister cells can grow rapidly if an adequate nutrient supply is available. Biofilm population that are rich in persister cells, therefore, have a greater potential to survive antimicrobial treatment and restore robustness to the biofilm once antimicrobial treatment has been terminated (Al Cunningham et al. 2011). In addition, bacteria that exist in biofilm can upregulate the expression of stress-response genes, shock proteins, and multidrug pumps (efflux pumps) and develop into a more antimicrobial-resistant phenotype (Al Cunningham et al. 2011). In general it can be stated that the nature of biofilm structure and the physiological characteristics of the resident bacteria offer the community a high degree of resistance to many types of antimicrobials (Costerton 1999; Rosan et al. 1999).

Endodontic disease, that is apical periodontitis, has recently been categorized as a biofilm-mediated infection (Ricucci and Siqueira 2010). The pattern of arrangement of bacterial community in root canal is consistent with the criteria that were originally used to classify periodontitis as a biofilm-mediated disease. Numerous studies have revealed a high prevalence of bacterial biofilms in the root canal and on the root surface of teeth with apical periodontitis (Ricucci and Siqueira 2010). A significant reduction of biofilm during endodontic treatment and prevention of its reestablishment subsequent to treatment are essential goods for root canal treatment. Clinical studies have shown that even after meticulous chemomechanical preparation and obturation of the root canal, a significant amount of biofilm is often present in the anatomical complexities of the root canal system and on the wall of main canal that was not engaged by the root canal file. (Ricucci and Siqueira 2010; Nair et al. 2005). While a favorable treatment outcomes using these protocols still occurs in 73–90 % of the teeth that are treated, the level of favorable outcomes has shown little change over the past 50 years (Friedman 2002). It appears that new antibiofilm strategies will have to be introduced into clinical practice if a rise in favorable treatment outcomes is to be realized. These novel strategies should routinely capable of not only disrupting biofilm but also doing so safely throughout the entire root canal system. At present several new strategies have been developed and are currently being tested in endodontics. These, as well as those in current use, will be discussed in the following sections.

2 Therapeutic Strategies for Endodontic Biofilm Management

Generally, the strategies designed to eliminate bacterial biofilms are directed at (1) inactivating resident bacteria within the biofilm structure or (2) disrupting biofilm structure and simultaneously killing resident bacteria (Fig. 2). These objectives can be met by employing a variety of antimicrobial agents and/or treatment strategies. Among the antimicrobials are agents that (1) produce slow destruction of biofilm structure, (2) destroy persister cells, (3) disrupt the quorum sensing signals that are necessary to maintain the biofilm, (4) diffuse into the biofilm structure to initiate killing of resident bacteria or are used in combination with other strategies that enhance diffusion and then initiate killing, and (5) destroy both biofilm matrix and resident bacteria (Prince 2002). Another approach in the biofilm management is

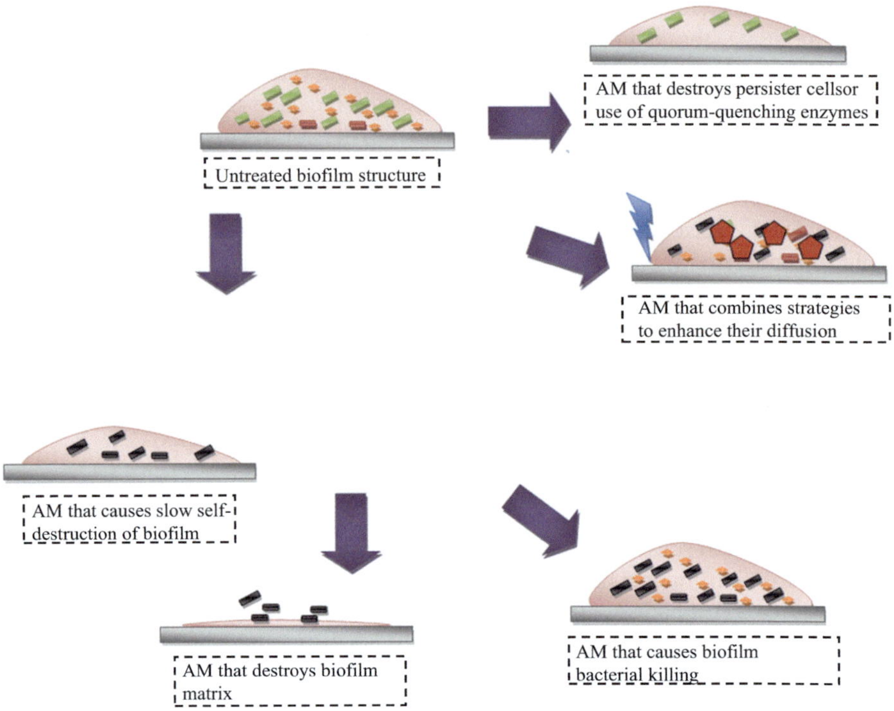

Fig. 2 Schematic diagram showing different antibiofilm strategies (Kishen 2010)

the prevention of bacterial cell adherence, which is the first step in biofilm formation. This particular strategy involves the use of chemically or ultrastructurally modified biomaterials that resist bacterial adherence (Kishen et al. 2008a; An et al. 1996).

The newer antimicrobial strategies adapted for endodontic should be directed at disrupting biofilm structure and destroying resident and persister bacteria in a highly irregular environment such as a root canal system. They have to do so in a manner that does not cause detrimental physical, mechanical, and/or chemical damages to the root dentin and periodontal tissues. In the next section, several of the new and advanced therapeutic approaches designed to manage endodontic biofilm will be reviewed and discussed.

2.1 *Antibacterial Nanoparticles*

Nanoparticles are microscopic particles with one or more dimensions in the range of 1–100 nm. They possess unique physico-chemical properties that are very different from their bulk or powder counterparts. Nanoparticles that are antimicrobial, for example, possess a broad spectrum of antimicrobial action and possess a much lower potential of inducing microbial resistance than conventional antibiotics. Magnesium oxide and calcium oxide slurries have demonstrated bactericidal action against both gram-positive and gram-negative bacteria (Sawai 2003). Zinc oxide slurry has been shown to be bacteriostatic, but more so against gram-positive bacteria than gram-negative bacteria (Yamamoto 2001). The antimicrobial properties of these metallic oxides arise from their ability to generate an active oxygen species (ROS), a radical form of oxygen that is cytotoxic for bacteria. Their antimicrobial potential further increases if they possess a high surface area, a high charge density, and a high potential for bacterial interaction (Sawai et al. 1998). The electrostatic interaction of a positively charged nanoparticle and a negatively charged bacterial cell allows a large number of nanoparticles to accumulate on the bacterial cell membrane when they interact. This produces an increase in membrane permeability and a rapid loss of membrane function that ultimately leads to bacterial cell death. Heavy metal ions can disrupt bacterial cell function in a variety of ways (Stohs and Bagchi 1995; Yoon et al. 2007; Reddy et al. 2007). Copper ions, for example, induce oxidative stresses (Cioffi et al. 2005) and a disruption in redox cycling, which causes damage to the bacterial cell membrane and the bacterial ribosomal DNA. Zinc ions, in concentrations above the essential threshold level, on the other hand, inhibit bacterial enzymes including dehydrogenase. This impedes cell metabolism and causes cell death (Beard et al. 1995). Silver ions inactivate proteins and inhibit DNA replication (Feng et al. 2000). The nanoparticles can be synthesized from powders of silver, copper oxide, and zinc oxide, for broad antimicrobial applications (Kim et al. 2007).

A reverse strategy in biofilm management is inhibiting its development. As previously mentioned, adherence of microorganisms to a substrate is the first step in biofilm formation. In the oral cavity, adherence to tooth structure enables bacteria to resist the normal flushing action of saliva and to survive under rather harsh growth conditions (Jefferson 2004; Busscher and van der Mei 1997; Busscher et al. 1995; An and Friedman 1998). Bacterial adherence experiments have highlighted that different endodontic irrigants can reduce adherence of *E. faecalis* to root dentin. The degree to which this occurs is dependent upon the irrigant and its concentration. Irrigation with EDTA (17 %, pH 7.3) following the use of sodium hypochlorite (5.2 %), for example, raises bacterial adherence to root dentin by 33 % (Kishen et al. 2008b). This most likely occurs because EDTA, a chelating agent, can demineralize dentin to a depth of 20–30 μm following 5 min of application (Marshall et al. 1995; Habelitz et al. 2002) and leave the collagen component of the matrix exposed on the dentin surface. Collagen is an excellent substrate for the adherence of many bacterial species including *E. faecalis*. If sodium hypochlorite is used after EDTA, exposed collagen is removed and the opportunity for bacterial adherence is negated (Kishen et al. 2008b; Basrani et al. 2007). Adherence to dentin may also be altered by cationic nanoparticles in an aqueous suspension if they are allowed to settle onto the negatively charged dentin surface (Kishen et al. 2008b). These are examples of how alterations in the physicochemical properties of dentin can change its potential as a substrate for biofilm formation.

Chitosan is a natural bioactive biopolymer derived by the deacetylation of chitin. It binds to a negatively charged surface and possesses excellent antimicrobial and antifungal properties. The exact mechanism by which chitosan and its derivatives act on bacteria is still not fully understood. It is believed that the electrostatic interaction between positively charged chitosan nanoparticles and a negatively charged bacterial cell membrane alters the membrane permeability to dysregulate cell function (Rabea et al. 2003). A recent study examined the antimicrobial properties of ZnO and resin-based root canal sealers that contained chitosan and ZnO nanoparticles. The study demonstrated that adding these nanoparticles to the root canal sealers improved their direct (based on a direct antibacterial assay) and diffusible antibacterial action (based on a membrane-restricted antibacterial assay) (Kishen et al. 2008c) without adversely altering their flow characteristics. Another study examined the ability of chitosan nanoparticles and ZnO nanoparticles to eliminate bacterial biofilm under normal and experimentally aged conditions (nanoparticles conditioned by tissue fluids). It was found that the rate of bacterial killing depended upon the concentration of the nanoparticles and the duration of their exposure to biofilm. Interestingly, both chitosan nanoparticles and ZnO nanoparticles reportedly retained their antibacterial potential after aging for 90 days (Shrestha et al. 2010).

Bioactive glass, has received some interest as a vehicle for root canal disinfection. Bioactive glass is antibacterial and is composed of SiO_2, Na_2O, CaO_2, and P_2O_5 in different concentrations. Its antibacterial properties have been attributed to several factors most notably its high pH, its osmotic effects (relative to cytosol pressure), and its ability to precipitate Ca/P (Stoor et al. 1998). The feasibility of using bioactive glass as a root canal disinfectant was tested in vitro (Zehnder et al. 2004, 2006; Gubler et al. 2008) and shown to be significantly less antibacterial than calcium hydroxide (Zehnder et al. 2006). It also did not effectively prevent recontamination of a previously cleaned root canal (Gubler et al. 2008). It was theorized that a suspensions/slurries of 45S5 bioactive glass would be an effective antibacterial agent in the root canal because of its high and sustainable pH (Waltimo et al. 2009). While it was shown that a nanometric slurry of bioactive glass has a 12-fold higher specific surface area than its micrometric counterpart, the latter demonstrated considerably higher alkalinity and antimicrobial efficacy. This was in contrast to a previous study reported by the same authors that claimed higher antibacterial efficacy when the materials shifted from micron- to nano-size (Waltimo et al. 2007). Bioactive glass was used in an attempt to promote mineral deposition within the root canal in another. It was thought that by doing so, the use of root canal sealers would not be required. A combination of polyisoprene or polycaprolactone and nanometric bioactive glass 45S5 (bioactive glass) was employed for this purpose. Incorporating bioactive glass fillers into the polyisoprene and polycaprolactone resulted in a bioactive composite material with improved mineralization characteristics (Mohn et al. 2010). Although the authors concluded that the composite of polyisoprene, polycaprolactone, and bioactive glass showed potential as a "single" root canal filling material, more rigorous investigations are warranted before adopting this approach for routine clinical use.

In summary, most of the nanoparticles that have been tested have shown a higher antibacterial potential than their powder counterpart. Their higher reactivity and their ability to resist aging have also proven to be a clinical advantage. Most cationic antibacterial particles show excellent interaction with biomaterials, bacteria, and biofilms. In root canal therapy, nanoparticles may be introduced into the root canal as slurry or in combination with a root canal sealer. Though they have the ability to diffuse deep into the dentin, additional research is needed to assess their ability to inactivate biofilm in the anatomical complexities of the root canal system. Their interaction with the host tissues/immune cells also requires further investigation. Their acceptance for routine clinical use will depend upon both their antimicrobial effectiveness and a simple and practical method of delivering them into the root canal.

2.2 *Antimicrobial Photodynamic Therapy*

Photodynamic therapy (PDT) is based on the concept that a nontoxic dye, termed a photosensitizer (PS), can be preferentially localized in a tissue and subsequently be activated by light of an appropriate wavelength to generate cytotoxic singlet oxygen and other reactive oxygen species to create a desired therapeutic effect. A successful outcome with PDT is dependent upon optimal interaction among three elements—light, PS, and oxygen. Light-activated disinfection (LAD) has been known by a variety of names such as photodynamic antimicrobial chemotherapy (PACT), photoactivated disinfection (PAD), light-activated therapy (LAT), and antimicrobial photodynamic therapy (APDT).

The emitted light should have a specific wavelength, one that corresponds to the maximum wavelength absorption property of a photosensitizer. A photosensitizer molecule in its ground state is at a spectroscopic singlet (S_0). After the absorption of a photon (light), it passes from the ground state to its first excited state (S_1). In this state, the photosensitizer can return again to a ground state, or it can pass into a triplet excited state (T_1) via intersystem crossing. The photosensitizer in the triplet state is extremely reactive and can interact along either one or both of the following pathways to destroy a bacterial cell: (1) type I reaction: the photosensitizer triplet state can react with a target, other than oxygen, through hydrogen or electron transfer, to form radical ions that can react with oxygen, to produce a cytotoxic species, such as hydrogen peroxide, superoxide anion, and/or lipid derived radicals; (2) type II reaction: the photosensitizer triplet state can transfer excitation energy to ground state molecular oxygen to produce excited state singlet oxygen ($^1O_2$) (Dai et al. 2009).

Singlet oxygen is a strong oxidizing agent. They are highly reactive within a radius of action of less than 0.02 μm and has a lifetime of less than 0.04 μs in a biological environment (Moan and Berg 1991). The reaction of singlet oxygen with a cellular target leads to cell death. Two basic mechanisms that have been proposed to account for the lethal injury sustained by bacteria exposed to APDT are DNA damage and cytoplasmic membrane damage (Bertoloni et al. 2000; Menezes et al. 1990). During lethal photosensitization, singlet oxygen may interact with photo-oxidizable amino acid residues, such as His, Cys, Trp, and Tyr, in one protein molecule, to produce reactive species that may in turn interact with residues or free amino groups in another protein, to form a cross-link (Shen et al. 1996). The photo-oxidative effect caused by a phenothiazinium photosensitizer on bacteria has damaged multiple bacterial targets such DNA (Menezes et al. 1990), cell membrane (Wakayama et al. 1980), protease activity, and lipopolysaccharide (LPS) (Kömerik et al. 2000). Functional impairment of cell wall, extensive damage to chromosomal DNA, and degradation of membrane proteins, following methylene blue-mediated APDT of *E. faecalis*, has also been reported (George and Kishen 2008a).

APDT has the potential to destroy mammalian cells as well as microbial cells. Yet several studies have shown the selective killing of microbial cells over host cells, especially at photosensitization periods, and light fluence, required for an antimicrobial action. Soukos et al. compared the effect of APDT using a combination of toluidine blue O (TBO) and red light against *S. sanguis* and human gingival keratinocytes and fibroblasts. They reported effective bacteria killing with no loss of human cell vitality (Soukos et al. 1996). Soncin et al. reported selective killing of *S. aureus* and no effect on human fibroblasts and keratinocytes (four to sixfold) when exposed to APDT and cationic phthalocyanine, at relatively low light fluencies (Soncin et al. 2002), and George and Kishen demonstrated a 97.7 % killing for *E. faecalis* and a 30 % human fibroblast dysfunction rate with APDT and methylene blue (George and Kishen 2007a). Tissue toxicity becomes a serious issue when a high concentration or volume of photosensitizer is applied in an attempt to improve the treatment outcome. Tissue toxicity can also occur with the use of initially nontoxic photosensitizers if they interact with target tissue in a way that gives rise to interaction by-products, which are toxic. This indirect form of tissue toxicity has yet to be fully explored, particularly within the framework of human applications of this technology. In spite of all these potential side effects, there are, at present, several photosensitizers useful in APDT that are in various stages of testing for FDA approval.

Compounds based on a phenothiazinium chromophore are emerging as promising candidates for use as an APDT photosensitizer (George and Kishen 2008a). Methylene blue and TBO, members of the phenothiazinium group of photosensitizers, have already gained acceptance for clinical use (Wainwright and Crossley 2002). Phenothiaziniums are generally cationic molecules, with a core structure composed of a planar tricyclic aromatic ring system that functions as the chromophore (Wainwright and Giddens 2003). In addition to phenothiaziniums, cationic porphyrins (Merchat et al. 1996), phthalocyanines (Minnock et al. 1996), and chlorins (Hamblin et al. 2002) have also gained acceptance as photosensitizers, because of their ability to inactivate gram-positive and gram-negative bacteria. Methylene blue, TBO, rose bengal, erythrosine, chlorin (e6), and hematoporphyrin are currently being investigated for use in the same way.

Soukos and co-workers proposed that by covalently conjugating a suitable photosensitizer to a *poly-L-lysine* chain, a bacteria-targeted photosensitizer delivery vehicle could be constructed that would effectively kill gram-positive and gram-negative microorganisms (Soukos et al. 1997). They tested their hypothesis against selected species of these microorganisms, by preparing a conjugate of chlorin (e6) and a *poly-L-lysine* chain (20 lysine residues) and illuminating them after 1 min of incubation with red light. This produced a killing rate of >99 % for gram-positive *Actinomyces viscosus* and gram-negative *Porphyromonas gingivalis* (Soukos et al. 1997). Later, Polo et al. used a conjugate of *poly-L-lysine* and porphycenes and reported a significant killing of gram-negative microorganisms (Polo et al. 2000) and Hamblin et al. used a *poly-L-lysine*-c_{e6} conjugate with a chain length of lysine, to kill selected gram-positive and gram-negative species (Hamblin et al. 2002). Nanoparticles are ideal carriers for photosensitizer molecules in APDT. The combination of nanoparticles and photosensitizer molecules has led to a new and expanding field of interdisciplinary research. Nanomaterials, such as TiO_2, ZnO, as well as fullerenes and their derivatives, can generate the singlet oxygen necessary for killing bacteria. Nanoparticles are also combined with photosensitizer molecules to enhance their action. Different strategies used with this technology are shown in Fig. 3. Basically they are (1) photosensitizer supplemented with nanoparticles; (2) photosensitizer encapsulated within nanoparticles, (3) photosensitizer bound or loaded to nanoparticles, and (4) nanoparticles that serve as a photosensitizer. The effectiveness of APDT against *E. faecalis* biofilm and human dental plaque bacteria has been investigated in vitro using methylene blue-loaded poly(lactic-co-glycolic) (PLGA) nanoparticles, activated with red light (665 nm). These cationic methylene blue-loaded nanoparticles exhibited bacterial phototoxicity against planktonic and biofilm forms of the bacteria. The nanoparticles were found to be concentrated on the bacterial cell walls at all test time points. It was concluded that the methylene blue-loaded PLGA nanoparticles displayed a potential for use in APDT disinfection of the root canal (Klepac-Ceraj et al. 2011; Pagonis et al. 2010).

Light sources used in APDT can be coherent (lasers) or noncoherent (lamps). The choice of light source is determined by the nature of the site, the required light dose, and the choice of photosensitizer. Lasers provide monochromatic, coherent, and collimated light, and a wide range of output power. Laser light can be easily coupled into a fiber optic cable, which can serve as a delivery system (probe) for irradiating complex anatomy, such as that found in the root canal. Nd:YAG, KTP, HeNe, GaAlAs, and diode lasers, light emitting diodes, and xenon arc lamps have all been employed in APDT. To date, none has shown superiority over the other (Prasad 2003).

Fig. 3 Schematic diagram showing different methods of combining nanoparticles and photosensitizers (Kishen 2010)

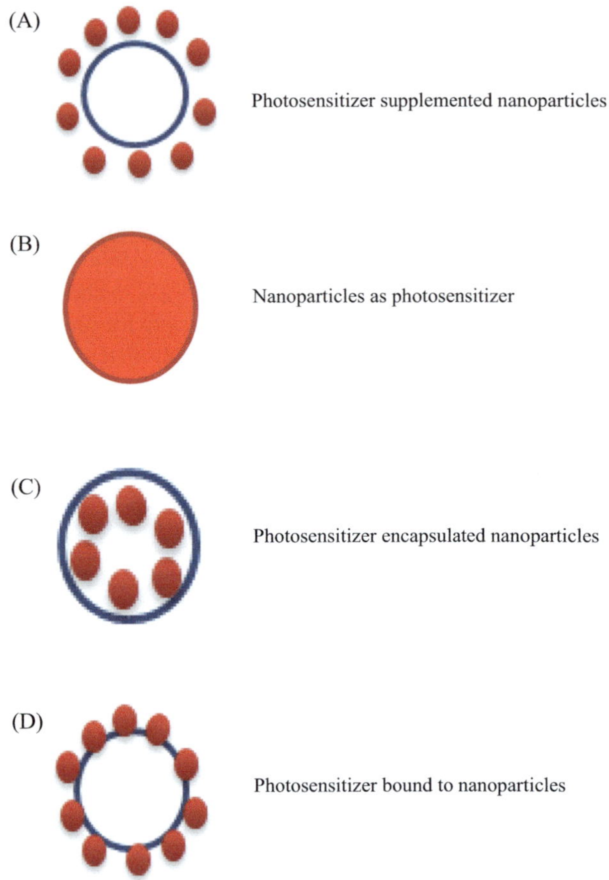

There have been numerous in vitro studies reported in the literature regarding the use of APDT in root canal disinfection. Meire et al. (2009) and George and Kishen (2007b, 2007b, 2008b) showed that *E. faecalis* could be effectively killed by APDT using methylene blue and TBO as a photosensitizer and red light as an activator. Soukos et al. tested APDT using methylene blue as a photosensitizer on a broad range of endodontic pathogens and reported it was effective in eliminating all the test organisms with the exception of *E. faecalis,* which in part survived (Soukos et al. 2006). In another study, Williams et al. used TBO and red light and reported a significant reduction in the bacterial load for suspensions of *S. intermedius*, *P. micros*, *P. intermedia*, and *F. nucleatum* (George and Kishen 2007b). The application of this technology in clinical endodontics still has to overcome several hurdles.

The effective use of APDT in root canal disinfection is limited by several *tissue-specific* factors. These factors are (1) the penetration of the activating light energy into the infected tissue, (2) the penetration of optimum photosensitizer concentration into the infected tissue, (3) the limited availability of environmental oxygen in the infected tissue, and (4) the ability of excess photosensitizer to produce dentin discoloration. Light propagation through tissue is influenced by processes dealing with reflection, absorption, scattering, and transmission. In general, approximately 4–6 % of light is lost to reflection. Light is also absorbed by tissue. This is mainly due to the presence of free water molecules, proteins, pigments, and other macromolecular constituents of the tissue. Different tissues have different absorption coefficients, which is a reflection of its transparency per unit of length. This creates a gradient, where intensity light at the surface is greater than in deeper areas of the tissue. Since different wavelengths of light possess different abilities to penetrate the tissue "transparency," is also influenced to a great extent by the wavelength of the incoming light/laser irradiation. Light is also scattered in tissue. Scattering has a marked effect on light intensity and directionality. Scattering and refraction causes widening of the light beam, a change in its directionality and a proportional loss in its fluence rate (power per unit area) (Prasad 2003). A loss of light energy in tissue such as dentin can compromise the antimicrobial action when APDT is used. In an attempt to overcome this, George and Kishen dissolved methylene blue in different solutions—water, 70 % glycerol, 70 % polyethylene glycol (PEG), and a mixture of glycerol, ethanol, and water (MIX) in a ratio of 30:20:50—and analyzed their photophysical, photochemical, and photobiological characteristics in a dentin substrate (George and Kishen 2007b). They found that aggregation of methylene blue molecules was significantly higher in water as compared to the other solutions. The MIX-based methylene blue formulation demonstrated effective penetration into dentinal tubules, enhanced singlet oxygen generation, and in turn improved overall bactericidal action. It also proved to be significantly better than methylene blue in causing bacterial cell wall impairment and chromosomal DNA damage (George and Kishen 2008a). George and Kishen also showed that an oxidizer and oxygen carrier was incorporated with the photosensitizer formulation to form an emulsion; it produced a significant increase in the photo-oxidation capabilities and facilitates the comprehensive disruption of a mature endodontic biofilm structure (George and Kishen 2008b).

Prokaryotic and eukaryotic cells possess a family of membrane proteins termed efflux pumps. The efflux pumps remove amphiphilic molecules from cell cytosol. Since many drugs, such as chlorpromazine, azithromycin, and tetracaine, are amphiphilic in nature, efflux pumps can effectively clear them from that location. Efflux is the process by which bacteria transport potentially toxic compounds from the cytosol to outside of the cell (Ryan et al. 2001). Many of these efflux systems have broad substrate profiles that allow structurally diverse drugs, chemicals, and compounds to be extruded to the surface making its expression a virulence characteristic. Efflux pump expression has been shown to be enhanced when bacteria exist as a biofilm bacteria as compared its planktonic state. This makes bacteria in a biofilm potentially more resistant to certain types of antimicrobial treatment. It was speculated that the use of an efflux pump inhibitor (EPI) could overcome this resistance. Tegos and Hamblin showed that phenothiazinium dyes, which are structurally characterized as amphipathic cations, are also substrates of multidrug efflux pumps (MEP). They showed that when inhibitors of bacterial MEP were used in combination with phenothiazinium dyes, the antimicrobial action of APDT could be potentiated (Tegos and Hamblin 2006; Tegos et al. 2008). Since efflux pumps are usually highly active in biofilm, they are strategic targets for antibiofilm treatment (Kvist et al. 2008; Zhang and Mah 2008). This was shown in a study by Kishen et al. who used an EPI in combination with phenothiazinium photosensitizer, to assist in the disruption and destruction of biofilm (Kishen et al. 2010).

The potential of APDT in root canal disinfection has been investigated in vitro by different studies. Results of these investigations are summarized in Table 1 (Bonsor et al. 2006a, b; Garcez et al. 2008, 2010). It was generally concluded that when combined with standard chemomechanical preparation of the root canal, APDT had the potential to significantly reduce the microbial load if it could be optimized for use in a root canal environment. Currently, APDT is not considered to be an optional protocol in root canal disinfection, only a supplement to them. Further research is still necessary in order to maximize its antibiofilm potential. This will involve improving its performance in the presence of tissue inhibitors, optimizing light delivery to the root canal, introducing new photosensitizers (and formulations) for use in the root canal, and defining a clinical approach that will produce consistent results. Research into ways to achieve this end is currently underway and involves combining the use of photodynamic therapy with bioactive micro- (Pagonis et al. 2010) and nanoparticles (Shrestha and Kishen 2011).

2.3 Laser-Assisted Root Canal Disinfection

A laser (**L**ight **A**mplification by **S**timulated **E**mission of **R**adiation) is a device that emits light through a process of optical amplification. The photons in a laser beam are emitted as a coherent, unidirectional, monochromatic light beam that can be collimated into an intensely focused ray of energy. The nature of laser–tissue interaction is influenced by the properties of laser (e.g., wavelength, energy density, and pulse duration) and the optical characteristics of the tissue (e.g., absorption, reflection, transmission, and scattering). Different types of lasers can produce different effects on the same tissue, and the same laser can have varying effects on different tissues. The nature of light absorption by a tissue, and transmission of light through a tissue as mentioned before, is wavelength dependent. It should be noted that light intensity does not remain constant as it passes through a defined volume of tissue; hence, its biological effect changes in accordance with its depth of penetration. Tissue response to laser energy is complex and influenced by the thermal properties of the target tissue as well as the type of laser. These thermal properties are *thermal diffusivity, thermal coefficient of expansion, heat capacity, phase transformation temperatures, and latent heat of transformation.* The clinician selects the laser type which generally has the ability to control the (a) applied power (power density), (b) total energy applied over a given area of tissue (energy

Table 1 Summary of relevant in vivo studies carried out using antimicrobial photodynamic therapy

No	Author/date	Objective and materials	Methodology	Conclusion
1	Bonsor et al. (2006a)	Aimed to evaluate the antimicrobial efficacy of root canal disinfection by combining PDT with conventional endodontic treatment Clinical study: 32 root canals from 14 patients	Irrigation with 20 % citric acid and 2.25 % sodium hypochlorite PDT with TBO and diode laser (12.7 mg/L-1, 100 mW, 120 s) Samples collected by filing	Cleaning and shaping resulted in complete bacterial killing in 86.7 % of samples Combination of cleaning and shaping + PDT resulted in complete bacterial killing in 96.7 % of samples
2	Bonsor et al. (2006b)	Aimed to compare the effect of a combination of 20 % citric acid + PDT with the use of 20 % citric acid + 25 % sodium hypochlorite on bacterial load in prepared root canals Clinical study: 64 patients	Procedure similar to previous study	Combination of 20 % citric acid + PDT resulted in complete bacterial killing in 91 % of samples 20 % citric acid + 2.25 % sodium hypochlorite resulted in complete bacterial killing in 82 % of samples
3	Garcez et al. (2008)	Aimed to analyze the antimicrobial effect of PDT in association with endodontic treatment Clinical study: 20 patients First session of cleaning/shaping + PDT; at the end of the first session, the root canal was filled with $Ca(OH)_2$, and after 1 week, PDT was also performed in the second session	Irrigation with 2.5 % sodium hypochlorite, 3 % hydrogen peroxide and 17 % EDTA PDT with polyethyleneimine (PEI) chlorin (e6 [ce6]) conjugate (2 min, 9.6 J, 240 s) Paper point sampling	First session produced 98.5 % bacterial reduction (1.83 log reduction) Second session produced 99.9 % bacterial reduction (1.14 log reduction) Second session PDT was more effective than the first session
4	Garcez et al. (2010)	Studied the antimicrobial effect of PDT combined with endodontic treatment in patients with necrotic pulp, infected with microflora resistant to antibiotic therapy Clinical study: 30 teeth from 21 patients with periapical lesions that	PDT used polyethylenimine-chlorin (e6) as a photosensitizer and a diode laser (40 mW, 4 min, 9.6 J)	Endodontic therapy alone produced a significant reduction in numbers of microbial species (only 3 teeth were free of bacteria) The combination of endodontic therapy with PDT eliminated all drug-resistant

(continued)

Table 1 (continued)

No	Author/date	Objective and materials	Methodology	Conclusion
		had been treated with conventional endodontic treatment and antibiotic therapy		species and all teeth were bacteria free

density), (c) rate and duration of laser irradiation (pulse repetition), and (d) mode of energy delivery (continuous/pulsed energy; direct/indirect tissue contact) (Miserendino and Robert 1995).

Lasers have been primarily used in root canal treatment to enhance elimination of microorganisms from the root canal system, subsequent to its conventional biomechanical enlargement, as the parameters of laser use for disinfection do not produce ablative effects on dentin (Miserendino and Robert 1995; Moshonov et al. 1995). Infrared lasers such as CO_2, Nd:YAG, diode, and erbium lasers have been the lasers of choice for disinfection. The bactericidal effect of laser depends upon the wavelength and energy level of the laser used, as well as its thermal effect. The laser-induced thermal effect produces an alteration in the bacterial cell wall that changes its osmotic gradient, causing swelling and cell death. It is interesting to also note that gram-negative bacteria have shown a higher resistance to destruction by laser irradiation than have gram-positive bacteria. This may be due to differences in the cell wall structure (Miserendino and Robert 1995).

When used in the root canal, the depth of penetration of the laser into the dentin will depend upon the other parameters, such as wavelength and power density. As previously mentioned, the depth of penetration generally decreases with an increase in the degree of tissue absorption. The delivery of laser energy throughout the root canal system and absorption of laser energy by the dentin are important considerations in laser root canal disinfection, since they influence not only the degree to which biofilm can be destroyed but also the degree to which dentin may be structurally and physically altered.

Laser killing of biofilm bacteria was investigated in an in vitro study by Rooney et al. in which an energy absorber, either black Indian ink or 38 % silver ammonium solution, was placed into an infected root canal before irradiating with it with a pulsed Nd:YAG laser (1064 nm). They reported bacterial killing rates in the main root canal in a range of 80–90 % (Rooney et al. 1994). Schoop et al. in another in vitro study compared the antibacterial potential of two lasers, the Nd:YAG and the Diode laser (810 nm). They found that the Nd:YAG laser reduced the bacterial load by 85 % to a depth of 1 mm in the dentin, while the diode laser (810 nm) reduced the load by only 63 % and to a depth of 750 μm in the dentin. The difference in laser penetration and bacterial killing was attributed to differences in the degree of absorption of the wavelengths emitted by the two laser types (Schoop et al. 2004). Bergmans et al. studied the effect of Nd:YAG laser irradiation on several endodontic pathogens in vitro. They found that biofilms containing these pathogens were difficult to eradicate, even with direct exposure to laser light. This led them to conclude that Nd:YAG laser irradiation should be a supplement and not an alternative to existing root canal disinfection protocols (Bergmans et al. 2006). Stabholz et al. studied the bactericidal potential of an erbium laser in a root canal model and found it to be inferior to that of Nd:YAG laser. Thermal energy produced by the erbium laser has a high affinity for water molecules; therefore, most of its energy is absorbed on the surface making it less active below it (Wang et al. 2007).

There are several limitations associated with the intracanal use of high-power laser energy. The emission of laser energy from the tip of an optical fiber or laser guide is directed downwards into the root canal and not laterally toward the root canal walls where biofilm is also present. Thus, it is almost impossible to obtain uniform killing throughout the entire root canal let alone in its anatomical complexities (Goodis et al. 2002; Stabholz et al. 2003). Safety is another limitation, particularly in regard to the potential for thermal damage to the dentin and the periapical tissues. This effect can be dangerous in teeth with close proximity to vital structures such as the mental foramen or the mandibular nerve (Stabholz et al. 2003). A modified beam delivery system has been tested for Er:YAG laser to deflect direct transmission of laser energy from the laser tip or fiber optic to the apex. This system consists of a hollow tube that permits lateral emission of Er:YAG laser radiation (side firing), rather than straight-line emission through the single opening at its terminal end, that was previously used (Stabholz et al. 2003). This new endodontic side-firing spiral tip was designed to comply with the shape and volume of root canals prepared by rotary nickel–titanium instrumentation. The tip is also sealed at its distal end, to prevent transmission of irradiation to and through the apical foramen (George and Walsh 2011). This design was introduced to enhance the ability of the laser to penetrate into and destroy microbes on the lateral walls of the root canal and in the dentinal tubules. Its clinical effectiveness is yet to be reported.

Noiri et al. investigated the antibiofilm effect of Er:YAG laser on in vitro monospecies biofilm of *A. naeslundii*, *E. faecalis*, *L. casei*, *P. acnes*, *F. nucleatum*, *P. gingivalis*, and *P. nigrescens* grown on hydroxyapatite disks for 21 days (aerobically for 7 days and anaerobically for 14 days). They reported that despite significant reductions in the numbers of viable cells in most of the biofilms tested, complete elimination of biofilm structure/bacteria did not occur with any of the Er:YAG laser irradiation energy densities tested. Interestingly, during growth, the *L. casei* decalcified the HA disks to a depth of approximately 200 μm and then invaded the porous decalcified layer, where they remained protected for the laser irradiation (Noiri et al. 2008). It has already been noted that the antibiofilm potential of the Er:YAG laser is influenced by the water content, components of extracellular matrix, cell density, and the absorption coefficient of the dentin. The temperature increase during irradiation can range from 7.3 °C at 20 mJ to 40.2 °C at 80 mJ.]. Yavari examined the ability of high-power settings of Er,Cr:YSGG laser irradiation (2 W and 3 W output powers for 16 s) to eradicate monospecies biofilm of *E. faecalis* (48 h) in vitro. He concluded that though 2- and 3-W of Er,Cr:YSGG laser could kill *E. faecalis* in root canal models, its effect was less remarkable than that produced by NaOCl (Yavari et al. 2010). It is important to realize that surface degradation, microbial penetration into the dentin, as well as the presence of biofilm in inaccessible site are common occurrences in root canal infection and that if this technology is to have clinical applicability, it must be safe and effective under those conditions.

Earlier studies have studied potential changes in the ultrastructure of root canal dentin as a concomitant/adverse effect to root canal disinfection using different lasers. It was noted that when lasers were used in a dry root canal, both near- and mid-infrared lasers produce characteristic thermal injury to the dentin. Human dentin has low absorption coefficients in the near-infrared range; nonetheless, Nd: YAG laser irradiation is still capable of melting the dentin surface (Fried et al. 1995). Moriyama et al. showed morphological changes could be produced on the dentin of endodontically prepared root canals by Nd:YAG laser irradiation. The changes were thermally related and appeared as cracks in the dentin, partial removal of smear layer, and sealing of many dentinal tubules due to melting of the dentin. Longer pulses of laser delivery caused more severe and more extensive melting. While the thought of eventually creating a more regular surface with sealed tubules by increasing the number of pulses appeared desirable, there were fears that the high number of thermal cycles necessary to achieve this effect could lead to deeper dentin cracks (Moriyama et al. 2004). During the photothermal interaction, tissue molecules absorb photons and generate heat that is dissipated through the tissue. Since tissue needs time to propagate the heat, longer laser pulses result in the production of higher temperatures in deeper regions of the tissue. With shorter pulses that use the same average energy, the higher temperatures develop at the surface (Armon and Laufer 1995). Thermal damage to tissue is a temperature/time-dependent process. The resultant confinement of thermal stress depends upon the laser pulse duration and the tissue absorption coefficient (μa). The use of longer pulses leads to longer interaction times and results in more evidence of a thermal effect (van Leeuwen et al. 1995). In separate studies, Marchesan et al. and Gurbuz et al. showed that the presence of water or irrigation solution in the root canal limits thermal interaction of laser beam from a diode laser (2.5 W, 15 Hz) and Nd:YAG laser (1.5 W, 100 mJ, 15 Hz) with the dentin on the root canal wall (Marchesan et al. 2008; Gurbuz et al. 2008). It has also been shown that the presence of water in root canal space prevents dentin damage when an erbium laser is used (Yamazaki et al. 2001). Several relevant clinical studies that examined the antimicrobial efficacy of high-power laser in endodontics are summarized in Table 2 (Koba et al. 1999; Dostálová et al. 2002; Leonardo et al. 2005). At present there is no strong evidence available to support the use of high-power lasers in endodontic disinfection.

Application of the fundamentals of liquid –laser interaction has created a new area of research. This principle is the basis for laser-activated irrigation (LAI) and photon-initiated photoacoustic streaming (PIPS) as an approach to root canal irrigation and disinfection (Kimura et al. 2011; Blanken et al. 2009; De Moor et al. 2009). The interaction of Er,Cr:YSGG laser with liquid irrigant in the root canal has been attributed to the efficient absorption of the mid-infrared wavelength light by water. This leads to the vaporization of irrigant and formation of vapor

Table 2 Summary of relevant in vivo studies carried out using laser-assisted disinfection

No:	Author/date	Objective and materials	Methodology	Conclusion
1	Koba et al. (1999)	Evaluated the postoperative symptoms and healing after root canal treatment with pulsed Nd:YAG laser 44 teeth from 38 patients Radiological evaluation used to assess the reduction of periapical lesions at 3 and 6 months	Nd:YAG (1 W, 15 pps for 1 s) 5 % NaOCl and 3 % H2O2 used to disinfect (control)	No significant differences was found between the groups regarding periapical healing
2	Dostálová et al. (2002)	Studied the ability of Er:YAG laser radiation with a movable waveguide to disinfect root canals Root canal of 44 premolars and molars were treated using a step-back technique; 10 teeth were then treated with calcium hydroxide, and 22 teeth were irradiated with the waveguide	5.25 % NaOCl used to disinfect (control) Er:YAG (100 mJ, 30 pulses, repetition rate 4 Hz) Before and after treatment, the colony-forming units were counted to determine 21 different microorganisms	Conventional treatment was effective in 60 % of the root canal Application of calcium hydroxide was effective in 80 % of the root canal Er:YAG laser irradiation via movable waveguide was effective in 100 % of the root canal
3	Leonardo et al. (2005)	Evaluated the antimicrobial effect of Er: YAG laser applied after cleaning and shaping of root canals of dog's teeth with apical periodontitis 40 root canals of dogs' premolar teeth with periapical lesions were used	Group I –cleaning and shaping only Group II –cleaning and shaping and Er:YAG laser application (63-mJ output/15 Hz) After coronal sealing, the root canals were left empty for 7 days before microbiological analysis	Er:YAG laser applied after cleaning and shaping did not reduce microorganisms in the root canal system

bubbles, which expand and implode with secondary cavitation effects that induce high-speed fluid motions into and out of the root canal. The thermal component of this interaction is moderate in intensity. The creation of bubbles is similar in both water and sodium hypochlorite solution. If the liquid does not absorb radiation, there are no bubbles, cavitation, pressure buildup, or fluid motion. At the beginning of the laser pulse, the laser energy is absorbed in a 2 mm-thick layer on liquid, which is instantly superheated to a boiling temperature at high pressure and converted into vapor. This vapor, at high pressure, expands at high speed and provides an opening in front of the fiber transmitting the laser light. As the laser continues to emit energy, the light passes through the bubble and evaporates the water surface in front of the bubble, driving a channel through the liquid, until the pulse ends (Kimura et al. 2011; Blanken et al. 2009).

The laser system used for PIPS/LAI is equipped with a novel 400 mm diameter radial and stripped tip. Subablative parameters (average power 0.3 W, 20 mJ at 15 Hz) are used to produce photomechanical effects, when light energy is pulsed in liquid. When activated in a limited volume of fluid, the high absorption of the Er: YAG wavelength in water, combined with the high peak power derived from the short pulse duration that is used (50 μs), results in a photomechanical phenomenon (De Moor et al. 2009). Earlier it was demonstrated that middle-infrared laser energy when delivered with conical modified fibers influences the configuration of the shock wave and subsequently improves the photomechanical effect in the root canal (George and Walsh 2011). Peters et al. studied the efficacy of laser-activated and ultrasonically activated root canal disinfection with conventional irrigation, specifically its ability to remove in vitro 3-week-old bacterial biofilm formed on root canal walls. This study showed that the laser-activated disinfection did not completely remove bacterial biofilms from the apical third of the root canal or from infected dentinal tubules. However, the fact that laser activation generated more bacteria-free samples and left less bacteria/biofilm in the apical canal than ultrasonic activation warrants further investigation (Peters et al. 2011).

Thus it could be concluded that the current evidence is insufficient to recommend laser therapies as an adjunct to chemomechanical disinfection of infected root canals. This does not imply that lasers should not be used as an adjunct to conventional chemomechanical preparation, but instead emphasize the need for future high-quality research in this field before it acts as a “frontline” protocol.

2.4 *Ozone*

Ozone (O_3) is an energized, unstable gaseous form of oxygen, which dissociates readily back into oxygen (O_2), liberating a reactive form of oxygen, called singlet oxygen (O_1). The singlet oxygen is highly reactive and is capable of oxidizing cells. Ozone has been suggested as a means of destroying microorganisms without promoting the development of drug resistance (Restaino et al. 1995; Paraskeva and Graham 2002). Ozone gas (HealOzone, KaVo, Biberach, Germany) is currently being used clinically for that purpose in endodontic treatment. However, studies have yielded inconsistent results regarding its efficacy in destroying endodontic pathogens. This inconsistency has been attributed to the lack of information about the optimum duration of application and concentration of the ozone to be used (Nagayoshi et al. 2004; Arita et al. 2005; Hems et al. 2005). An ozone gas with a concentration of 4 g m^3 is currently being used in endodontics for root canal disinfection. At that concentration it is slightly less cytotoxic than NaOCl (2.5 %) to periapical tissue and oral mucosa. In vitro studies have shown that aqueous ozone (up to 20 μg mL-1) is essentially nontoxic to oral cells (Estrela et al. 2007; Ebensberger et al. 2002; Noguchi et al. 2009).

Hems et al. evaluated the antibacterial potential of ozone against *E. faecalis*, in planktonic and biofilm cultures (48-h-old biofilm grown on cellulose nitrate membrane filter). Interaction times that ranged from 30 s to 240 s were used in both instances. It was concluded that ozone had an antibacterial effect on the planktonic form of *E. faecalis* and those suspended in fluid, but little effect when it was present as a biofilm. Its overall antibacterial efficacy was less than that of sodium hypochlorite, under the test conditions of the study (Hems et al. 2005; Estrela et al. 2007). Huth et al. assessed the antimicrobial efficacy of aqueous (1.25–20 μg mL^{-1}) and gaseous ozone (1 53 g m^{-3}) as an antiseptic against endodontic pathogens in suspension and in a biofilm. *E. faecalis*, *Candida albicans*, *Peptostreptococcus micros*, and *Pseudomonas aeruginosa* were grown for 3 weeks in planktonic culture and as monospecies biofilm in root canals of extracted teeth and then exposed to the two forms of ozone. It was concluded that high concentrations of gaseous and aqueous ozone had a dose-, strain-, and time-dependent effect in destroying the test microorganisms in suspension and biofilm form (Huth et al. 2006). In another study, the antimicrobial efficacy of dissolved ozone was tested against planktonic and biofilm models of *Pseudomonas fluorescens*. It was observed that even at low concentrations, ozone (0.1 ± 0.3 ppm) could completely kill the test bacteria in planktonic form after 15 or 30 min of contact time. Killing in the biofilm models was less effective and resulted only in a decrease of approximately two orders of magnitude, when compared to the planktonic model. No significant increase in the antibiofilm action was observed with an increase in the contact time (Viera et al. 1999). Kuştarci et al. evaluated the antimicrobial potential of potassium titanyl phosphate (KTP) laser and gaseous ozone in experimentally infected root canals. They reported that both KTP laser and gaseous ozone demonstrated significant antibacterial action. The gaseous ozone proved to be more effective than the KTP laser, but neither was as effective as 2.5 % NaOCl (Kuştarci et al. 2009). A summary of relevant antimicrobial studies carried out to using ozone is shown in Table 3.

The reduced effectiveness of ozone against sessile bacteria when compared to planktonic bacteria appears to be attributable to several different factors (Viera et al. 1999): (1) The EPS in biofilm structure may form a physical/chemical barrier that prevents deep penetration of the dissolved ozone into the biofilm (Stoodley et al. 1994). (2) Biofilm has microbial colonies surrounded by water channels through which liquid movement is controlled by convective flow (Stoodley et al. 1994; Rasmussen and Lewandowski 1998). Blockage of these biofilm channels by oxidation products of ozone can impede the penetration of ozone into the inner layers of the biofilm (Lawrence et al. 1994). (3) There is the presence of phenotypically altered microbial communities that possess enhanced resistance to antimicrobials in the deeper aspects of the biofilm. A recent study has also suggested that ozone dissolved in oil can be used as root canal medicament during treatment (Silveira et al. 2007). However, questions on the flow characteristics of ozonized oil, the chemical stability of ozonized oil, and the nature, if any, of its

Table 3 Summary of relevant antimicrobial studies carried out to using ozone

No	Author/date	Objective/methodology	Authors/conclusion
1	Nagayoshi et al. (2004)	Effect of ozonated water against *E. faecalis*- and *S. mutans*-infected dentin of bovine incisors	Ozonated water application might be useful for root canal irrigation
2	Hems et al. (2005)	To evaluate the potential of ozone as an antibacterial agent using *E. faecalis*. The antibacterial efficacy of ozone was tested against both broth and biofilm cultures. Ozone was sparged for 30, 60, 120, and 240 s	NaOCl was found to be superior to ozonated water in killing *E. faecalis* in broth culture and in biofilms
3	Estrela et al. (2007)	To determine the antimicrobial efficacy of ozonated water, gaseous ozone, 2.5 % sodium hypochlorite, and 2 % chlorhexidine in human root canals infected by *E. faecalis* for 60 days	The irrigation of infected human root canals with ozonated water, 2.5 % NaOCl, and 2 % chlorhexidine and the application of gaseous ozone for 20 min was not sufficient to inactivate *E. faecalis*
4	Kuştarci et al. (2009)	Evaluated the antimicrobial activity of potassium titanyl phosphate (KTP) laser and gaseous ozone in experimentally infected root canals (*E. faecalis* for 24 h)	2.5 % NaOCl was superior in its antimicrobial abilities compared with KTP laser and gaseous ozone

interaction with root dentin/obturating materials have to be answered before advocating its routine clinical use (Ng 2004; Guinesi et al. 2011). A systematic review by Azarpazhooh and Limeback highlighted conflicting evidence on the antimicrobial efficacy, making the routine use of ozone questionable (Azarpazhooh and Limeback 2008).

2.5 Herbal and Enzyme Alternatives

Recent trends in antibiofilm research point toward the possible use of natural extracts from plants in the treatment of biofilm-mediated infection. Polyphenols are naturally present in a variety of plants (Bravo 1998; Duthie and Crozier 2000). They are characterized by the presence of more than one phenol units per molecule (Bravo 1998). These substances are well known for their antimicrobial activity. For example, anacardic acid found in the liquid extract of cashew nut shell has been shown to exhibit antimicrobial activities against *Streptococcus mutans* and *Staphylococcus aureus* (Kubo et al. 1992, 2003). Other polyphenols have been used for the preservation of food against microbial contaminants. Many factors such as bacterial species/strains, type of polyphenol, concentration of polyphenol, microbial cells density, temperature, and synergistic and inhibitive effect when used with other antimicrobials can influence their antimicrobial potential.

Morinda citrifolia (MCJ) is a herb, which has a broad range of antibacterial, antiviral, antifungal, analgesic, anti-inflammatory, and immune-enhancing effects (Li et al. 2003; Wang et al. 2002). MCJ contains the antibacterial compounds L-asperuloside and alizarin. An in vitro study that investigated the antimicrobial activity of 2 % chlorhexidine gel, propolis, MCJ, 2 % povidone-iodine (POV-I), and calcium hydroxide on *E. faecalis*-infected root dentin reported that chlorhexidine gluconate produced the best antimicrobial effect (100 %), followed by 2 % POV-I (87 %), propolis (71 %), MCJ (69 %), and calcium hydroxide (55 %) (Kandaswamy et al. 2010). Another in vitro study compared the effectiveness of MCJ with sodium hypochlorite and chlorhexidine gluconate to remove the smear layer from the canal walls of endodontically instrumented teeth and reported that its efficacy was comparable to that of NaOCl in conjunction with EDTA (Murray et al. 2008).

Turmeric (*Curcuma longa*) has been used extensively as a food preservative in Southeast Asia and has been used in traditional medicine for the treatment of numerous diseases. Curcumin (diferuloylmethane), the main bioactive component of turmeric, has been shown to have a wide spectrum of biological actions, including antisepsis, anti-inflammatory, and antioxidative (Cowan 1999). A recent report suggests that curcumin in aqueous form exhibits phototoxic effect against gram-positive and gram-negative bacteria (Rukkumani et al. 2003). Triphala, another traditional preparation, consists of dried and powdered fruits of three medicinal plants: *Terminalia bellerica*, *Terminalia chebula*, and *Emblica officinalis* in equal proportions. Triphala can reportedly kill 100 % of *E. faecalis* in culture in 6 min. In formulations such as this, different plants can enhance the potency of the compound, through an additive or synergistic effect (Prabhakar et al. 2010). Green tea polyphenols are derived from the young shoots of the tea plant *Camellia sinensis*. Green tea polyphenols have also shown statistically significant antibacterial action against *E. faecalis* biofilm present on tooth substrate (Prabhakar et al. 2010). Phenol and natural phenolic compounds except ethyl linoleate and tocopherol have also shown a significant ability to reduce biofilm formation by *P. aeruginosa* but not eradicate it, at the concentration level used in the study. However, the exact mechanism by which it accomplished this was not reported (Jagani et al. 2009).

The major advantages of using herbal products are that they are readily available, are cost-effective, have long shelf life, have low toxicity, and do not initiate microbial resistance. Combination of natural polyphenol with nanoparticles and photodynamic therapy may open newer vistas in bacteria-specific killing (targeted bacterial killing) without undue effects on healthy tissues and mammalian cells. At the moment however, that remains speculative. Table 4 summarizes the major class of antimicrobial compounds extracted from plants (Cowan 1999).

Enzymes have been suggested for the removal of biofilm components, especially EPS from inanimate surfaces such as orthopedic implants (Donelli et al. 2007). Two carbohydrate-containing moieties of staphylococcal biofilms, a linear poly-b-(1–6)-N-acetyl-D-glucosamine (PNAG) and teichoic acid, have been targeted using enzymes such as dispersin B and proteinase K (Donelli et al. 2007; Chaignon

Table 4 Summary of the major class of antimicrobial compounds from plants [164]

No	Class	Example(S)	Mechanism
1	Phenolics	Catechol	Membrane disruption
		Hypericin	Bind to adhesins, complex with cell wall, inactivate enzymes
		Warfarin	Bind to adhesins
			Interaction with eukaryotic DNA (antiviral activity)
2	Terpenoids, essential oils	Capsaicin	Membrane disruption
3	Alkaloids	Berberine	Intercalate into cell wall and/or DNA
		Piperine	
4	Lectins and polypeptides	Mannose-specific agglutinin	Block viral fusion or adsorption

et al. 2007; Sadovskaya et al. 2006). These studies have shown that rinsing the implant surface with these enzymes can prevent formation of staphylococcal biofilms. Yet, applying these enzymes in an in vivo environment that is home to multispecies pathogenic biofilms may pose some challenges due to the specificity of their action. Further studies are required to evaluate whether these enzymes work synergistically or additively with other antibiofilm strategies. In addition, the effect of these enzymes on a biological substrate and their mechanism of action have to be better understood before implementing such treatment strategies clinically.

3 Concluding Remarks

Current understanding emphasizes that endodontic disease is a biofilm-mediated infection and that elimination of bacterial biofilm from the root canal system and the exterior root surface may be necessary to maximize the prospects for a favorable treatment outcome in the management of this disease. Unfortunately, the root canal environment is a challenging locale to accomplish this goal. It has been for this reason that different protocols ranging from antimicrobial root canal irrigation to advanced methods that incorporate lasers, photoactivated disinfection, and nanoparticles have been tried. While several of the advanced antimicrobial protocols have shown a significant inhibitory effect against several types of microbial biofilms under in vitro conditions, more in vivo studies are required to evaluate whether they are clinically practical and effective. These studies will also have to verify their safety and their biocompatibility with dental and dentally related tissue. The most promising direction of this research appears to be the development of strategies aimed at the disruption of biofilm and prevention of biofilm formation. In bringing this new technology to the marketplace, it is important to combine potent antibiofilm strategies with an effective and easy-to-use delivery system, if it is to enjoy broad acceptance. Those engaged in research in this field feel confident that the coming years will herald a new and effective approach to treat biofilms in endodontics.

References

Al Cunningham, RR, Stewart P, Camper A, Stoodley P, Lennox J, Anderson V (2011) Center for Biofilm Engineering, Montana State University, Biofilms: hypertext book. http://www.hypertextbookshop.com/biofilmbook/v004/r003/2011

An YH, Friedman RJ (1998) Concise review of mechanisms of bacterial adhesion to biomaterial surfaces. J Biomed Mater Res 43:338–348

An YH, Stuart GW, McDowell SJ, McDaniel SE, Kang Q, Friedman RJ (1996) Prevention of bacterial adherence to implant surfaces with a crosslinked albumin coating in vitro. J Orthop Res 14:846–849

Anderl JN, Franklin MJ, Stewart PS (2000) Role of antibiotic penetration limitation in Klebsiella pneumoniae biofilm resistance to ampicillin and ciprofloxacin. Antimicrob Agents Chemother 44:1818–1824

Arita M, Nagayoshi M, Fukuizumi T, Okinaga T, Masumi S, Morikawa M, Kakinoki Y, Nishihara T (2005) Microbicidal efficacy of ozonated water against Candida albicans adhering to acrylic denture plates. Oral Microbiol Immunol 20:206–210

Armon E, Laufer G (1995) Analysis to determine the beam parameters which yield the most extensive cut with the least secondary damage. J Biomech Eng 107:286–290

Azarpazhooh A, Limeback H (2008) The application of ozone in dentistry: a systematic review of literature. J Dent 36:104–116

Basrani BR, Manek S, Sodhi RN, Fillery E, Manzur A (2007) Interaction between sodium hypochlorite and chlorhexidine gluconate. J Endod 33:966–969

Baumgartner JC, Siqueira JF, Sedgley CM, Anil K (2008) Microbiology of endodontic disease. In: Ingle JI, Bakland LK, Baumgartner JC (eds) Ingle's endodontics, 6th edn. BC Decker, Ontario, pp 221–222

Beard SJ, Hughes MN, Poole RK (1995) Inhibition of the cytochrome bd-terminated NADH oxidase system in Escherichia coli K-12 by divalent metal cations. FEMS Microbiol Lett 131:205–210

Bergmans L, Moisiadis P, Teughels W, Van Meerbeek B, Quirynen M, Lambrechts P (2006) Bactericidal effect of Nd:YAG laser irradiation on some endodontic pathogens ex vivo. Int Endod J 39:547–557

Bertoloni G, Lauro FM, Cortella G, Merchat M (2000) Photosensitizing activity of hematoporphyrin on Staphylococcus aureus cells. Biochim Biophys Acta 1475:169–174

Blanken J, De Moor RJ, Meire M, Verdaasdonk R (2009) Laser induced explosive vapor and cavitation resulting in effective irrigation of the root canal. Part 1: a visualization study. Lasers Surg Med 41:514–519

Bonsor SJ, Nichol R, Reid TM, Pearson GJ (2006a) Microbiological evaluation of photo-activated disinfection in endodontics (an in vivo study). Br Dent J 25:337–341

Bonsor SJ, Nichol R, Reid TM, Pearson GJ (2006b) An alternative regimen for root canal disinfection. Br Dent J 201:101–105

Bravo L (1998) Polyphenols: chemistry, dietary sources, metabolism, and nutritional significance. Nutr Rev 56:317–333

Busscher HJ, van der Mei HC (1997) Physico-chemical interactions in initial microbial adhesion and relevance for biofilm formation. Adv Dent Res 11:24–32

Busscher HJ, Bos R, van der Mei HC (1995) Initial microbial adhesion is a determinant for the strength of biofilm adhesion. FEMS Microbiol Lett 128:229–234

Chaignon P, Sadovskaya I, Ragunah C, Ramasubbu N, Kaplan JB, Jabbouri S (2007) Susceptibility of staphylococcal biofilms to enzymatic treatments depends on their chemical composition. Appl Microbiol Biotechnol 75:125–132

Cioffi N, Ditaranto N, Torsi L, Picca RA, Sabbatini L, Valentini A, Novello L, Tantillo G, Bleve-Zacheo T, Zambonin PG (2005) Analytical characterization of bioactive fluoropolymer ultra-thin coatings modified by copper nanoparticles. Anal Bioanal Chem 381:607–616

Costerton JW (1999) Introduction to biofilm. Int J Antimicrob Agents 11:217–221

Costerton JWLZ, DeBeer D, Caldwell D, Korber D, James G (1994) Biofilms, the customized microniche. J Bacteriol 176:2137–2142

Cowan MM (1999) Plant products as antimicrobial agents. Clin Microbiol Rev 12:564–582

Dai T, Huang YY, Hamblin MR (2009) Photodynamic therapy for localized infections–state of the art. Photodiagnosis Photodyn Ther 6:170–188

De Moor RJ, Blanken J, Meire M, Verdaasdonk R (2009) Laser induced explosive vapor and cavitation resulting in effective irrigation of the root canal. Part 2: evaluation of the efficacy. Lasers Surg Med 41:520–523

del Pozo JL, Patel R (2007) The challenge of treating biofilm-associated bacterial infections. Clin Pharmacol Ther 82:204–209

Denyer SP, Maillard JY (2002) Cellular impermeability and uptake of biocides and antibiotics in Gram-negative bacteria. J Appl Microbiol 92:35S–45S

Donelli G, Francolini I, Romoli D, Guaglianone E, Piozzi A, Ragunath C, Kaplan JB (2007) Synergistic activity of dispersin B and cefamandole nafate in inhibition of staphylococcal biofilm growth on polyurethanes. Antimicrob Agents Chemother 51:2733–2740

Dostálová T, Jelínková H, Housová D, Sulc J, Nemeć M, Dusková J, Miyagi M, Krátky M (2002) Endodontic treatment with application of Er:YAG laser waveguide radiation disinfection. J Clin Laser Med Surg 20:135–139

Dunne WM Jr, Mason EO Jr, Kaplan SL (1993) Diffusion of rifampin and vancomycin through a Staphylococcus epidermidis biofilm. Antimicrob Agents Chemother 37:2522–2526

Duthie G, Crozier A (2000) Plant-derived phenolic antioxidants. Curr Opin Clin Nutr Metab Care 3:447–451

Ebensberger U, Pohl Y, Filippi A (2002) PCNA-expression of cementoblasts and fibroblasts on the root surface after extraoral rinsing for decontamination. Dent Traumatol 18:262–266

Estrela C, Estrela CR, Decurcio DA, Hollanda AC, Silva JA (2007) Antimicrobial efficacy of ozonated water, gaseous ozone, sodium hypochlorite and chlorhexidine in infected human root canals. Int Endod J 40(2):85–93

Feng QL, Wu J, Chen GQ, Cui FZ, Kim TN, Kim JO (2000) A mechanistic study of the antibacterial effect of silver ions on Escherichia coli and Staphylococcus aureus. J Biomed Mater Res 52:662–668

Fried D, Glena RE, Featherstone JD, Seka W (1995) Nature of light scattering in dental enamel and dentin at visible and near-infrared wavelengths. Appl Opt 34:1278–1285

Friedman S (2002) Prognosis of initial endodontic therapy. Endod Top 2:59–88

Garcez AS, Nuñez SC, Hamblin MR, Ribeiro MS (2008) Antimicrobial effects of photodynamic therapy on patients with necrotic pulps and periapical lesion. J Endod 34:138–142

Garcez AS, Nuñez SC, Hamblim MR, Suzuki H, Ribeiro MS (2010) Photodynamic therapy associated with conventional endodontic treatment in patients with antibiotic-resistant microflora: a preliminary report. J Endod 36(9):1463–1466

George S, Kishen A (2007a) Advanced noninvasive light-activated disinfection: assessment of cytotoxicity on fibroblast versus antimicrobial activity against Enterococcus faecalis. J Endod 33:599–602

George S, Kishen A (2007b) Photophysical, photochemical, and photobiological characterization of methylene blue formulations for light-activated root canal disinfection. J Biomed Opt 12:034029

George S, Kishen A (2008a) Influence of photosensitizer solvent on the mechanisms of photoactivated killing of Enterococcus faecalis. Photochem Photobiol 84:734–740

George S, Kishen A (2008b) Augmenting the anti-biofilm efficacy of Advanced Noninvasive Light Activated Disinfection with emulsified oxidizer and oxygen carrier. J Endod 34:1119–1123

George R, Walsh LJ (2011) Performance assessment of novel side firing safe tips for endodontic applications. J Biomed Opt 16:048004

Gilbert P, Allison DG, McBain AJ (2002) Biofilms in vitro and in vivo: do singular mechanisms imply cross-resistance? J Appl Microbiol 92:98S–110S

Goodis HE, Pashley D, Stabholz A (2002) Pulpal effects of thermal and mechanical irritant. In: Hargreaves KM, Goodis HE (eds) Seltzer and Benderís dental pulp. Quintessence, Hanover Park, IL, pp 371–410

Gubler M, Brunner TJ, Zehnder M, Waltimo T, Sener B, Stark WJ (2008) Do bioactive glasses convey a disinfecting mechanism beyond a mere increase in pH? Int Endod J 41:670–678

Guinesi AS, Andolfatto C, Bonetti Filho I, Cardoso AA, Passaretti Filho J, Farac RV (2011) Ozonized oils: a qualitative and quantitative analysis. Braz Dent J 22:37–40

Gurbuz T, Ozdemir Y, Kara N, Zehir C, Kurudirek M (2008) Evaluation of root canal dentin after Nd:YAG laser irradiation and treatment with five different irrigation solutions: a preliminary study. J Endod 34:318–321

Habelitz S, Balooch M, Marshall SJ, Balooch G, Marshall GW Jr (2002) In situ atomic force microscopy of partially demineralized human dentin collagen fibrils. J Struct Biol 138:227–236

Hamblin MR, O'Donnell DA, Murthy N, Rajagopalan K, Michaud N, Sherwood ME, Hasan T (2002) Polycationic photosensitizer conjugates: effects of chain length and Gram classification on the photodynamic inactivation of bacteria. J Antimicrob Chemother 49:941–951

Hems RS, Gulabivala K, Ng YL, Ready D, Spratt DA (2005) An in vitro evaluation of the ability of ozone to kill a strain of Enterococcus faecalis. Int Endod J 38:22–29

Huth KC, Jakob FM, Saugel B, Cappello C, Paschos E, Hollweck R, Hickel R, Brand K (2006) Effect of ozone on oral cells compared with established antimicrobials. Eur J Oral Sci 114:435–440

Jagani S, Chelikani R, Kim DS (2009) Effects of phenol and natural phenolic compounds on biofilm formation by Pseudomonas aeruginosa. Biofouling 25:321–324

Jefferson KK (2004) What drives bacteria to produce a biofilm? FEMS Microbiol Lett 236:163–173

Kandaswamy D, Venkateshbabu N, Gogulnath D, Kindo AJ (2010) Dentinal tubule disinfection with 2 % chlorhexidine gel, propolis, morinda citrifolia juice, 2 % povidone iodine, and calcium hydroxide. Int Endod J 43:419–423

Kim JS, Kuk E, Yu KN, Kim JH, Park SJ, Lee HJ, Kim SH, Park YK, Park YH, Hwang CY, Kim YK, Lee YS, Jeong DH, Cho MH (2007) Antimicrobial effects of silver nanoparticles. Nanomedicine 3:95–101

Kimura Y, Tanabe M, Imai H, Amano Y, Masuda Y, Yamada Y (2011) Histological examination of experimentally infected root canals after preparation by Er:YAG laser irradiation. Lasers Med Sci 26:749–754

Kishen A (2010) Advanced therapeutic options for endodontic biofilms. Endod Top 22(1):99–123

Kishen A, Shi Z, Neoh KG (2008a) Antibacterial nanoparticulates to prevent post-treatment endodontic infection. J Endod 34:15–20

Kishen A, Sum CP, Mathew S, Lim CT (2008b) Influence of irrigation regimens on the adherence of Enterococcus faecalis to root canal dentin. J Endod 34:850–854

Kishen A, Shi Z, Shrestha A, Neoh KG (2008c) An investigation on the antibacterial and antibiofilm efficacy of cationic nanoparticulates for root canal disinfection. J Endod 34:1515–1520

Kishen A, Upadya M, Tegos GP, Hamblin MR (2010) Efflux pump inhibitor potentiates antimicrobial photodynamic inactivation of Enterococcus faecalis biofilm. Photochem Photobiol 86:1343–1349

Klepac-Ceraj V, Patel N, Song X, Holewa C, Patel C, Kent R, Amiji MM, Soukos NS (2011) Photodynamic effects of methylene blue-loaded polymeric nanoparticles on dental plaque bacteria. Lasers Surg Med 43:600–606

Koba K, Kimura Y, Matsumoto K, Gomyoh H, Komi S, Harada S, Tsuzuki N, Shimada Y (1999) A clinical study on the effects of pulsed Nd:YAG laser irradiation at root canals immediately after pulpectomy and shaping. J Clin Laser Med Surg 17:53–56

Kömerik N, Wilson M, Poole S (2000) The effect of photodynamic action on two virulence factors of gram-negative bacteria. Photochem Photobiol 72:676–680

Kubo I, Muroi H, Himejima M (1992) Antibacterial activity of totarol and its potentiation. J Nat Prod 55:1436–1440

Kubo I, Nihei K, Tsujimoto K (2003) Antibacterial action of anacardic acids against methicillin resistant Staphylococcus aureus (MRSA). J Agric Food Chem 51:7624–7628

Kuştarci A, Sümer Z, Altunbaş D, Koşum S (2009) Bactericidal effect of KTP laser irradiation against Enterococcus faecalis compared with gaseous ozone: an ex vivo study. Oral Surg Oral Med Oral Pathol Oral Radiol Endod 107:e73–e79

Kvist M, Hancock V, Klemm P (2008) Inactivation of efflux pumps abolishes bacterial biofilm formation. Appl Environ Microbiol 74:7376–7382

Lawrence JR, Wolfaardt GM, Korber DR (1994) Determination of diffusion coefficients in biofilms by confocal laser microscopy. Appl Environ Microbiol 60:1166–1173

Leonardo MR, Guillén-Carías MG, Pécora JD, Ito IY, Silva LA (2005) Er:YAG laser: antimicrobial effects in the root canals of dogs' teeth with pulp necrosis and chronic periapical lesions. Photomed Laser Surg 23:295–299

Lewis K (2005) Persister cells and the riddle of biofilm survival. Biochem Mosc 70:267–274

Li RW, Myers SP, Leach DN, Lin GD, Leach G (2003) A cross-cultural study: anti-inflammatory activity of Australian and Chinese plants. J Ethnopharmacol 85(1):25–32

Marchesan MA, Brugnera-Junior A, Souza-Gabriel AE, Correa-Silva SR, Sousa-Neto MD (2008) Ultrastructural analysis of root canal dentine irradiated with 980-nm diode laser energy at different parameters. Photomed Laser Surg 26:235–240

Marshall GW Jr, Balooch M, Kinney JH, Marshall SJ (1995) Atomic force microscopy of conditioning agents on dentin. J Biomed Mater Res 29:1381–1387

Meire MA, De Prijck K, Coenye T, Nelis HJ, De Moor RJ (2009) Effectiveness of different laser systems to kill Enterococcus faecalis in aqueous suspension and in an infected tooth model. Int Endod J 42:351–359

Menezes S, Capella MA, Caldas LR (1990) Photodynamic action of methylene blue: repair and mutation in Escherichia coli. J Photochem Photobiol B Biol 5:505–517

Merchat M, Bertolini G, Giacomini P, Villanueva A, Jori G (1996) Meso-substituted cationic porphyrins as efficient photosensitizers of gram-positive and gram-negative bacteria. J Photochem Photobiol B Biol 32:153–157

Minnock A, Vernon DI, Schofield J, Griffiths J, Parish JH, Brown ST (1996) Photoinactivation of bacteria. Use of a cationic water-soluble zinc phthalocyanine to photoinactivate both gram-negative and gram-positive bacteria. J Photochem Photobiol B Biol 32:159–164

Miserendino L, Robert PM (1995) Lasers in Dentistry. Quintessence, Hanover Park, IL

Moan J, Berg K (1991) The photodegradation of porphyrins in cells can be used to estimate the lifetime of singlet oxygen. Photochem Photobiol 53:549–553

Mohn D, Bruhin C, Luechinger NA, Stark WJ, Imfeld T, Zehnder M (2010) Composites made of flame-sprayed bioactive glass 45S5 and polymers: bioactivity and immediate sealing properties. Int Endod J 43:1037–1046

Moriyama EH, Zângaro RA, Villaverde AB, Lobo PD, Munin E, Watanabe IS, Júnior DR, Pacheco MT (2004) Dentin evaluation after Nd:YAG laser irradiation using short and long pulses. J Clin Laser Med Surg 22:43–50

Moshonov J, Orstavik D, Yamauchi S, Pettiette M, Trope M (1995) Nd:YAG laser irradiation in root canal disinfection. Endod Dent Traumatol 11:220–224

Murray PE, Farber RM, Namerow KM, Kuttler S, Godoy FG (2008) Evaluation of *Morinda citrifolia* as an endodontic irrigant. J Endod 34:66–70

Nagayoshi M, Kitamura C, Fukuizumi T, Nishihara T, Terashita M (2004) Antimicrobial effect of ozonated water on bacteria invading dentinal tubules. J Endod 30:778–781

Nair PNR (1987) Light and electron microscopic studies of root canal flora and periapical lesions. J Endod 13:29–39

Nair PNR (2006) On the causes of persistent apical periodontitis: a review. Int Endod J 39:249–281

Nair PNR, Sjögren U, Krey G, Kahnberg K-E, Sundqvist G (1990) Intraradicular bacteria and fungi in root-filled, asymptomatic human teeth with therapy-resistant periapical lesions: a long-term light and electron microscopic follow-up study. J Endod 16:580–588

Nair PN, Henry S, Cano V, Vera J (2005) Microbial status of apical root canal system of human mandibular first molars with primary apical periodontitis after "one-visit" endodontic treatment. Oral Surg Oral Med Oral Pathol Oral Radiol Endod 99:231–252

Ng RPY (2004) Sterilization in root canal treatment: current advances. Hong Kong Dent J 1:52–57

Nichols WW, Evans MJ, Slack MPE, Walmsley HL (1989) The penetration of antibiotics into aggregates of mucoid and non-mucoid Pseudomonas aeruginosa. J Gen Microbiol 135:1291–1303

Noguchi F, Kitamura C, Nagayoshi M, Chen KK, Terashita M, Nishihara T (2009) Ozonated water improves lipopolysaccharide-induced responses of an odontoblast-like cell line. J Endod 35:668–672

Noiri Y, Katsumoto T, Azakami H, Ebisu S (2008) Effects of Er:YAG laser irradiation on biofilm-forming bacteria associated with endodontic pathogens in vitro. J Endod 34:826–829

Pagonis TC, Chen J, Fontana CR, Devalapally H, Ruggiero K, Song X, Foschi F, Dunham J, Skobe Z, Yamazaki H, Kent R, Tanner AC, Amiji MM, Soukos NS (2010) Nanoparticle-based endodontic antimicrobial photodynamic therapy. J Endod 36:322–328

Paraskeva P, Graham NJ (2002) Ozonation of municipal wastewater effluents. Water Environ Res 74:569–581

Peters OA, Bardsley S, Fong J, Pandher G, Divito E (2011) Disinfection of root canals with photon-initiated photoacoustic streaming. J Endod 37:1008–1012

Polo L, Segalla A, Bertoloni G, Jori G, Schaffner K, Reddi E (2000) Polylysine-porphycene conjugates as efficient photosensitizers for the inactivation of microbial pathogens. J Photochem Photobiol B Biol 59:152–158

Prabhakar J, Senthilkumar M, Priya MS, Mahalakshmi K, Sehgal PK, Sukumaran VG (2010) Evaluation of antimicrobial efficacy of herbal alternatives (*Triphala* and green tea polyphenols), MTAD, and 5 % sodium hypochlorite against *Enterococcus faecalis* biofilm formed on tooth substrate: an in vitro study. J Endod 36:83–86

Prasad PN (2003) Introduction to biophotonics. John Wiley & Sons, New York

Prince AS (2002) Biofilms, antimicrobial resistance, and airway infection. N Engl J Med 347:1110–1111

Rabea EI, Badawy ME, Stevens CV, Smagghe G, Steurbaut W (2003) Chitosan as antimicrobial agent: applications and mode of action. Biomacromolecules 4:1457–1465

Rasmussen K, Lewandowski Z (1998) Microelectrode measurements of local mass transport rates in heterogeneous biofilms. Biotechnol Bioeng 59:302–309

Reddy KM, Feris K, Bell J, Wingett DG, Hanley C, Punnoose A (2007) Selective toxicity of zinc oxide nanoparticles to prokaryotic and eukaryotic systems. Appl Phys Lett 24:2139021–2139023

Restaino L, Frampton EW, Hemphill JB, Palnikar P (1995) Efficacy of ozonated water against various food-related microorganisms. Appl Environ Microbiol 61:3471–3475

Ricucci D, Siqueira JF Jr (2010) Biofilms and apical periodontitis: study of prevalence and association with clinical and histopathologic findings. J Endod 36:1277–1288

Rooney J, Midda M, Leeming J (1994) A laboratory investigation of the bactericidal effect of a Nd:YAG laser. Br Dent J 17:61–64

Rosan B, Correeia FF, DiRienzo JM (1999) Corncobs: a model for oral microbial biofilms. In: Busscher HJ, Evans LV (eds) Oral biofilms and plaque control: concepts in dental plaque formation, Harwood Academic Publishers, The Netherlands, pp 145–162
Rukkumani R, Sri Balasubashini M, Menon VP (2003) Protective effects of curcumin and photo-irradiated curcumin on circulatory lipids and lipid peroxidation products in alcohol and polyunsaturated fatty acid-induced toxicity. Phytother Res 17:925–929
Ryan BM, Dougherty TJ, Beaulieu D, Chuang J, Dougherty BA, Barrett JF (2001) Efflux in bacteria: what do we really know about it? Expert Opin Investig Drugs 10:1409–1422
Sadovskaya I, Chaignon P, Kogan G, Chokr A, Vinogradov E, Jabbouri S (2006) Carbohydrate-containing components of biofilms produced in vitro by some staphylococcal strains related to orthopaedic prosthesis infections. FEMS Immunol Med Microbiol 47:75–82
Sawai J (2003) Quantitative evaluation of antibacterial activities of metallic oxide powders (ZnO, MgO and CaO) by conductimetric assay. J Microbiol Methods 54:177–182
Sawai J, Shoji S, Igarashi H, Hashimoto A, Kokugan T, Shimizu M, Kojima H (1998) Hydrogen peroxide as an antibacterial factor in zinc oxide powder slurry. J Ferment Bioengin 86:521–522
Schoop U, Kluger W, Moritz A, Nedjelik N, Georgopoulos A, Sperr W (2004) Bactericidal effect of different laser systems in the deep layers of dentin. Lasers Surg Med 35:111–116
Shen HR, Spikes JD, Kopecekova P, Kopecek J (1996) Photodynamic crosslinking of proteins I. Model studies using histidine- and lysine-containing N-(2-hydroxypropyl) methacrylamide co-polymers. J Photochem Photobiol B Biol 34:203–210
Shrestha A, Kishen A (2011) Polycationic chitosan conjugated photosensitizer for antibacterial photodynamic therapy. Photochem Photobiol. doi:10.1111/j.1751-1097.2011.01026.x
Shrestha A, Shi Z, Neoh KG, Kishen A (2010) Nanoparticulates for antibiofilm treatment and effect of aging on its antibacterial activity. J Endod 36:1030–1035
Silveira AM, Lopes HP, Siqueira JF Jr, Macedo SB, Consolaro A (2007) Periradicular repair after two-visit endodontic treatment using two different intracanal medications compared to single-visit endodontic treatment. Braz Dent J 18:299–304
Soncin M, Fabris C, Busetti A, Dei D, Nistri D, Roncucci G, Jori G (2002) Approaches to selectivity in the Zn(II)–phthalocyanine photosensitized inactivation of wild-type and antibiotic-resistant Staphylococcus aureus. Photochem Photobiol Sci 1:815–819
Soukos NS, Wilson M, Burns T, Speight PM (1996) Photodynamic effects of toluidine blue on human oral keratinocytes and fibroblasts and Streptococcus sanguis evaluated in vitro. Lasers Surg Med 18:253–259
Soukos NS, Hamblin MR, Hasan T (1997) The effect of charge on cellular uptake and phototoxicity of polylysine chlorin e6 conjugates. Photochem Photobiol 65:723–729
Soukos NS, Chen PS, Morris JT, Ruggiero K, Abernethy AD, Som S, Foschi F, Doucette S, Bammann LL, Fontana CR, Doukas AG, Stashenko PP (2006) Photodynamic therapy for endodontic disinfection. J Endod 32:979–984
Stabholz A, Zeltser R, Sela M, Peretz B, Moshonov J, Ziskind D, Stabholz A (2003) The use of lasers in dentistry: principles of operation and clinical applications. Compend Contin Educ Dent 24:935–948
Stohs SJ, Bagchi D (1995) Oxidative mechanisms in the toxicity of metal ions. Free Radic Biol Med 18:321–336
Stoodley P, Debeer D, Lewandowski Z (1994) Liquid flow in biofilm systems. Appl Environ Microbiol 60:2711–2716
Stoor P, Söderling E, Salonen JI (1998) Antibacterial effects of a bioactive glass paste on oral microorganisms. Acta Odontol Scand 56:161–165
Sundqvist G, Figdor D (2003) Life as an endodontic pathogen: ecological differences between the untreated and root-filled root canals. Endod Topics 6:3–28
Tegos G, Hamblin MR (2006) Phenothiazinium antimicrobial photosensitizers are substrates of bacterial multidrug resistance pumps. Antimicrob Agents Chemother 50:196–203

Tegos GP, Masago K, Aziz F, Higginbotham A, Stermitz FR, Hamblin MR (2008) Inhibitors of bacterial multidrug efflux pumps potentiate antimicrobial photoinactivation. Antimicrob Agents Chemother 52:3202–3209
van Leeuwen TG, Jansen ED, Motamedi M, Borst C, Welch AJ (1995) Pulsed laser ablation of soft tissue. In: Welch AJ, van Gemert MJC (eds) Optical-thermal response of laser-irradiated tissue. Plenum, New York
Viera MR, Guiamet PS, de Mele MFL, Videla HA (1999) Use of dissolved ozone for controlling planktonic and sessile bacteria in industrial cooling systems. Int Biodeter Biodegrad 44:201–207
Vrany JD, Stewart PS, Suci PA (1997) Comparison of recalcitrance to ciprofloxacin and levofloxacin exhibited by Pseudomonas aeruginosa biofilms displaying rapid-transport characteristics. Antimicrob Agents Chemother 41:1352–1358
Wainwright M, Crossley KB (2002) Methylene blue a therapeutic dye for all seasons? J Chemother 14:431–443
Wainwright M, Giddens RM (2003) Phenothiazinium photosensitisers: choices in synthesis and application. Dyes Pigm 57:245–257
Wakayama Y, Takagi M, Yano K (1980) Photosensitized inactivation of E. coli cells in toluidine blue-light system. Photochem Photobiol 32:601–605
Waltimo T, Brunner TJ, Vollenweider M, Stark WJ, Zehnder M (2007) Antimicrobial effect of nanometric bioactive glass 45S5. J Dent Res 86:754–757
Waltimo T, Mohn D, Paqué F, Brunner TJ, Stark WJ, Imfeld T, Schätzle M, Zehnder M (2009) Fine-tuning of bioactive glass for root canal disinfection. J Dent Res 88:235–238
Wang MY, West BJ, Jensen CJ, Nowicki D, Su C, Palu AK et al (2002) *Morinda citrifolia* (Noni): a literature review and recent advances in Noni research. Acta Pharmacol Sin 23:1127–1141
Wang QQ, Zhang CF, Yin XZ (2007) Evaluation of the bactericidal effect of Er, Cr:YSGG, and Nd:YAG lasers in experimentally infected root canals. J Endod 33:830–832
Yamamoto O (2001) Influence of particle size on the antibacterial activity of zinc oxide. Int J Inorg Mater 3:643–646
Yamazaki R, Goya C, Yu DG, Kimura Y, Matsumoto K (2001) Effects of erbium, chromium: YSGG laser irradiation on root canal walls: a scanning electron microscopic and thermographic study. J Endod 27:9–12
Yavari HR, Rahimi S, Shahi S, Lotfi M, Barhaghi MH, Fatemi A, Abdolrahimi M (2010) Effect of Er, Cr: YSGG laser irradiation on Enterococcus faecalis in infected root canals. Photomed Laser Surg 28:S91–S96
Yoon KY, Hoon Byeon J, Park JH, Hwang J (2007) Susceptibility constants of Escherichia coli and Bacillus subtilis to silver and copper nanoparticles. Sci Total Environ 15:572–575
Zehnder M, Söderling E, Salonen J, Waltimo T (2004) Preliminary evaluation of bioactive glass S53P4 as an endodontic medication in vitro. J Endod 30:220–224
Zehnder M, Luder HU, Schätzle M, Kerosuo E, Waltimo T (2006) A comparative study on the disinfection potentials of bioactive glass S53P4 and calcium hydroxide in contra-lateral human premolars ex vivo. Int Endod J 39:952–958
Zhang L, Mah TF (2008) Involvement of a novel efflux system in biofilm-specific resistance to antibiotics. J Bacteriol 190:4447–4452

Index

L.E. Chávez de Paz et al. (eds.), *The Root Canal Biofilm*, Springer Series on Biofilms 9, DOI 10.1007/978-3-662-47415-0

E

MIX
Papier aus verantwortungsvollen Quellen
Paper from responsible sources
FSC® C105338

If you have any concerns about our products,
you can contact us on
ProductSafety@springernature.com

In case Publisher is established outside the EU,
the EU authorized representative is:
Springer Nature Customer Service Center GmbH
Europaplatz 3, 69115 Heidelberg, Germany

Printed by Libri Plureos GmbH
in Hamburg, Germany